Atlas of
Breeding B
Of the Maritime Provinces

Cartography by Linda Ann Payzant and L. Peter M. Payzant

Drawings by Azor Vienneau

Cover painting by Donald Curley

Anthony J. Erskine

Produced as part of the Nova Scotia Museum Program
of the Department of Education, Province of Nova
Scotia

Minister The Hon. Guy J. Le Blanc
Deputy Minister Armand F. Pinard

Co-published by Nimbus Publishing Limited and the
Nova Scotia Museum with the generous assistance of
the Department of Natural Resources

A product of the Nova Scotia Government
Co-publishing Program

Printed by McCurdy Printing, Halifax
Design by Graphic Design Associates, Halifax

COVER PAINTING by Don Curley: Adult Red-tailed
Hawk *(Buteo jamaicensis)* preparing to feed its young.

Canadian Cataloguing in Publication Data
   Erskine, Anthony J., 1931 -
      Atlas of Breeding Birds of the Maritime Provinces

      Co-published by: Nova Scotia Museum.
      ISBN 1-55109-010-4

1. Bird populations – Maritime Provinces
2. Birds – Maritime Provinces – Breeding distribution.
I. Vienneau, Azor. II. Nova Scotia Museum. III. Title.

QL685.5.M37E77 1992   598.29715   C92-098513-0

Contents

List of Tables

List of Figures

Foreword

One of the best things about a foreword is that you can say pretty well whatever you want to say in it, within broad limits of relevance.

This Atlas owes its very existence to Linda and Peter Payzant. Because so many aspects of their contribution unfortunately will be invisible in the final product, I must try to convey some sense of what they did for it. Back in 1984 they were the prime movers in focusing the interests of a handful of people to the point where an *ad hoc* committee (later to become the Steering Committee) was formed. Over Christmas of that year Peter and Linda drafted a funding proposal for the project, which was finally accepted in August 1985. Since then the Atlas has been a labour of love and the centre of gravity in their lives. They always endeavoured (and sometimes had to fight) to ensure that the highest possible standards were maintained in everything from the handling of raw data to the production of the final maps. They wrote the software, did all the computing and virtually all the map production for the project. On top of all this, they were the RCs for Region 18 (one of the first to be completed), and did a great deal of square-bashing in other regions in the last two field seasons. There were many periods when for weeks at a stretch, every single hour of their free time was spent on Atlas matters.

The Atlas has been lucky both in the people it has attracted as volunteers, and in those it has searched out and hired. Our volunteers gave both time and money, many repeatedly and very generously. Our two Atlas Co-ordinators each brought to the job special skills we happened to need at that time. Judith Kennedy came with a thorough knowledge of how an atlas was run and had forms designed, procedures in place, RCs trained and atlassers out in the field without a hitch. In our last two field seasons, Brian Dalzell, in addition to superb atlassing skills, brought an apparently inexhaustible capacity to suffer heat, rain, blackflies and cold canned food while living out of a rented van. Without him there would have been huge blank areas in northern and central New Brunswick.

Our Atlas breaks significant new ground in presenting actual population estimates derived from the species abundance codes recorded by many atlassers. A few in the North American atlassing community have yet to be convinced of the value of such estimates but I predict that well before the current generation of North American atlasses is completely published, such estimates will be standard.

My involvement in this atlas has been one of the most profoundly satisfying of my life. It has enlarged my perspective on birds, as well as my circle of friends. And it has given me indelible memories: a patch of bark on a fir tree suddenly turning into a vibrating clump of fledglings as a Brown Creeper arrives with a moth in its beak; a female Northern Harrier coming for me at eye level, six feet away, and swooping up only at the last second when I raise my arms; the howling of coyotes at four in the morning outside the old farmhouse where I stayed in Cape Breton Island; and the sudden goose-bumps when I realized that a Boreal Owl had just responded to my tape-recording of its call.

If I'm around for the next atlas, I'll be out there.

Fred W. Scott
Chairman, Steering Committee
11 February 1992

Acknowledgements

Like all such projects, the Maritimes Atlas is the result of efforts by many people and organizations, and we thank all of you. Without your support, in time, money, enthusiasm, facilities, and advice, provided willingly and without stinting, we would not have had the opportunity to produce this book. The persons who provided data, or made financial donations, or provided advice or other help, are thanked herein. We wish we could thank you all personally.

The evolution of the Atlas is outlined in the Introduction, and the committee structures are set out in Appendix A. We thank, individually and collectively, the members of the Maritimes Bird Atlas Trust, the Steering Committee, the Treasurer, the Co-ordinators, and the Regional Co-ordinators (all listed alphabetically below):

Chris Adam, Ford Alward, Peter Austin-Smith, Vicky Bunbury, Dan Busby, Bill Caudle, Hilaire Chiasson, David Christie, David Clark, Rosemary Curley, Dave Currie, Brian Dalzell, Jerome D'Eon, Ted D'Eon, Bob Dickie, Tony Erskine, Roger Foxall, Al Gibbs, Don Gibson, Gay Hansen, Dave Harris, Judith Kennedy, Erwin Landauer, Fulton Lavender, Don MacNeill, Roslyn MacPhee, Mary Majka, Mike Malone, Blake Maybank, Ian McLaren, Eric Mills, Linda and Peter Payzant, Peter Pearce, Ian and Christine Ross, Fred Scott, Jean Timpa, Harry Walker, Rob Walker, Jim Wilson.

Special thanks are due the Nova Scotia Museum, which provided the Atlas with a home from the very beginning, and whose support ran right through and included the publication of this book. Office support makes a project run smoothly, and its lack could stall everything. We thank the Museum director and staff who ensured that the Atlas did not stall.

Along with a home, a project must have money to function. For this we thank Supply and Services Canada, which provided start-up funding (the first year and a half), and the Canadian Wildlife Service, Environment Canada, which carried the major financial load the rest of the way. With the money from these agencies, we hired a co-ordinator for the project to keep it moving despite all obstacles. It moved, and we thank DSS and CWS for making this possible.

Other major direct financial support came from the Province of Prince Edward Island (Fish and Wildlife Branch, Department of the Environment), Bowater Mersey Paper Co. Ltd., the James L. Baillie Memorial Fund, Stora Forest Industries, the New Brunswick Museum, the Themadel Foundation, the New Brunswick Department of Natural Resources and Energy, and from special fund-raising activities, as well as from the many organizations and individuals listed below. A direct appeal to Atlas participants in each of the last two years provided funding to support the Co-ordinator in Atlas field work during the summers of 1989 and 1990, thus ensuring coverage of many areas poorly worked until then.

The College of Geographic Sciences at Lawrencetown, N.S., was especially helpful in assisting us with mapping our first year's data, and their map of the Maritimes appears on all of the full-page species treatments. We thank Bob Maher and David Colville of the College for their help. Gerald Black of the Department of Fisheries and Oceans provided essential expertise in the PostScript language, thus freeing us from the need to plot our maps on a pen plotter.

The Nova Scotia Department of Lands and Forests (now part of Natural Resources) provided us with a complete set of topographic maps for use in the Atlas office. In the final field season, they donated crucial air transport for two field teams to reach some very remote squares. Parks Canada supplied accommodations to a team owling in Cape Breton Highlands National Park. The legal firm of Patterson, Kitz donated their services.

One area of specialized, and essential, support deserves special mention. Two major North American atlasses provided very welcome expertise when our project was in its early stages. The New York Breeding Bird Atlas supplied us with complete listings of their software, which were very helpful to us in planning our data processing. The Ontario Breeding Bird Atlas helped us with organizational matters, including allowing us to adopt many of their forms and practices virtually unchanged. We thank the Ontario and New York atlas committees for these courtesies, which saved us much time and effort.

Peter and Linda Payzant were responsible for all our computer programming (see Appendix C), including production of the final digital map files and overlay maps. We thank them for these special efforts, over and above their many other roles in the project, and also for the many special listings produced to ease the task of condensing the Atlas data into the species accounts and interpreting them.

Most of the population estimates were derived from observers' abundance indices by a statistical program based on an algorithm (see Appendix D) developed by Chris Field, whose professional expertise was essential in making these numbers believable. Thanks again, Chris.

Rudy Stocek, Peter Austin-Smith and Tony Lock supplied valuable numerical population data for many species, enabling me to confirm estimates derived from observers' data or to generate estimates directly in cases where data from the field were too sparse for use in our computer program.

Illustrations are, by now, traditional in atlasses, and they make all the difference in the visual appeal of the page. The delightful bird drawings in this atlas are by Azor Vienneau, who somehow found the time to produce them between his regular duties as an artist at the Nova Scotia Museum and his many hours of atlassing.

The text of this book was drafted using the facilities of the Canadian Wildlife Service, Atlantic Region, partly while I was employed by CWS under a Senior Assignment Position and partly on my own time, before and after my retirement. The Atlas project and I personally thank CWS for providing me with this opportunity.

Portions of the manuscript were reviewed by Tony Lock, Peter and Linda Payzant, Peter Austin-Smith, Rudy Stocek and Ron Weir, and I thank them for their assistance. I particularly wish to thank David Christie, who was especially helpful in reviewing virtually the whole book.

The field effort, involving more than 1,100 persons, was of course the largest contribution in kind, comprising some 43,000 hours over the five field seasons. This could not have been bought for money. The efforts of many persons went far beyond what any job would demand, and could only be justified as "a fun thing," for all of us! Each of us would like to single out for special thanks some individual whose efforts really stood out, but there is no fair way to thank a few without snubbing the rest. You know who you were, and we thank all of you. This book is the memorial to your efforts.

The following persons contributed to Atlas field work in 1986–90. Their efforts made the Atlas.

Chris Adam, Marc Adam, Thom Adams, Jeanne U. Addelson, Marika Ainley, Roger Albert, Mr. Alder, Betty Allard, Charlie Allen, Chris Allen, Garry Allen, George & Margaret Alliston, Michael Almon, Ford Alward, Paul Alward, David Ambridge, Daryl Amirault, Diane Amirault, Robert W. Anderson, Lenore & Roger Andrew, Robert F. Andrle, Chris Antle, Doug Archibald, Anne-Marie Arseneault, Kurt Arseneault, Ronald Arsenault, William O. Astle, Brian Atkinson, Lise & Normand Aubut, Peter J. Austin-Smith.

Kempton Baird, Ross Baker, Don Baldwin, Bernie & Betty Ball, George Ball, Bob Bancroft, Dan Banks, Stephen Barbour, Wayne Barchard, Anne Bardou, Peter Barkhouse, Gordon Barron, Bill & Eleanor Barrow, Myrtle Bateman, Mr. & Mrs. J. Bates, Kevin Bayort, Harry Beach, A. M. Beaulieu, Yvon Beaulieu, Bill Beaverbank, Gwen A. Beck, Charles L. Bell, Greg Bell, Tom Bellis, Craig Benkman, Gerard & Denise Benoit, David Bernard, Ross Bernard, Charles & Betty Berry, Stephen Bettles, Ronald Betts, R. G. S. Bidwell, Clare Birch, Chip Bird, Donald J. Bird, Dan Birt, G. Bishop, Archie Black, Bruce Blackwell, Roy Blakeburn, Sherman Bleakney, J. Blenis, Mary & Owen Bloise, Blomidon Naturalists Society, Paul Bogaard, Larry & Alison Bogan, Barb Boiduk, Janet Boss, Yves Bosse, Bryce R. Boswell, Simon Bouchard, Dan Bourque, Jean & Ronald Bourque, Stephen Bourque, Bowater Mersey Paper Co., Bill Bowerbank, Thelma Bowers, Martyn & Sandy Bowler, Bob Bowles, George Boyd, J. Bray, Calvin Brennan, Harry & Jean Brennan, James Bridgland, Florence Britton, Joan Bromley, Mr. & Mrs. Carl Brown, Jennifer Brown, Jim Brown, M. Brown, Marge Brown, John Brownlie, Nico Bruinooge, Arthur & Marie Bryant, Brenda Brydon,

Phyllis J. Bryson, Mr. & Mrs. E. Bull, Vicky Bunbury, Rick Burger, Beulah Burman, Sandy Burnett, Bruce Burns, Roger Burrows, Jamie Burton, Brenda & Tony Burzynski, Michael Burzynski, Dan Busby, Lee S. Bushell, Stephen Bushell, Mr. & Mrs. J. Buyting, Barbara A. Byrd.

Cape Breton Highlands National Park Staff, Dorice Caissie, Marcel Caissie, Pat & Bob Caldwell, Raymond R. Calhoun, Roger Calkins, Mrs. W. Calvert, Bill Cameron, Eric Cameron, Ian Cameron, Mary & Doug Cameron, Anne Camozzi, Ansel A. Campbell, Mrs. Ben Campbell, D. Campbell, David Campbell, Duff & Kay Campbell, Mr. & Mrs. Jim Campbell, Malcolm Campbell, Moira Campbell, Jean Carmichael, Eric R. Carr, Danielle Carrier, Clara Carter, David Cartwright, Karen Casselman, W. G. Caudle, Roland D. Chaisson, Mr. Chapman, Mary Alice Chapman, Christine Chasse, Hilaire Chiasson, Raymond Chiasson, Rose-Aline Chiasson, Ed & Linda Chirico, Robert Chivers, Tom Chmiel, David Christie, Edward Christie, Sheila Christie, J. Churchill, Lana Churchill, Margaret Churchill, Ed Claridge, David Clark, Diane Clark, Margaret Clark, Paul Clark, Will & Caryl Clark, Alice Clements, Pat Clifford, R. Clifford, Club d'Ornithologie de Moncton, Jean Cochrane, Shirley Cohrs, Cyril Coldwell, Larry & Lynn Coldwell, Karl & Donna Cole, Percy Cole, Chad Coles, Harold Collete, Paul Collins, Clayton & Jessie Colpitts, Constance Colpitts, Lucy Colpitts, Pearl Colpitts, David Colville, Paul & Ruth Colville, Peter Comeau, Vivian Comeau, Lynda Conrad, Weldon & Winnifred Conrad, Ray Cook, Sandra Cooke, Cathy Coombes, Arthur D. Cooper, Enid Cooper, Gerald Corbett, Mr. & Mrs. Robert Cormack, Donald Cormier, J. Albert Cormier, Bob & Mary Cotsworth, Alan Covert, Kevin Craig, Lorna Creamer, Ann Crocker, Peter & Marion Croft, Donna Crossland, Wayne & Cindy Crouse, Phyllis Crowe, Eleanor Crowley, Gary Cullins, Dorothy Curley, Georgie Curley, Rosemary Curley, Wanda Curley, Barb Currie, David A. Currie, S. Currie, Ted & Linda Currie, Elinor Curry, Grant A. Curtis, Joan Czapalay.

Linda Dakin, Louis Daley, J. Allison Dalton, Brian Dalzell, Halton Dalzell Jr., Jeff Dalziel, Ken Dance, George & Vivian Daniels, Harry N. Darrow, David Dauphinee, Marcel David, Heather Davidson, Cheryl Davis, R. Davis, Steve Davis, Mr. & Mrs. John Dawson, Tracey Dean, Richard DeBow, Hank Deichmann, Kyle DeLeavey, Queenie & Wayne DeLeavey, F. P. F. DeLong, Frank Delorey, Joe Delorey, Tom DeMarco, Peter DeMarsh, Andre d'Entremont, Delisle D'Entremont, Jacqueline D'Entremont, Raymond D'Entremont, Roland D'Entremont, Brenda E. D'Eon, Jerome D'Eon, Ted C. D'Eon, Con Desplanque, Vera & Paul DeWitt, Jan deWitte, Faith & Jane DeWolfe, Randy Dibblee, Don Dickey, Bob & Helen Dickie, G. E. Dickie, Andy & Mary Kate Didyk, Sabine Dietz, Maureen Dixon, Ben K. Doane, Mr. & Mrs. Fred Dobson, James A. Dobson, Jean-Claude Doiron, Robert Doiron, Joe Dolphin, Raymond Dort, Mr. Douglas, Nancy Dowd, Pamela Doyle, Guylaine Drolet, Larry Drost, Cliff Drysdale, Ken Dubberke, Carmon Dube, Ducks Unlimited Canada, Tom Duffy, John Dulanty, Lewis & Elizabeth Dumont, Charles D. Duncan, Gordon Dunphy, Margaret Dunphy, Todd Dupuis, Ann Dutton, Paul Duval, Lucy Dyer.

Jim Edsall, Clem & Pat Egolf, Phyllis Ehler, Bill & Nickey Eisenhauer, Mark Elderkin, Paul Elderkin, Ruth Eldridge, Chris Ellingwood, James Elliott, Alex & Darlene Ellis, Dale Ellis, Helen Ellis, Rebecca Ellis, Rebecca & Robert Ellis Jr., Marcel Emond, Patrice Emond, A. J. & Janet Erskine, Florence Erskine, Dorothy Everett, Joe & Faye Everett.

Alonza E. Fahey, Paul Fairclough, Fred Fairley, Stephen Farmer, Joseph Farquhar, Audrey Faulkner, Cam Fenton, Ernest Ferguson, Sylvia Ferguson, Eric Fiander, Chris Field, Brian Fields, Jakko Finne, George Finney, Bill & Ruby Fisher, Geraldine Fitzgerald, Mary Ann Fitzpatrick, Stephen Flemming, David Fletcher, Michael J. Fletcher, Anna & John Foley, Harriet Folkins, Noel Fontaine, H. D. Ford, Shirley Forrest, Bernard Forsythe, Fred Forsythe, George Foster, Ron Fournier, Patricia Fox, Roger Foxall, Beatrice Fralick, Don Fraser, Cliff Friesen, Sylvia J. Fullerton, Fundy National Park Staff.

George Gagnon, Mr. & Mrs. J. R. Gallagher, Freeman Gallant, Jean-Raymond Gallien, Wayne Garden, T. Rod Gardner, Neville Garrity, G. Garron, Tom Gatz, Leonard Gaudet, Terry Gauthier, Jocelyne Gauvin, Sharon Gay, Mary Geddes, Diana & Lloyd Geil, Ellis Gertridge, Al & Wendy Gibbs, Donald G. Gibson, Margaret Gibson, John Gilhen, Mary Gill, Barbara

& Gordon Gilliland, Scott Gilliland, Mike Gillis, Kenneth Gilmour, Lionel Girouard, Bill & Elizabeth Glen, Alan P. Godfrey, Tommy Godfrey, Audard Godin, Gilles Godin, Virgile Godin, Andre Goguen, Jim Goltz, Vernon Goodfellow, Minola Goodwin, Irene Gorham, R. Allen Gorham, A. Graham, H. Graham, Mr. & Mrs. Arthur Grant, Scott Grasman, June Graves, Ben & Anna Gray, Tom Greathouse, Henry H. Green, Lorena Green, Polly Greene, Mrs. Ronald Greene, Gary Greer, Ken Gregoire, Diane Griffin, F. Grondin, Greg Guidry.

Jean-Claude & Lily Hachey, Jane & Eric Hadley, Bernice Hafner, Halifax Field Naturalists, Eddy E. Hall, Helen & Hubert Hall, Ross Hall, D. Halliday, Gay Hansen, M. Hardy, Hinrich Harries, Dave Harris, Don Harris, Griffin & Dianne Harris, Sharon Harris, Tom Harrison, Lorna Hart, Edward Hartt, Frank Hatheway, Harold Hatheway, David & Betty Hatt, Robert & Shirley Hauth, Gordon Havens, Sharon Hawboldt, Nick & Gabriel Healy, Ralph Hemming, Elwin Hemphill, Harold Hicken, Peter Hicklin, Randy Hicks, Shawn Hicks, Ivy Higgins, Barb Hildebrand, Maxine & Ken Hill, Neil Hill, Marilyn Hiltz, Tammy Hiltz, Frank Himsl, Barbara A. Hinds, Harold Hinds, Eric Hiscock, Sabine Hitzelberger, J.A. Hoffman, Geoff Hogan, Vern Hollis, Mr. Hollis, John W. Hollway, Les Homans, Peter Hope, Sean Hope, Philip B. Hoppin, Jeff Horne, Harold Horwood, Ron Hounsell, Charles Hubbard, Etta Hudgins, Pat Hudson, Lisa C. Huff, Jack Hughes, Theresa Hughes, Marilyn Hunt, Earlene Hunter, Gordon Hunter, Mark Hunter, Pam Hunter, Wendy Hunter, Charles E. Huntington, Nelson G. Hurry, Derek & Glennys Hutton, Gisa Hynes.

Stephen Illsley, Roger & Connie Ince, Enid M. Inch, Spencer & Helen Inch, Michael Inkpen, Mr. & Mrs. Robert Inkster, Nancy Ironside, Sylvia Irvine.

Ross James, Ian Jamieson, Joan V. Jarvis, June Jarvis, L. Jarvis, Sam K. Jarvis, Brian Jenkins, Roger A. Jenkins, Maude Jodrey, Linda Johns, Eric Johnsen, Bruce Johnson, Russ Johnson, Sheldon Johnson, Cecil Johnston, Delbert Johnston, H. Winston Johnston, Leroy Johnston, Maude Joudrey, Robert Joudrey.

John Kearney, Donald Keith, Heather Kelley, Doug Kelly, D.J. Kennedy, Joe Kennedy, Judith Kennedy, Fred & Margaret Kenney, Gerry Kennific, Mr. Ketchum, Ed Kettela, Meika Keunicke, A. Keuning, Ken L. Kierstead, Vernon N. Kierstead, Don Kimball, Richard L. Kingston, Franklin & Linda Kinney, Leslie Klapstein, Joyce Knapp, Janos Kovacs.

Pat Lacey, Todd LaFrance, Rick Lair, Victor Lamkey, Ralph & Joan Lamrock, Erwin Landauer, Blaise Landry, Joel Landry, Nicole Landry, Donald E. Langille, Judy Langille, Rosita Lanteigne, Louis LaPierre, James W. LaPointe, Ejnar Larsen, Lars Larsen, Randolph Lauff, Charles T. Laugher, Fulton Lavender, Lance Laviolette, Florida LaVoie, Madeleine LaVoie, David Lawley, Ken Lawson, Alfred Laybolt, Betty Learmouth, Rheal LeBlanc, Mel & Ginny Lee, V. Lee, Linda Leeman, Len LeGard, Jim & Carol Legge, Luc Lemieux, Charles S. Lennox, Mary Ann Lidstone, Diane Lindsey, Leslie Linkletter, Daryl Linton, Doug Linzey, Robert Lisk, Joan Lloyd, Anthony Lock, Mr. & Mrs. John Lockhart, Randy Loft, George Long, Frank Longstaff, David Lounsbury, Sara Lounsbury, Kaye & George Love, Helen Lovely, Viola Lovitt, Allen Lowe, Zoe Lucas, Gerry Lunn, Gwendolen C. Lunn, Mike Lushington, Mary Lynyak.

Anne L. MacDonald, Eugene MacDonald, Gerald MacDonald, Graham MacDonald, Maria MacDonald, Pat MacDonald, Paul MacDonald, Ralph & Marie MacDonald, R.J. & Dianne MacDonald, Robert MacDonald, Rod MacDonald, John MacDonell, Don MacDougall, Evelyn MacDougall, G. & A. MacDougall, Ralph MacDougall, Doug MacEachern, Gerri MacEachern, Harold & Joyce MacEachern, Nancy MacFarlane, Regan MacGillivray, Arlene & Paul MacGuigan, Andrew MacInnis, C.J. MacInnis, Cyril J. MacInnis, George MacInnis, John MacInnis, Gerard MacIntyre, Andy MacKay, Heather E. MacKay, Morgan MacKay, Heather MacKinley, Blair MacKinnon, Colin MacKinnon, Dave MacKinnon, Howard MacKinnon, Wade MacKinnon, Floyd MacKnight, Angus MacLean, Alan MacLeod, Dave MacLeod, Gordon MacLeod, John MacLeod, Joyce MacLeod, Kay MacLeod, Peter MacLeod, Carol MacMillan, Harriet MacMillan, G. M. MacNaughton, David & Kathleen MacNearney, Dan MacNeil, Doug MacNeil, John MacNeil, John A. MacNeil, Don & Carol MacNeill, Rob & Jenny MacNeill, Scott MacNeill, Harold D. MacPhee, Roslyn MacPhee, Kent MacRae, Bruce

Mactavish, Christine MacWilliams, Alan Madden, Lorna Maddox, Ray & Joan Mahabir, Heather Maher, R. V. Maher, Christopher Majka, Mary Majka, Scott Makepeace, Jill Malins, Margaret Mallett, Michael Malone, Mrs. Glen Manthorne, George E. Manuel, Maritimes Nest Record Scheme, Abby Marshall, F. Kathryn Marshall, Joe T. Marshall, Alex Martin, Barry Martin, Chris Martin, Eldric Martin, Frances Martin, Gilles & Michel Martin, Kathy Martin, Leo Martin, Kathleen Masui, H. Matthews, Pam Matthews, Roscoe Mault, Blake Maybank, M. B. Mayne, Dennis Mazerolle, Jean-Marie Mazerolle, Charles F. McAleenan, Keith McAloney, Donald McAlpine, A. McArthur, Dan McAskill, Scott McBurney, Sandra McCartney, Patricia McCleave, Allan McCormick, Mr. & Mrs. Everett McCormick, Mr. & Mrs. J. R. McCormick, Jeff McCormick, Roland McCormick, Mr. & Mrs. R. McCullough, Robert & Wendy McDonald, Breeda McDonnel, Geraldine McGill, Shannon & Laurie McGowan, Ernest & Marg McGrath, Elsie McIntosh, Sally McIntosh, Bob G. McIsaac, Perry McIsaac, John & Gwen McKenzie, Bridget McKeough, Ruth McLagan, Ian A. McLaren, Reid McManus, Douglas B. McNair, Lloyd & Pearl McNair, Antoine Melanson, Reg Melanson, George Mercer, Valerie Meredith, Richard Merriam, Nelson Merritt, Ken Meyer, Gert Michaud, Pierrette Michaud, Beulah Michelin, Ruth Miller, Wilma Miller, Eric L. Mills, Allan Millward, A. Milner, Jean & Myers Milner, Randy Milton, Gary Moore, Mr. & Mrs. Ron Moore, Bernice A. Moores, Rene Moreau, Mr. & Mrs. S. Morehouse, Anthony & Diana Morris, Phillip Morris, Christopher Morrison, Doug Morrison, Hal Morrison, Dave Morrow, James Morrow, Bill & Jean Morse, Gail Morse, Paul Mortimer, Roger Mosher, Farley Mowat, Kerstin Mueller, Eric Mullen, Kirk Munn, Anne Muntz, Erich Muntz, Philip & Clarice Muntz, Thelma R. Murchison, Dan Murnaghan, Bill Murphy, Ethelda E. Murphy, Frederick Murphy, Pat Murphy, Richard Murphy, William Murphy, Cathy & Allan Murrant, Mr. & Mrs. W. A. Murray, Lorne & Mary Myers, Sandra Myers, Dave Myles.

Danielle Nadeau, Chris Naugler, Dorothy Neilson, Bill Nelson, Reg & Ruth Newell, G. & M. Nickerson, Glen H. Niles, Peter Northcott, Howard Norton, Nova Scotia Department of Lands & Forests.

Janice Oakes, Michael O'Brien, Rahn O'Connell, Kevin O'Donnell, John C. Oland, Maxine Oldham, Margie Olive, Phil & Mary Oliver, Michael Olsen, Laurie Orr, Elizabeth Otter, Luke Otter, Jean Ouellette, Roger Outhouse.

Larry & Cheryl Page, Stephen & Sheila Palmer, Peter Papoulidis, Jean Paquin, Gerry Parker, Richard Parry, Thomas S. Parsons, Warren Parsons, C. Alan Pater, Lisa Patrice, W. D. Paul, Jean-Yves & Yolande Paulin, Linda & Peter Payzant, Winnie Peach, Peter Pearce, Brian Pellerin, Ron Pellerin, Mr. & Mrs. George Pelletier, Brenda L. Penak, Ruth Penner, Donald Pentland, Lillian Perry, Frank & Geraldine Peters, Jill Peters, Kenneth & Doris Peters, Warren & Nancy Peters, Murray Peterson, Thomas J. Pettigrew, Mabel & Bruce Peveril, Mark Phinney, Fran Piercey, Roger & Pat Pocklington, Jeannette Poirier, Vincent Poirier, H. & K. Popma, Arthur Porter, Ghyslain Pothier, Richard Poulin, Terry Power, Mary Pratt, Mark Prendergast, P.E.I. Fish & Wildlife Division, Gordon Pringle, Dean Prior, Roy Proctor, Charles & Lori Prosser, Lloyd Prosser, Gini Proulx, Lisa Proulx, Peter Puleston, Grace Pulley, Mark Pulsifer, Don and Joyce Purchase, Mr. & Mrs. Ivor Quaggin, Derek Quann.

Laverne Rabatich, Mike Rae, Sue Rankin, Mary Rawlinson, Mr. & Mrs. Kelsey Raymond, Janice Reede, Don Reeves, Nancy Reid, R. N. Renouf, Tim Reynolds, Pam Rhyno, Vincent & Lou Rice, Leonel Richard, Sandy Richards, Ruth Richman, Elta Rideout, Stan Riggs, Lloyd Ripley, J. D. Rising, Roger Rittmaster, Twila Robar-DeCoste, Donald C. Roberson, Peter Roberts, Joe Robertson, Paul-Emile Robichaud, Richard Robichaud, Roland Robichaud, J. Robinson, Sylvain Rochefort, Susan Rodda, Barbara & Terje Rogers, Bill Rogers, Ruth Rogers, Clarence Rose, Ian & Christine Ross, Margaret & Bill Ross, Doreen & Willis Rossiter, Michael W. P. Runtz, L. Rutherford.

Dwayne Sabine, Susan Sanderson, Marie Sappier, Bev Sarty, Kim Saunders, Eric Savoie, Mike Savoie, Mary Jane Savoy, Ernest Sawatzky, Bryan Scallion, Gary Schneider, Trevis Schriver, Julie Schroeder, Fred W. Scott, Mr. & Mrs. John Scott, Stanley Scott, Danny Sears, Norm Seymour, Tom Sheppard, Ian & Robin Sherman, Dave Shutler, Heather Siliker, Mark

Simpson, George Sinclair, Scott Sinclair, Sedgewick Sinclair, Julie Singleton, Karen Sippeley, Louis Sippley, Al Smith, Bruce Smith, Bruce Smith, Craig Smith, David F. Smith, Eleanor Smith, Mary E. Smith, Shirley Smith, Mabel B. Smythe, Nellie Snyder, Colin Somers, Ruth Somers, Dusan Soudek, Fran Spalding, Kathleen Spicer, Maddy & Bernie Spicer, Esther M. Sporle, Sandra Stafford, Cindy Staicer, Cecil Stairs, Peter Van Stam, Mrs. Frank Stanton, Annette Stark, Brian Starzomski, Stephen Starzomski, Grenville Steeves, Jamie Steeves, John E. Steeves, Sue Stephenson, Richard Stern, Sylvia Sterns, Brian Stevens, Clarence Stevens, Donald Stevens, Clayton Stewart, John Stewart, Norman Stewart, Paul Stewart, Walter Stewart, D. Stiles, Rudy Stocek, Murray Stockley, Ralph Stopps, John Stub, Sunbury Shores Arts & Nature Centre, David Sutherland, Rick Swain, Leo W. Sweeney, David Sweet, Jane Symmes.

Miriam Tams, Ron R. Tasker, Derek Tay, Mr. & Mrs. Karl Tay, Harold Taylor, Ilda Taylor, Jim & Bernice Taylor, Kenneth & Linda Taylor, Pat Taylor, Roger Taylor, Gerald Teasdale, Gaetane Theriault, Michel Therriault, Bill & Brenda Thexton, Bob Thexton, Georgette Thibodeau, Martin & Gisele Thibodeau, R. J. Thibodeau, Tony & Morina Thomas, Bob Thompson, Dale Thompson, Randy Thompson, Reg & Jean Thompson, Alison Thomson, Harry Thurston, Jean M. Timpa, Sean Timpa, Stuart Tingley, Pamela Tompkins, Robert Tordon, Julie Towers, Marcella Towle, Elizabeth Townsend, Sandy Trenholm, Fred Tribe, Gordon & Judy Tufts, Lillian B. Tufts, Donald & Donna Turner, Donna Turner, Jack & Julie Turner, J. David Tweedie, Lorne & Audrey Tyler.

Cheryl Uhlman, Charles & Joanne Upton, Eva Urban, Walter Urban, Peter Urquhart, John vanderMeer, Maureitus VanZutphen, I. W. Varty, Stan Vass, Charlene Veinotte, Azor Vienneau, Steve Vines, Deanna Vogelsanger.

Sandy Wagner, David & Winifred Wake, Eleanor Waldron, Harry E. Walker, Ian R. Walker, Paul Walker, Philip Walker, Rob Walker, Elaine Wallace, Ronald Walsh, Brian Walters, Hugh Warburton, Marjorie Warburton, Mary Jane Ward, John Warwick, Owen Washburn, Bob Watt, Rheanna Watt, Murray Watters, Sid Watts, Dorothy Weatherby, Lorne Weaver, E. A. Webber, Barbara Weir, Ron D. Weir, Gary & Dottie Welch, John Henry Wells, Marguerite Wheatley, Dorothy Whipple, Alice White, Don & Alma White, Doug White, Joy & Peter White, Karl White, Mark White, Melvin White, Ralph White, Jan Whitelaw, C. Whitman, Doug Whitman, George Whitman, Stewart Whitman, Robert Whitney, Richard Wilbur, Dave Wilder, John R. Wile, Gene Wilhelm Jr., Trevor Wilkie, Audrey Wilkinson, Donald Wilkinson, Hugh Williams, Rhys & Pixie Williams, Mary Willms, Peter Wilshaw, Erika Wilson, James G. Wilson, Kevin Wilson, Mel Wilson, Valerie Wilson, Ajo & Liz Wissink, Arthur Witham, James Wolford, Blair Wood, Whyman Wood, Stephen Woodley, Terrie Woodrow, Gary Woodworth, Frank Woolaver, John & Mary Wright.

Paul Yates, Anna Yeo, Barry Yoell, Layton Yorke, Dave & Joan Young, Marian Zinck.

The following individuals and organizations made financial contributions toward the operation of the Atlas project:

Christopher Adam, Marc Adam, Mary K. Akerland, Rev. Ford Alward, David Ambridge, Daryl Amirault, Stephen Archibald, L. Armstrong, Jean-Paul Arsenault, Peter J. Austin-Smith.

Barbara Bain, Robert Bancroft, Stephen Barbour, Harry Beach, Judith E. Beaton, Yvon Beaulieu, Rev. Tom Bellis, Nancy Blair, Blomidon Naturalists' Society, Paul Bogaard, Larry Bogan, Bowater Mersey Paper Co. Ltd., Thelma Bowers, Catherine Bradshaw, Calvin Brennan, Brian Brown, Marge Brown, Jennifer Brown & Gary Woodworth, Phyllis J. Bryson, Robert R. Buchanan, Sandy Burnett, Bruce Burns, Roger Burrows.

Roger Calkins, Ansel A. Campbell, David Campbell, Moira Campbell, Lloyd & Elsie Cannon, Eric R. Carr, John Cartwright, W. G. Caudle, Els Cawthorn, Hilaire Chiasson, C. H. Chipman, Robert Chivers, Barbara Christie, David Christie, Sheila Christie, Sandra E. Clarke, Alice Clements, Club d'Ornithologie de Moncton, Jean Cochrane, Karl & Donna Cole, Donald Cormier, Bob Cotsworth, Alan Covert, Kevin Craig, Ann Crocker, Rosemary Curley, Grant A. Curtis, Joan Czapalay.

Halton Dalzell, Marcel David, Con Desplanque, Vera & Paul DeWitt, Robert & Helen Dickie, Dorothy Doak, Dr. & Mrs. Fred Dobson, Pamela Doyle, Lewis & Elizabeth Dumont, Gordon Dunphy, Lucy Dyer.

Clem & Pat Egolf, Ann Eldridge, Dr. Jim Elliott, A. J. Erskine, Joseph Farquhar, Gwen J. Fichaud, David Flemming, Harriet Folkins, Bernard Forsythe, Harold L. Forsyth, Susan & Gary Foshay, Roger Foxall, Sylvia J. Fullerton.

Michael Gillis, Franklin C. Gilmore, Sherman Glazebrook, Maud Godfrey, Tommy Godfrey, Lorette Goquer, Irene Gorham, Arthur Grant, Elizabeth M. Grant, Tom Greathouse, Henry H. Green, Milton Gregg, Diane Griffin, Deborah Hearn, Dr. C. W. Helleiner, The Halifax Field Naturalists, Peter Hicklin, Barbara A. Hinds, Harold Hinds, James R. Hirtle, John W. Hollway, Peter Hope, Cathy & Weldon Horne, David Hussell & Erica Dunn.

Enid M. Inch, Mr. & Mrs. Robert Inkster, James L. Baillie Memorial Fund, Joan V. Jarvis, Brian Jenkins, Bruce Johnson, Cecil Johnston, Dr. H. Winston Johnston, Keith Keddy, Judith Kennedy, Fred & Margaret Kenney.

Charles T. Laugher, Fulton Lavender, Betty Learmouth, Dr. Mel & Ginny Lee, Charles S. Lennox, Barbara Lock, Tony Lock, Alice Lockhart, Dr. & Mrs. John Lockhart, Georges Long, George & Kaye Love, Viola Lovitt, Zoe Lucas.

Regina Maass, Clive MacDonald, Gerald MacDonald, R. B. MacDonald, Harold & Joyce MacEachern, Gerard MacIntyre, David MacKinnon, Walter E. MacKinnon, Angus MacLean, J. A. MacLeod, Kay MacLeod, David & Kathleen MacNearney, Jack MacNeil, John MacNeil, Don & Carol MacNeill, Chris Mackie, Alan Madden, Ray & Joan Mahabir, Mary Majka, Michel Màrtin, Pam Matthews, Donald McAlpine, Alison McArthur, J. Dan McAskill, Patricia McCleave, Robert & Wendy McDonald, Shannon & Laurie McGowan, Elsie McIntosh, John & Gwen McKenzie, Ruth McLagan, Ian A. McLaren, Douglas McNicol, Ruth Miller, James Milne, Lorna Mitchell, William Modeen Jr., Josephine Smith Monkman, Nancy More, Mrs. M. C. Morehouse, Jean Morine, Jean & William Morse, Roger Mosher, Thelma R. Murchison, Sandra Myers.

Natural History Society of Prince Edward Island, New Brunswick Federation of Naturalists, New Brunswick Museum, Judge John R. Nichols, Mrs. Carol Nicholson, G. & M. Nickerson, Doris E. Niles, Glen H. Niles, Nova Scotia Bird Society, Nova Scotia Bird Society Sanctuary & Scholarship Trust Fund.

John C. Oland, Maxine Oldham, Michael Olsen, Elizabeth Otter, Stephen & Sheila Palmer, Thomas S. Parsons, Warren Parsons, Paul Raymond Photography, Linda & Peter Payzant, P. A. Pearce, Bruce Pellerin, Ronald J. Pellerin, Lillian Perry, Dr. Stuart S. Peters, Warren & Nancy Peters, Mark Phinney, Arthur Porter, Roy Proctor, Lori Prosser, Lisa Proulx, Margaret Pugsley, Grace Pulley, Mr. & Mrs. D. W. Purchase, Derek Quann.

Ruth Richman & Gary Schneider, J. D. Rising, Peter Roberts, Ruth Rogers, Sally Ross, Rural Delivery Magazine, Les Rutherford, Saint John Naturalists Society, Bryan Scallion, Trevis Schriver, Fred W. Scott, Barbara L. Shaw, David Shutler, Scott C. Sinclair, J. Singleton & R. Whitney, David F. Smith, Mary E. Smith, Ruth E. N. Smith, Dr. Dusan Soudek, Esther M. Sporle, Cindy Staicer, John E. Steeves, Dr. Richard Stern, Mary Stirling, Rudy Stocek, Stora Forest Industries, John Stub, David Sweet.

James W. Taylor, Roger Taylor, The Themadel Foundation, The Wild Bird Company, Gaetane Theriault, Bill & Brenda Thexton, Georgette Thibodeau, Tony & Morina Thomas, Alison Thomson, Jean M. Timpa, Stuart Tingley, Elizabeth Townsend, Marion & Ralph Townsend, Gordon & Judy Tufts, Julie Turner.

Cheryl Uhlman, Walter Urban, Maureitus VanZutphen, Azor Vienneau, Richard Vinson, David & Winifred Wake, Mrs. Eleanor Waldron, Rob Walker, Timothy P. Walker, Lorne Weaver, Ron D. Weir, Gary & Dottie Welch, Dr. Dan Welsh, Marguerite Wheatley, Don & Alma White, Donald White, Louise White, Doug Whitman, Mary Williams, Rhys & Pixie Williams, Mary Willms, Peter Wilshaw, James G. Wilson, Mel Wilson, James Wolford, Terrie Woodrow, Frank Woolaver, John & Mary Wright, Dave & Joan Young, and many anonymous donors.

How to Read the Atlas Maps

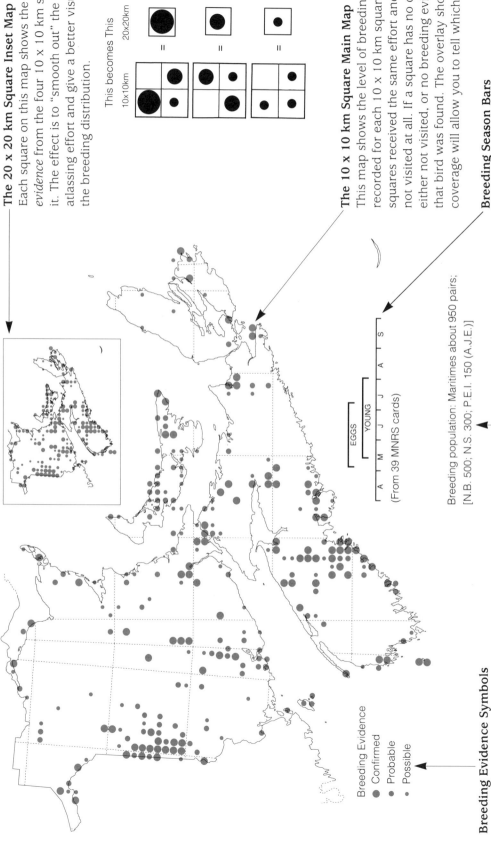

The 20 x 20 km Square Inset Map

Each square on this map shows the *best breeding evidence* from the four 10 x 10 km squares inside it. The effect is to "smooth out" the variation in atlassing effort and give a better visual picture of the breeding distribution.

This becomes This

10x10km   20x20km

The 10 x 10 km Square Main Map

This map shows the level of breeding evidence recorded for each 10 x 10 km square. Not all squares received the same effort and some were not visited at all. If a square has no dot, it was either not visited, or no breeding evidence for that bird was found. The overlay showing square coverage will allow you to tell which is the case.

Breeding Season Bars

These show the earliest and latest dates for eggs and unfledged young, based on cards sent into the Maritimes Nest Records Scheme. The number of cards for that species is given in parentheses below the bars.

EGGS

YOUNG

A M J J A S

(From 39 MNRS cards)

Breeding population: Maritimes about 950 pairs; [N.B. 500; N.S. 300; P.E.I. 150 (A.J.E.)]

Breeding Population Estimates

Where a range is given (for example, 25,000 ± 5000), it is derived statistically from abundance estimates supplied by atlassers. Otherwise it is from another source, indicated by initials (AJE = the author; CWS = Canadian Wildlife Service, etc.).

Breeding Evidence Symbols

The 10 x 10 km square boundaries are not shown but dots are always in the centres of the squares in which they occur.

Breeding Evidence
- Confirmed
- Probable
- Possible

I
Introduction

Bird atlases describe, with maps and text, where birds occur in an area, in one season or at all times of year. Early bird books outlined the distribution of each species in words, in general terms, or in varying detail depending on available information. Range maps appeared first for well-studied groups such as waterfowl (e.g., Phillips 1923–26). Only since 1950 has enough information become available to allow detailed distribution maps to be drawn for all species, even in well-studied areas. The maps in Godfrey's *The Birds of Canada* (1st ed., 1966) were the first published covering all species in many parts of this country. (The Golden field guide *Birds of North America*, also published in 1966, included range maps, but at a much smaller scale.) Maps based on available data usually showed the peripheral locations joined by a line, and breeding or other occurrence was indicated, implicitly but often incorrectly, to occur in all suitable areas within that boundary.

Sharrock (1976), in introducing *The Atlas of Breeding Birds in Britain and Ireland*, the first bird atlas of the computer age, reviewed experiments in the shift from simple outline maps to showing bird distribution found by sampling all units within a uniform grid. The British/Irish atlas provided the model for subsequent grid-based bird atlasses all over the globe. The Ontario (1987) and New York State (1988) bird atlas publications are noteworthy examples from North America. Both followed closely the British pattern.

A Maritime Provinces breeding bird atlas was suggested first in 1970. The Maritimes are the only provinces in Canada without extensive, remote, road-less hinterlands; so this was the one part of our country in which all areas of a province might be surveyed by volunteer effort using grid-based sampling. The time had not yet come for an atlas project here in 1970, and a subsequent proposal in 1981–82 also failed to gather momentum when the scale of organization and funding needed became apparent.

In 1984, an atlas project was again suggested as one way that volunteer bird observers in Nova Scotia might make a worthwhile contribution to knowledge, and thus to bird conservation. The Nova Scotia Museum supported the concept whole-heartedly. The Canadian Wildlife Service, Atlantic Region, was also supportive and urged the inclusion of all three Maritime Provinces. The organization needed for Nova Scotia alone could serve all three provinces with a relatively minor increase in scale, and the distributional picture provided by a Maritimes atlas would be more widely useful than one from Nova Scotia by itself. Funding for a co-ordinator was the first major hurdle, surmounted by start-up contributions from the federal Department of Supply and Services through an unsolicited proposal, with ongoing support from the Canadian Wildlife Service, Environment Canada, carrying much of the subsequent load. The Nova Scotia Museum provided office space and a multitude of associated services. Details of the organization and operation are outlined in Appendices A. and B.

The largest part of the overall effort, in the collecting of field records and in summarizing the data for future use, was donated by the 1,100 + volunteer observers and regional co-ordinators. Many of these people also made generous cash donations in the last two years to further the project.

We announced the project, did practice field-work, and hired our first co-ordinator in 1985, and the project ran, with only an expectable quota of snags, through 1986–90. As in most atlas projects, about half the total dataset was assembled in the last two years, following a slow start. Publication of the results was anticipated from the start, and serious planning for this book began in 1988. Most of the text, both for the chapters and the species accounts, was drafted by the summer of 1990, using preliminary data. The maps and tabulations of data needed for deriving distribution patterns, population estimates, and other "number-crunching" reached definitive form only after most 1990 data were incorporated in January 1991, when final revisions and editing began.

This book deals with bird distribution and numbers. Details of methods used are summarized in Appendices. They followed closely the patterns of other breeding bird atlasses, and we benefited greatly by the work done earlier by others, especially in the Ontario and British atlasses. The background information for understanding the distribution of birds in the Maritimes is assembled in the next two chapters: one is mostly geographical, setting out the factors that influence bird distribution and our detection of the birds; the other is historical, summarizing what we can infer about changes in bird distribution and numbers, mainly resulting from human actions, since Europeans first settled the Maritimes soon after 1600. The next chapter summarizes some general results of the Atlas project. The final introductory section gives a general exposition of the content of the species accounts and provides cautions on their use, with a discussion of the effects of varying coverage and varying detectability of species on the mapped data.

The 214 distribution maps and accompanying species accounts summarize most of the information arising from the Atlas project. They include, for all regularly breeding species, population estimates by province, a review of population changes over the period since the start of European settlement, and a forecast of each species' prospects for the future. The species accounts also summarize general distributional and natural history information on each species, from existing publications and from the Maritimes Nest Records Scheme. We highlight any distribution patterns evident for species that are not found everywhere in the Maritimes, and we note habitat use or nesting habits that are more characteristic of a species here than elsewhere.

Three further chapters make a start on using the Atlas data. One looks for geographic patterns that are shared by groups of species and correlates some of these with bio-physical aspects of the Maritimes that may underlie the patterns. The next explains how the population estimates came to be included in the Atlas and discusses some of their strengths and weaknesses. The last chapter looks at the distributional and population data, for individual species or groups of species, as bases for recommendations on environmental conservation. Many other uses will emerge; in fact, requests for use of the Atlas data began to arrive long before the end of data collection. We hope that the published Atlas will satisfy most needs for the data. Copies of the computerized data files will be deposited at the provincial museums in the Maritimes, and at the Canadian Wildlife Service, Atlantic Region, for consultation when more details are needed.

Abbreviations used in the text

The traditional abbreviations for provinces and states are used, not the two-letter codes introduced by the U.S. Postal Service.

BBS—Breeding Bird Survey, a co-operative project organized by the Canadian Wildlife Service and the U.S. Fish and Wildlife Service, for monitoring bird numbers through volunteer surveys.

COSEWIC—Committee on Status of Endangered Wildlife in Canada, a federal/provincial group responsible for designating species as endangered, threatened, etc.

CWS—Canadian Wildlife Service, Environment Canada, Atlantic Region headquarters in Sackville, N.B.

DDT—the well-known pesticide, used widely in forest spraying in New Brunswick 1952–67, and elsewhere.

DUC—Ducks Unlimited Canada, Maritimes office in Amherst, N.S.

MNRS—Maritimes Nest Records Scheme, files at CWS, Sackville, N.B.

II

The geographic background

The Maritime Provinces as an environment for breeding birds

Various Canadian geography books and atlasses have treated the Maritime Provinces (hereafter referred to as the Maritimes) mainly as an environment for people (e.g., D.S. Erskine, *in* Warkentin, 1968). The biological and physical factors that affect the distribution and activities of people, as mapped in other publications, also affect the distribution of birds. The economic activities of people also affect the surrounding areas and the other living things that dwell there, including birds. We must include several maps based on other studies to illustrate these factors and activities. These will be introduced in this chapter, in the context of factors that affect bird distribution.

The Maritimes (Figure 1) include a peninsula and nearby islands extending eastward from the North American continent into the Atlantic Ocean. They are geologically complex (see Roland, 1982), comprising the ancient remains of several successive continental margins which were fragmented when earlier "proto-Atlantic" oceans closed and re-opened. The island of Newfoundland and the Gaspé Peninsula are parts of the same physiographic region, which is distinct both from the still older continental rocks of the Canadian Shield to the northwest and from the more recent coastal plain to the southwest.

The ancient rocks of the Maritimes were worn down by the natural action of water, wind, sun, and ice over the ages, and the remaining land is low, attaining 750 m (2,500 ft) above present sea-level only at a few summits in north-central New Brunswick. The land was entirely covered by the Pleistocene ice-sheets, which melted from our region about 15,000 years ago, but the lowering of the sea level in that epoch exposed parts of the continental shelf seaward of the ice, including the area around Sable Island, N.S. Subsequent melting of the glaciers temporarily inundated at least one-third of the present land. The present map of the Maritimes represents the distribution of land and sea only over the last five thousand years, an instant in geological time, and only a brief moment even in the time that our present bird species have existed.

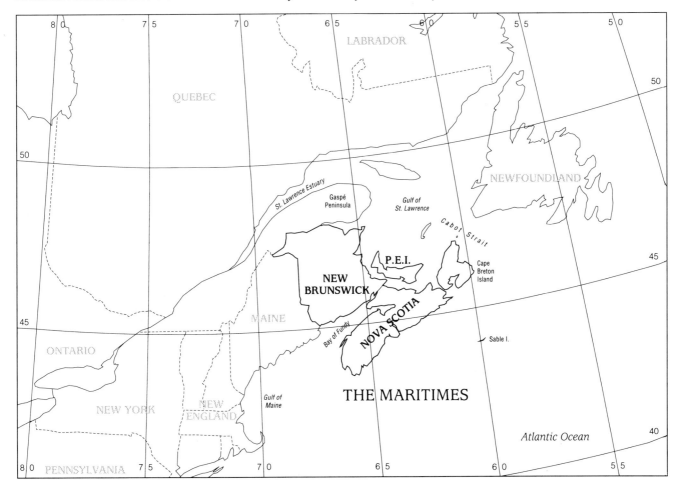

Figure 1. *Geographic location of the Maritime Provinces of Canada.*

In physiographic terms, the Maritimes include three easily recognizable regions (Figure 2):

- the uplands of the northwestern third of New Brunswick, which are parts of the eastern edge of the Appalachian mountain chain extending from Georgia to the Gaspé Peninsula;
- the lowlands of central and eastern New Brunswick, Prince Edward Island, and northern Nova Scotia, which are made up of much younger sedimentary rocks, interposed around its margins between harder rocks of former continental edges;
- the uplands of the Atlantic slope of Nova Scotia, which are made up of rocks of varying age and hardness ground down to a relatively level peneplain and covered with a thin layer of earth and gravel left by the glaciers.

The Bay of Fundy/Gulf of Maine to the southwest and the St. Lawrence estuary/Cabot Strait to the northeast are breaks of long-standing significance in the landmass and biogeography of the region.

Factors affecting distribution of breeding birds in the Maritimes

Our Atlas is about bird distribution—where the birds are breeding and where they aren't. This section outlines the geographic and biological framework within which birds breed in the Maritimes and discusses those factors that may influence where individual species or species-groups breed.

Distribution is related to abundance; the edge of the breeding range of a species is where its breeding density falls to zero. The region where average annual reproduction fails to replace average annual losses in the population is another aspect of the range limit. This last point helps to explain why, in all breeding categories, a few birds were found in some years in places where they had not succeeded in maintaining a breeding population, either before or subsequently. Data collection for the Atlas was not planned as an experiment to explore the reasons why the distribution limits of a species lie where they do. Our inferences about why all birds do not breed everywhere in the Maritimes are based largely on correlations of bird distribution with various environmental influences.

The distributions of all living things on this planet, including birds, depend on habitats that meet their basic needs of food and cover, the latter including both shelter from the elements and security from predators. Birds are high up in the food web, and they play relatively minor roles in the energy flows of most ecosystems. Like humans, they are consumers rather than

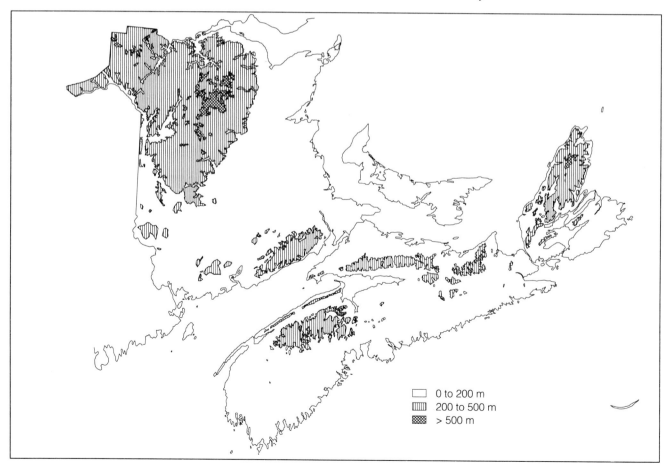

Figure 2. *The 200m and 500m elevation contours.*

4

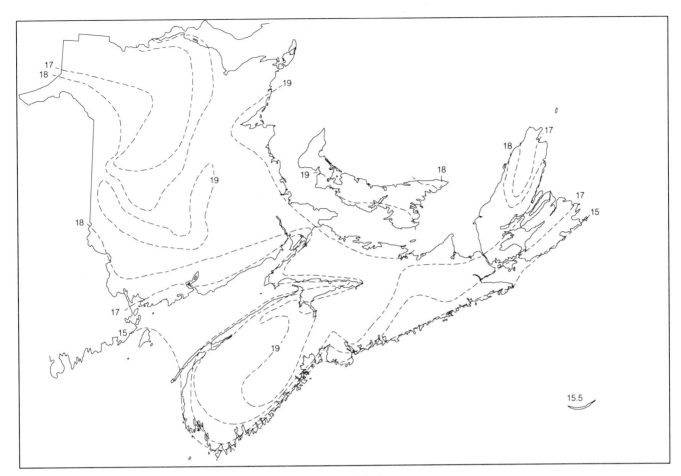

Figure 3. *Isotherms of mean July temperatures (°C) (after Gates 1975).*

producers, and thus are limited by the distribution and abundance of what they consume, unless other factors impose more pressing restrictions. This section reviews some of the possible influences in nature, and also some of those arising from human actions, that affect where birds occur in our region.

(a) Biophysical factors.

Bird distributions on land, at any season, are most easily correlated with vegetation types, and especially with extent and type of forest cover or lack of it. The distribution of plant communities depends on climate and also on soils, which ultimately are derived from bedrock under the influence of climate, with more limited effects of fire and animals. Freshwater birds depend for food and cover on plants and other animals growing in the waters they use, which in turn depend on dissolved nutrients derived from soils and bedrock adjoining both upstream and local parts of the drainage basin. Marine birds rely on the effects of currents and tides in concentrating nutrients and food organisms in the oceans. A concise discussion of factors influencing bird distribution involves ignoring most interactions of causative factors, about which many books have been written, and thus risks serious over-simplification.

Climate, expressed most familiarly in varying temperature and precipitation, influences the vegetation of an area, and through it the birds that breed there. The east-facing coasts of Canada, like the east coasts of Asia at similar latitudes, have cooler climates in summer than those of western Europe and western America. This relative coolness derives partly from cool oceans nearby, as the east-facing coasts of America and Asia are washed by south-flowing cold currents whereas the west coasts of these continents are influenced by currents flowing from the tropics. The Nova Scotian Current, moving alongshore from northeast to southwest, passes on the cooling effects of the ice-laden Labrador Current as well as carrying (in spring) ice originating in the Gulf of St. Lawrence. This cools most of Nova Scotia somewhat, but especially a rather narrow strip along its outer (Atlantic) coasts, extending also to extreme southwest New Brunswick (Figure 3) (see also Gates, 1975)

Tidal currents from the southwest, striking underwater reefs at the mouth of the Bay of Fundy, bring to the surface cold bottom waters, which keep that inlet and its coasts in both New Brunswick and Nova Scotia cool throughout the summer. At that season, the inland

5

portions of New Brunswick experience the heat and humidity characteristic of summers in the eastern interior of North America, from which the prevailing winds blow to the Maritimes. Contact between warm, humid air from inland regions and the cold waters alongshore produces fog, which extends the cooling effect inland in areas such as the New Brunswick-Nova Scotia border region around the head of the Bay of Fundy, where onshore winds are prevalent.

Proximity to the outer, cool coasts of the Maritimes often has more effect on distribution of living things than does latitude or elevation. North-to-south climatic zonation, of vegetation or birds, is poorly developed here. Southwestern Nova Scotia, though several degrees of latitude farther south, has few of the distinctive breeding birds of the eastern broad-leafed forest region which extends into interior New Brunswick—Great Crested Flycatcher or Scarlet Tanager, for example. Conversely, Blackpoll Warblers and Fox Sparrows, birds of the northern boreal forest biome, breed along and near the Atlantic coasts of Nova Scotia and around the mouth of the Bay of Fundy, but are not found elsewhere in the Maritimes except in the higher, cooler parts of northern New Brunswick and northern Cape Breton Island.

Other climatic influences that affect the forest zonation of the Maritimes, and with it the bird distribution, are inter-linked and correlated with temperatures.

- Winter temperatures here are too low for various southern plants and trees and for some bird species permanently resident in the warmer climes where these grow (e.g., Red-bellied Woodpecker, Tufted Titmouse).

- Often it is the more or less continuous snow cover in winter, which accompanies and results from lower temperatures, that limits bird distribution by limiting the opportunities for foraging on the ground. When artificial feeding is provided, southern bird species such as Wild Turkey or Northern Cardinal may survive our winters in those parts of the Maritimes with the mildest temperatures.

- Precipitation also varies within the Maritimes to some extent, especially where moist winds from the sea strike the steep slopes of the Cape Breton Highlands. The varying ratios of rainfall vs. evaporation, the latter depending on temperature during the growing season, influence the vegetation types to some extent. Inclusion of precipitation seems not to explain plant distribution and bird distribution here better than does temperature alone, so it was not mapped here. Temperature is correlated with duration of sunshine and thus with cloud cover, which is linked to precipitation.

- The Maritimes is one of the windier areas of Canada, but the effects of wind on biotic distribution are restricted to cooling where the prevailing winds blow from cool seas alongshore and warming where they blow from the warm interior of the continent. Thus, wind also acts on bird distribution through the medium of temperature.

The distinction between cool and warm areas seems to be the single most important climatic factor acting on bird distribution in the Maritimes.

The effects of bedrock, the soils evolved from it, and the climate combine to determine the *climax vegetation* of an area. Early outlines of biogeographic zonation (Merriam, 1894), based on fauna as well as flora, placed the Maritimes, along with much of southern Quebec and southern Ontario, in a Transition zone between the Canadian (i.e., boreal) zone to the north and the Upper Austral (i.e., temperate) zone to the southwest. Subsequent forest classifications (Halliday, 1937, modified by Rowe, 1959) recognized that the Maritimes forests in the Acadian Forest Region differed from those of the Great Lakes–St. Lawrence Forest Region in that the Maritimes' cooler climate gives our forests a more boreal nature than others at similar latitudes.

We recognized four major forest types within the Maritimes forests, by combining zones identified by Rowe (1959) and Loucks (1962) (Figure 4). There also is great variability within short distances, arising from variations in soils and bedrock geology and from variations in the cultural history of the area.

Most differences in the forest cover, and in the birds, are quantitative rather than qualitative. The forests of southern inland New Brunswick tended, before disturbance, to be more often broad-leafed, with higher proportions of maples, beech, yellow birch, white pine, and hemlock, as in the New England and Great Lakes regions. However, the more boreal forest types featuring spruces, balsam fir, white birch, and poplars are major components of the forest cover throughout, even in southern New Brunswick. The northern tree species predominate at high elevations in central and northern New Brunswick and along the outer coasts of Nova Scotia, but south-facing slopes and sheltered valleys everywhere include trees of the temperate forest. Mixed forests and inter-mixed stands of different types are frequent, even without the disturbing influences of recent cutting or fire. Many forest birds breed all over the Maritimes, as most of the mixed stands include their various preferred habitats. The Maritimes is often viewed as a single biotic region with many local variations towards the adjacent but distinct boreal and temperate forest regions.

Forests in the Maritimes were always affected by fire, although lightning, the only recurrent natural cause of fires, is infrequent here compared to more continental areas, and our forests are seldom dry enough to allow uncontrolled wildfires to cover large areas. There seems not to have been any general tradition in the Maritimes of aboriginal peoples starting wildfires to encourage grazing by wild herbivores or to clear land for agriculture, as native farming here was minimal. After European settlement in the Maritimes became general, human actions caused far more forest fires than did lightning, but during the past century, people also worked to an increasing degree to prevent forest fires from spreading. The effect of fire is mainly to return forest cover, temporarily, to earlier stages of successional growth. Repeated burning may remove seed-sources for some trees and other plant species. Also, burning often destroys the organic matter in surface layers of the soil, thus leading to impoverished plant cover in an area, which favours some plants and birds and places others at a disadvantage.

Outbreaks of insects, usually in larval stages, often lead to defoliation and sometimes to death of forest trees. Like fire, such outbreaks result in reversion of forests to earlier successional stages. Insect infestations, however, do not have fire's adverse effects on the ground cover and soil humus.

The concept of "climatic climax vegetation" is a useful aid to understanding plant and animal distribution, but it is an over-simplification. Very little area in a forested region would succeed to the climatic climax type even in the lasting absence of fire, insects, and logging. Some variations arise from the hills and valleys that make up the landscape (physiography). Both water and cool air drain from upland to lowland areas. Thus, the cooler, damper lowlands feature conifer forests, swamps, and wetlands, contrasting with the hardwood ridges, scrublands, and barrens on warmer, drier upland sites. Although wetland succession progresses gradually from open water through marsh and bog to lowland forest, this process is slow, requiring centuries or millenia, compared with the 100 years or less for upland forest succession after fire or cutting. Beaver dams on brooks drowned out most woody plants of the floodplain, but after the dams are abandoned these moist treeless areas support grass and forbs for some years, even decades. Dry upland sites with thin soils such as jack pine stands and blueberry barrens are susceptible to fires, which destroy humus, thus keeping vegetation sparse and perpetuating local habitat differences. Even in a region such as the Maritimes, where the climatic climax vegetation has been forest for several thousand years since the glacial epoch, there always has been a mosaic of woodland, shrubland, and temporarily open areas, interspersed with wetlands, and the birds were similarly varied on a local scale. Forest cover types thus influence bird distribution mostly

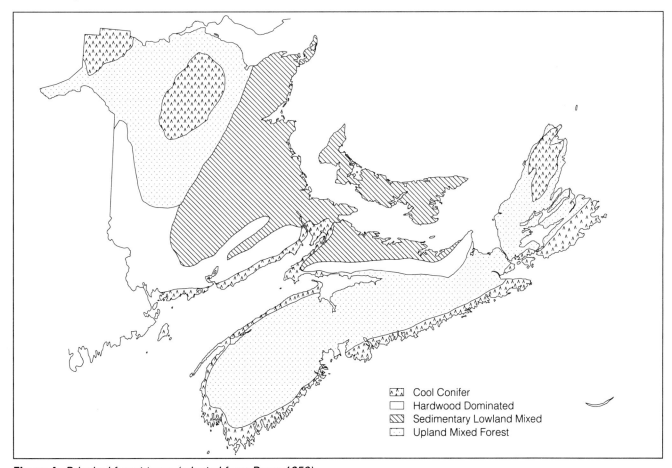

Figure 4. *Principal forest types (adapted from Rowe 1959).*

Legend:
- Cool Conifer
- Hardwood Dominated
- Sedimentary Lowland Mixed
- Upland Mixed Forest

on a rather local scale, except for the transitions to cooler and warmer forests near the edges of our region.

Bedrock underlies all life on Earth, but its influence on distribution tends to be indirect, acting through physiography, soils, water, or vegetation. The main geological distinctions within the Maritimes are between the hard-rock areas (Pre-Cambrian to Devonian, older rocks) of southern Nova Scotia and northwestern New Brunswick, and the sedimentary areas (Carboniferous to Triassic, more recent rocks) of Prince Edward Island and nearby parts of the other provinces.

The harder rocks underlie the highland regions of the Maritimes (see Figure 2), which have eroded more slowly than the sandstones and mudstones of the lowlands. The major river valleys and lake basins (Figure 5) show the erosion patterns even better than do the contours.

The hard-rock areas are more rugged, and their irregular terrain includes far more lakes, but lakes in these areas tend to be relatively infertile and unvegetated (oligotrophic). This aquatic habitat is characterized by Common Loons, with few other water birds. Lakes and marshes in the more level sedimentary areas, and particularly those where limestone or gypsum occur, are more fertile (eutrophic), and the main waterfowl and marsh bird areas are found there.

A related effect is that the coasts of areas underlain by sedimentary rocks have few outlying islands and reefs, whereas sand bars and eroding cliffs are frequent. These structural features affect the distribution of some seabirds. Common Eiders and our few primary seabirds breed mainly along hard-rock coasts featuring offshore islets, reefs, and cliffs, as they depend for food on resources of the marine ecosystem rather than on terrestrial fertility. Piping Plovers and Common Terns are found mainly along the flatter shores of sedimentary areas. A third type of Maritimes coastline, steep and straight with few inlets or offshore islands, as in the middle part of the Bay of Fundy and around northern Cape Breton Island, is little used by breeding seabirds of any kind.

The main agricultural areas, and the birds that frequent farmlands, are on the more fertile sedimentary areas in Prince Edward Island, eastern New Brunswick, and northern Nova Scotia, plus the Annapolis Valley, N.S. A possible exception is the potato country of the upper St. John and Restigouche valleys in New Brunswick, on Ordovician rocks. All agriculture, of course, is a result of human action, although ultimately it is restricted by soils, bedrock and climate.

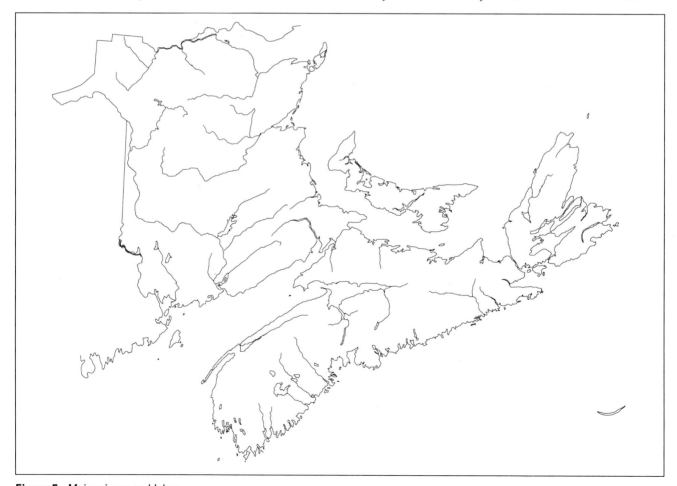

Figure 5. *Major rivers and lakes.*

(b) Cultural factors.

Cultural changes to the zonal habitats account for much of the Maritimes' present diversity of birds. Chapter III is devoted to an attempt to reconstruct the effects on the birds of these and other changes over the 400 years since the start of European settlement in the Maritimes. In this section we summarize the inferred changes in the region's habitats that human activities have made and that helped to determine the present distribution and numbers of the birds.

Settlement of the Maritimes, except for a brief period in the late 1700s, was undertaken for exploitation of natural resources, with the military presence thought necessary to secure the resources for one nation's use. *Fishing* was the first economic activity here, and was under way even before the beginnings of formal exploration searching for the riches of Cathay or El Dorado. Fishing villages grew up in coves along the outer coasts and islands, with a little minimal agriculture on adjacent rocky uplands. Most clearing of land was either a by-product of firewood-cutting or a deliberate removal of forest cover from which enemies, on four legs or two, might attack. The effect was a strip of open land along much of the Maritimes coastline, later kept open by wind and salt spray as well as grazing animals, providing greater opportunities for land birds of grasslands and edges. Early fishing activities affected marine bird distribution much more through disturbance and direct exploitation of colonial birds nesting on the islands from which many fishermen worked. Only in the last half-century has fishing reached levels at which it limits stocks of some fish species, and thus influences the seabirds that form parts of the same food webs.

Of the early fortified centres, only Port Royal near Annapolis Royal, N.S., (intermittently from 1605), and Fort Beausejour near Sackville, N.B., (constructed 1750), were in areas with appreciable farming potential, adjoining natural grasslands on the upper salt marsh. *Farming* by settlers from France, whether as freeholders or servants, was not actively encouraged, but was tolerated as it provided additional workers and helped to secure the land for national ends. The fur trade in the Maritimes never achieved the importance it had in Quebec and westward and it had little environmental impact except on the exploited species.

Settlement, and farming, became more general after 1760, although strategic/political considerations continued to be prominent for another half-century or more. The Maritimes remained, in government thinking, a site for hewing of wood and drawing of fish-nets, but many immigrants came to farm, more often in search of an independent way of life rather than in expectation of profits (Brookes 1972).

Clearing of forests for farming, by land-hungry settlers from Britain and briefly (1760–1783) from New England, continued into the early 1900s in some areas (Figure 6). Clearing for agriculture involved relatively small proportions, probably less than 10 per cent overall, of the forest area in New Brunswick and Nova Scotia, but in Prince Edward Island as much as 85 per cent of the land area may have been farmed at some time. Conversion of forests to farmlands involved a nearly complete change in the bird community of many local areas. The area under cultivation is now reduced everywhere, often by 50 per cent or more, from its maximum extent in the early part of this century.

Agricultural practices continue to affect bird use of the artificial open habitats created by settlement. Croplands, the most intensively worked type (except urban areas), are little used by nesting birds, and still less since toxic pesticides and herbicides came into general use after World War II. Mowing of hay fields affects their use by grassland species. Spring grass burning often ensures that reversion to woodland of abandoned farmland is deferred for many years after cultivation ceases.

The built-up parts of human settlements occupy even smaller areas than farms and other cleared lands, despite continuing urban sprawl since World War II (Figure 7). The juxtaposition of buildings ("artificial cliffs") with short grass, shrubbery, and shade trees nearby provides opportunities around human habitations for various birds that did not occur together in the wild before European settlement. Domestic animals such as cats and dogs pose threats to birds around most human settlements, but the buildings to house farm animals, and the waste grain and manure around farms with livestock, give greatly increased opportunities for birds able to exploit these situations.

More important in its effects on bird habitats than clearing of woodland for farming, because much more extensive, has been the *conversion of forest lands to earlier successional stages* by recurring use of the timber growing there. Firewood and house construction used forest trees from the start of settlement, and both of these uses continue to this day. The first commercial use of the forests was for ship-building; the Maritime Provinces were of world importance in marine shipping in the days of "wooden ships and iron men," from 1800 to 1870. Timber also was one cargo carried in ships leaving the Maritimes then. After iron began to replace wood for construction of larger ships in the 1870s, cutting of saw-logs took over as the main economic use of Maritimes forests. This in its turn was replaced in the 20th century by cutting for pulp and paper manufacture, which remains the major forest industry here (Black and Maxwell, 1972).

The effect of repeated cutting and occasional burning, with little serious replacement planting until the 1980s, has been to reduce much of the accessible forests of the Maritimes to second-growth woodland or scrub of limited commercial value. The use of such lands by birds has changed also, but the change in birdlife has been in species composition more than in overall density, as discussed in the next chapter. This has favoured birds of early successional stages over forest species, many of those benefiting being widespread generalists.

Alteration of marshes and other wetlands, by dyking, draining, or filling, to make use of the "reclaimed" lands for agriculture or settlement, continues to affect another group of birds in the Maritimes. Many "reclaimed" lands proved unusable for their intended purposes, as well as for the birds that formerly occupied them. Dyked grassland often remained too wet for mechanized agriculture until still further drainage under government subsidies became available in the 1970s. Some former wetlands were allowed to revert to marsh, or were re-flooded in the interests of waterfowl management. The latter is the only human influence on Maritimes wetlands that has generally benefited more wetland birds than it affected adversely.

Another, but less extensive, alteration of bird habitats for cultural purposes accompanied the *extraction of minerals from the ground*, the third major exploitative industry of the region (Brookes 1972). Many rather

small areas have been affected. The largest were those affected by strip-mining, for coal around Grand Lake, N.B., and gypsum around Windsor, N.S.; by mine tailings, especially in coal mining areas of eastern Cape Breton Island and northern Nova Scotia; and by settling ponds associated with metallurgical separation or smelting, mostly in northern New Brunswick. Much more numerous, and collectively more extensive, are excavations for sand, gravel, and fill, which occur in all regions. Most of these activities converted second-growth woodlands into largely unvegetated areas of little value to most birds, although pools of water in abandoned gravel pits sometimes prove attractive to shorebirds after vegetation has re-appeared there. Much of the excavation for rock, gravel, and other aggregates in recent years has been associated with road construction, which also involves the creation of earth and rock "cuttings" and embankments. Banks, whether in cuttings or around quarries or excavations, provide increased opportunities for birds that burrow in the earth for nesting, such as Belted Kingfishers and Bank Swallows.

Less immediately obvious, but even more pervasive, are the changes to bird habitats effected by *pollution*. Dumping of wastes has been a feature of all human set-

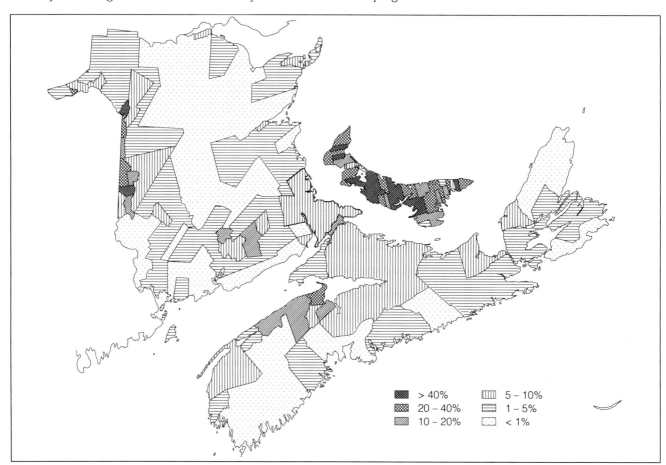

Figure 6. *"Improved" farmland as percentage of total land area, by census division (adapted from Statistics Canada, 1986 census of agriculture).*

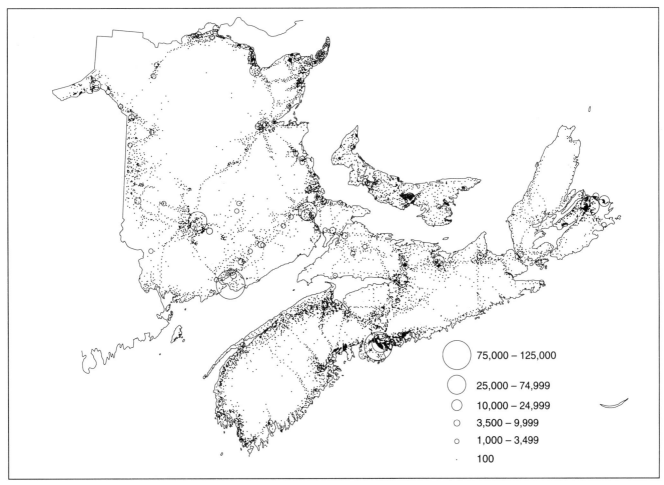

◯	75,000 – 125,000
◯	25,000 – 74,999
○	10,000 – 24,999
○	3,500 – 9,999
○	1,000 – 3,499
·	100

Figure 7. *Distribution of human population (after McCalla, R.J. 1988.* The Maritime Provinces Atlas, *Maritext Ltd., Halifax., Canada. 90 pp. By permission).*

tlements, from prehistoric kitchen-middens to the present landfills and sewers. Only in the 20th century has dumping of toxic wastes and other deliberate release of chemicals into the environment become recognized as a habitat problem. The application of persistent pesticides, such as DDT and related substances, affected the food chains of entire ecosystems. The pollution of waters by a vast array of substances, not all intrinsically toxic, has caused a variety of effects: siltation, by soil run-off after rain from agricultural lands and road construction; choking of ponds with vegetation through excessive farm effluents or sewage; poisoning of rivers by mine-wastes; or smothering of harbours by the accumulated debris from centuries of human settlement (Figure 7). Release of crude petroleum and bunker oils at sea makes a temporary but important change in the habitat of the birds that forage on or beneath the ocean's surface.

The habitat changes very briefly summarized above apply in varying degrees to different areas of the Maritimes. The next chapter discusses the effects on numbers or distribution of particular bird species that may be attributed to some of these changes.

III

The historical background: Changes in distribution and numbers of Maritimes birds

Our Atlas provides the baseline against which distributional changes of birds in the Maritimes will be measured for decades. It also provides the baseline on numbers, as no quantitative studies have extended widely enough to allow previous population estimates for most species. These new baselines start from the present time, 1986–90.

Earlier changes in bird populations of the Maritimes have been summarized several times, with varying thoroughness. The most extensive summary of earlier bird population changes covered only the period of detailed written records, 1878–1978 (Christie, 1979). But bird numbers here were changing long before people began recording changes, even before the Pleistocene glaciers covered the Maritimes—in fact ever since birds first evolved and occupied this region. Most changes were from natural causes, and thus part of the "normal" evolutionary process, up until European settlement of this region began. In this chapter, I summarize my own impressions of the status of birds in the Maritimes around the time of the first European settlements, roughly 1600 AD, so as to allow an outline of subsequent changes. The historic record of this period includes very few details on the status of birds before the late 1800s, and not a great deal more on changes in the vegetative cover of the area. Most of my perspective on former bird numbers is based on inference from imperfectly known environmental and cultural changes in other regions around the north Atlantic Ocean, with a few fragmentary references from the early writers. It should be taken as plausible rather than proven. [In the interests of readability, I prefer to let this warning suffice, rather than peppering the text with "probably," "presumably," or "must have been" in every other sentence of this book.]

The influences that affected bird numbers in the Maritimes since 1600 were of three types, all of which are still operating:
- direct and indirect impacts by people on the birds themselves;
- human alteration of the natural environments used by birds; and
- continuing natural changes, especially in climate, affecting both habitats and birds.

Direct and indirect impacts affecting bird numbers

People have used birds and their eggs and young for food whenever they were available, although few anywhere depended on birds for a major portion of their meat. Birds were important for food mainly in situations where the supply was large and predictable. The Faroe Islands (Europe) seabird economy and the James Bay (Ont.) spring goose hunt were among the rare cases where bird meat was of major continuing importance to indigenous peoples. Such use was seasonal, as storage of bird meat for subsequent use in winter or in other areas, by salting or drying, rarely emerged except where agriculture had made possible year-round settlements with permanent structures, usually at a later stage in the evolution of societies. This had not occurred in the Maritimes up to the time when Europeans first settled here. Uses of bird feathers for ritual, for decoration, or for insulation were minor in our area and were by-products of the search for food rather than primary reasons for killing birds.

Wild birds in the Maritimes in 1600 were important as human food only where their numbers were sufficiently large to allow predictable exploitation, i.e., at seabird colonies, migration assemblages of waterfowl, and roosts of Passenger Pigeons. The early European settlers in the Maritimes followed the example of the aboriginal inhabitants in exploiting all these food sources, and they also killed and ate any other birds large enough to be worth the effort needed to secure them.

Commercial hunting for birds within the Maritimes in the settlement era was limited by the small local markets. Seabird colonies, here and as far as the north shore of the Gulf of St. Lawrence, were decimated by commercial egging and shooting for the much larger New England markets, as described by Audubon and Nuttall in the 1830s. Bird hunting as "sport" emerged gradually as hunting for food became less necessary. Originally it was confined to the wealthy and those with time on their hands, e.g., soldiers whose garrisons were not under immediate threat of attack. Sport hunting perpetuated human killing of wild birds long after this had ceased to be an economic way of securing meat. Increased leisure time available for recreation and more disposable income led to a renewed increase in sport hunting pressure on game birds in easily accessible areas during the 20th century, and especially since World War II. The overall impacts on breeding bird populations of hunting activities have never been documented adequately anywhere in eastern North America (but see Phillips, 1923–26), and only some broad generalizations are possible here.

The effects of former exploitation of seabirds in the

Maritimes are intertwined with those of unplanned human disturbance of these birds, including predation by dogs, cats, and rats. All of the latter arose from the presence of Europeans in and around settlements on the islands where colonial seabirds bred along our coasts. It is futile to try to separate the effects of these adverse influences. Seabird breeding populations were greatly reduced from the levels prevailing in 1600, usually to far below present numbers, and often to extirpation. Among the pelagic seabirds, Leach's Storm-Petrels and Atlantic Puffins certainly bred, and Common Murres and Razorbills probably bred, in substantial numbers on many islands around the Maritimes; breeding birds of these species certainly numbered in thousands, and some in tens of thousands, around the Maritimes in 1600. All of these were nearly or completely extirpated by human activity before 1900. The minor peripheral colonies of Northern Gannets here were also lost. There is no conclusive evidence that Great Auks ever nested here, and their final extinction by 1844 did not come about from losses in the Maritimes. Recovery of pelagic seabird numbers here since 1900 has been slow and incomplete, although Leach's Storm-Petrels again breed in tens of thousands.

Inshore seabirds such as gulls and Common Eiders were hit as hard or harder, as access to their colonies was easier. Breeding numbers of these birds and of cormorants, all formerly in tens of thousands, were down to remnant levels by 1900. Terns, being smaller, were less attractive as food and were not exploited for decoration of hats until the late 1800s. They probably also benefited by reduction of gulls, but suffered from the introduction of dogs, cats, and rats to their nesting islands; the net effects may not have reduced their overall numbers substantially, as estimates suggested up to one million terns breeding on Sable Island around 1900. The vast increases in numbers of large gulls since these species were protected by treaty in 1916 have affected many other birds, and especially terns, whose numbers now are at all-time lows. Cormorants and their eggs seem never to have been taken regularly for human food except by the native peoples, owing to their fishy flavour; these birds were simply destroyed everywhere on the excuse that by eating fishes cormorants are harmful to human interests. They were reduced to a few relict colonies in the northeast by 1900 and have increased significantly only since World War II, but they have not yet regained their primaeval numbers.

Past effects of excessive exploitation of waterfowl in the Maritimes were also overlapped by concurrent human efforts to alter the wetland habitats used by these birds. This is not the place to try to tease out the relative contributions of these negative influences. The anecdotal reports from the past, when "the skies and marshes were black with ducks and geese," probably involved some imagination, but are too frequent to be dismissed as totally unfounded. Waterfowl numbers in 1600 were greater, probably very much greater, than at any time in the past century. Few species were lost for all time, but an unknown number may have been extirpated all over eastern North America.

The Labrador Duck disappeared completely during the settlement era, but apparently not as a result of hunting, and human actions in the Maritimes deserve no more than a small share of the blame for this extinction. The Bufflehead's status as a former breeding bird in the northeast is uncertain but probable; it was gone before 1900, when its overall numbers throughout eastern North America were much lower than at present. Several ducks were discovered as newly breeding in the northeast in the 1930s, most notably Ring-necked Ducks, with smaller numbers of Northern Pintails, Mallards, and later American Wigeons and Northern Shovelers. These may have been recent immigrants, as then supposed. Alternatively, the accessible northeastern populations of some or all of these species may have been decimated or extirpated by market hunting and other unregulated shooting prior to 1900, but had recovered to levels allowing re-settlement of the Maritimes by the 1930s. Wood Ducks were reduced to remnant numbers throughout eastern North America, but have recovered well, in the Maritimes and elsewhere. The apparently small former breeding populations of Red-necked and Horned Grebes were eliminated everywhere in the east by 1900, and they have not returned. The idea that hunting, at recent levels, can have any influence on numbers of game birds still meets resistance, notwithstanding all evidence.

Market hunting of migrant shorebirds up to 1916 decimated many of the larger species and brought Eskimo Curlews, Hudsonian Godwits, and Lesser Golden-Plovers close to extinction. Among our breeding birds in this group, only the Willet, always restricted in habitat to the seacoasts, was seriously reduced by hunting. The dyking of salt marshes also reduced its feeding areas, and the overall effect was a major reduction in numbers. Willets persisted as breeding birds in western Nova Scotia while they were extirpated elsewhere north of Virginia on the Atlantic coast. Their survival here was fortuitous rather than a result of planned conservation, and their recovery of former numbers and range may still be continuing.

Land birds exploited regularly for food were fewer and less concentrated, and we do not recognize adverse effects on most of them. The one obvious exception was the Passenger Pigeon, extinct in the wild by 1900, of which Maritimes hunters took a share while Maritimes lumbermen were eliminating the formerly more extensive mature hardwood forests on which this species depended. The pigeons were mostly gone from Nova Scotia and Prince Edward Island by 1850, but persisted in New Brunswick until the 1870s, not long before they became very scarce everywhere.

The direct effects of domestic animals on land birds seem likely to have been localized around settlements, and minor overall. Dogs and cats as predators can hardly have had more influence on numbers of wild birds, except in the newly created open areas around settlements, than had the more generally distributed native carnivores previously; the latter were greatly reduced during the settlement period by trapping and hunting. Other direct mortality factors introduced by human settlement, such as power lines, picture windows, and especially high-rise buildings, have proliferated enormously in the present century, but cannot have been important before 1900. Road kills, also a recent phenomenon, account for bird losses far greater than is usually recognized, as dead birds, unlike porcupines, are reduced to dust by subsequent traffic on major roads within hours or even minutes after the fatal impact (Stewart, 1971).

Impacts on birds through altered habitats

Changes in breeding bird habitats were even more sweeping, as well as rather better documented, than the effects of direct exploitation of birds. The early explorers found the Maritimes a forested land, and some visitors from other parts of Canada and the U.S.A. are surprised to find how much is still woodland, after 400 years of settlement and relatively unregulated logging.

The primaeval forests were rather different from those we see today. Mature closed forests, including large old trees, occupied much larger areas, notwithstanding the effects of wildfires, whose early occurrence here is preserved in place-names such as Brule and Burntland Brook. The forests then had higher proportions of broad-leafed trees, particularly of the "northern hardwoods"—sugar maple, beech, and yellow birch. The continued occurrence of forest plants characteristic of such habitats, and the lumber and former shipbuilding industries that depended on the wood of large trees, are evidence that the solid second-growth balsam fir and spruce forests were a feature mainly of the recent past.

Birds characteristic of mature, closed, mixed forests were more abundant in 1600 than at present, and those associated with more open, younger, and more coniferous habitats were less numerous when European settlement began. However, few forest bird species have disappeared from or appeared in the Maritimes as a result of forest changes over recent centuries. Relative numbers of different species were affected much more. Pileated Woodpecker and Barred Owl, which nest in

cavities in large, mature hardwood trees, presumably declined more than most smaller species. The birds of solid conifer stands, especially budworm-followers such as Bay-breasted and Cape May Warblers, increased, as did forest-edge species such as Chestnut-sided Warbler, which was considered a rare bird in Audubon's time, and of course the Robin. However, the somewhat cooler period, or "Little Ice Age," centred around 1600 may have discouraged some birds characteristic of more southern hardwood forests from breeding north into the Maritimes during the early settlement period. Whippoor-will, Red-headed Woodpecker, House Wren, Brown Thrasher, Warbling Vireo, Pine Warbler, and Indigo Bunting are species more characteristic farther south which may have been no more abundant here in the 17th and 18th centuries than now.

The early stages in forest succession, with shrubs, saplings, and open or patchy tree cover, occurred naturally after fire and wind-throw, but usually only for relatively brief periods on any site. Since 1600, these habitats, and the birds that use them, have proliferated and expanded enormously, and they have become permanent features of all edges around settled lands and along roadsides. Because of the temporary existence of such habitats in the past, the birds using them had to be adaptable generalists; these have increased more than any other group, some even colonizing the more vegetated parts of human settlements. Robin, Song Sparrow, Alder Flycatcher, and Common Yellowthroat are familiar examples.

Farmland, whether in intensive cultivation, hayland, pasture, or abandoned fields, is all new since 1600. Most forests in the Maritimes were cut over for firewood or timber, or burned, at some time in recent centuries. But only on Prince Edward Island did active farmland become the major habitat, replacing forests: 85 per cent of the Island's area was claimed to be arable land by the early 1900s. Dependence on wood, for building, fuel, and a multiplicity of other purposes, and a scarcity of good farmland in most areas, ensured that forest, though not virgin forest, remained the major habitat throughout New Brunswick and Nova Scotia. The greatest extent of cleared land was reached between 1880 and 1930 in different parts of the Maritimes, and the infiltration of birds of open habitats from the southwest continued even longer. The appearance and subsequent increase of Horned Larks, Grackles, Cowbirds, and, more recently, of Killdeers as breeding species in the present century exemplify this trend. Species indigenous to open lands here were few. Savannah and Sharp-tailed Sparrows were always present in saltmarsh grassland, and Bobolinks may have used floodplain meadows long before clearing for settlement began. Vesper Sparrows were originally scarce, only found in burns and sandy barrens, but they thrived in the early stages of land clearance. The scarcity of Eastern Meadowlarks and Upland Sandpipers, even since forest clearing, remains unexplained.

Towns and villages came later, and Canada still has no native species that are characteristically urban birds, although Purple Martin is restricted here to settled areas. Introduction of semi-domestic Rock Doves, for winter food, began before 1700, but was not well documented anywhere in Canada. House Sparrows were released in Nova Scotia in the 1850s, but became established widely only in the 1880s, when they spread in from introductions in the U.S.A. Starlings arrived here in 1915–25, also having spread from introductions in the eastern states; unlike our other introduced species which are permanent residents, some Starlings migrate from the Maritimes in winter, although more remain. House Finches are just starting to reach our urban areas, from introductions in the eastern U.S.A. in 1940. Buildings and other urban structures of human origin provide cavities and artificial cliffs, increasing availability of nest-sites which earlier had limited regional numbers of Barn and Cliff Swallows and Chimney Swifts.

Wetlands suffered more from human actions since 1600 than most other habitats, although only dumping and sewage were widespread influences on them before 1900. Dyking of saltmarsh for hayland and pasture began around the upper Bay of Fundy in the late 1600s. However, much of the dyked land was re-flooded with seawater every few years to keep up its fertility, so this habitat was not completely lost to its original birds until after World War I. Only the Sharp-tailed and Savannah Sparrows and Willet, with some ducks, were much affected as breeding birds by loss of saltmarsh. The drainage of fresh marshes, ponds, and lakes, in the name of agriculture, was done locally and on a small scale earlier, but became widespread only with mechanization, particularly after World War II. No wetland species is known to have disappeared as a result of marsh drainage, and it is impossible to sort out the relative influences of wetland destruction and of unrestricted hunting on waterfowl and marsh bird numbers in the years leading up to the Migratory Birds Convention of 1916. Dyking and draining of wetlands continued unabated until about 1960, with a slowing or even a slight reversal of that trend since then. Development of impounded marshes (impoundments), especially by Ducks Unlimited Canada, has allowed small numbers of various species of wetland birds not known here previously to become established on these artificial "prairie sloughs," as well as increasing opportunities for indigenous species; Redhead, Gadwall, and Moorhen are some recent immigrants.

Natural changes in the environment

Since 1600, the North Atlantic region, including the Maritimes, emerged from the Little Ice Age, a period of climatic cooling from the 1300s to the 1800s. Opinions differ as to whether subsequent warming ended around 1960 or has been overtaken by the effects of worldwide atmospheric pollution. Although concerns over "greenhouse effects" of CO_2 and other gases have emerged only recently, the Maritimes climate during much of the last 40 years seems to have been more generally warmer and drier than at any time since before the Little Ice Age. These climatic changes are all relatively small, 1–2°C warming being sufficient to advance the spring break-up and snow-melt by several weeks. Milder winters in particular may have helped the establishment of recent immigrant species that are permanent residents wherever they occur, Mockingbird and Cardinal being examples. Overall warming also reduced ice-cover along the coasts in spring, allowing earlier breeding by seabirds and perhaps also breeding by southern species. Recent breeding on coastal islands in the Maritimes by "southern herons," such as Black-crowned Night-Heron, Glossy Ibis, and perhaps Snowy Egret, may be results of the warming trend and of increasing populations not far to the south. Although long-term climatic changes would lead to alteration in the climax vegetation of the region, the changes in the past century are unlikely to have had important effects on our forests, certainly none distinguishable from the more sweeping changes caused by logging and clearing.

IV

What the Atlas project achieved

Raw statistics of the Atlas effort are summarized in Appendix E. We see there that the 1,120 atlassers spent 43,093 party-hours in assembling 144,642 records of 224 "species," among which 53,163 records carried abundance indices. Half of the records appeared on cards submitted by just 25 atlassers or teams. The picture conveyed by these numbers is, in the main, correct; a lot of people put in a lot of time, and they documented the status of all the species they could find across the Maritimes, and some people went many extra miles to ensure that the project attained its goals. Inevitably, these statistics tell only part of the truth. In this section we summarize briefly what was achieved.

Results

Most people who took part in the Atlas project want to know how their own findings compared to the rest. Basic kinds of information that were collected in the Atlas project were summarized for each square on the following maps, allowing comparisons with equivalent maps in the Ontario atlas:

party-hours (Fig. 8)
total number of species recorded (Fig. 9)
percentage of species with confirmed records (Fig. 10)
number of species with abundance indices (Fig. 11)

The patterns in the Maritimes were much more uneven than in Ontario, because our coverage was less complete. We did not use the same classes as in Ontario for grouping the data, because, with our less complete coverage, so many records would have been in the lowest class that some useful distinctions would have been concealed.

Where most species were found

The number of bird species detected is the first index to an area sought by modern birders, to whom the more species found the better the area. In the framework of the Atlas grid, however, more species detected may also arise from more time spent or more effective observers deployed in an area. The squares with the most species found (Table 1) are a rather surprising sub-sample of the Maritimes as the best places to look for birds!

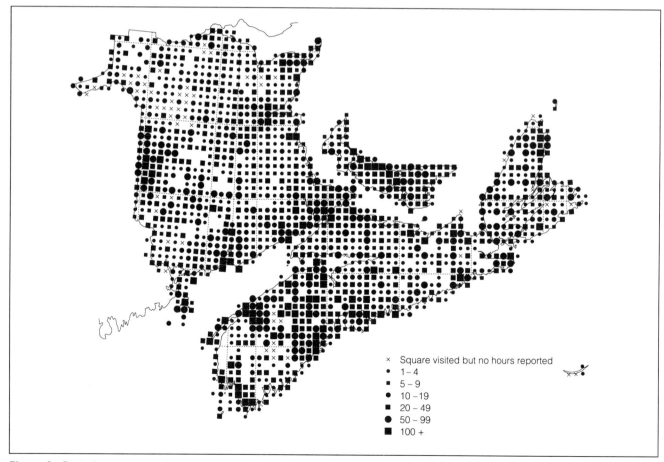

Figure 8. *Party hours reported per square.*

Table 1.

The 21 squares in which the most species were reported, with the party-hours registered in each.

Square no.	Name	Species detected	Species confirmed	Party-hours
20TNF75	Antigonish	134	89	153
19TFV75	Grand Manan	134	88	184
19TFA30	St. Stephen	131	76	43
20TMF07	Amherst	131	56	254
20TLH56	Tracadie	129	78	116
20TLF66	Riverside-Albert	126	76	226
19TFA60	Bonny River	124	61	40
19TFA44	Brockway	123	58	35
20TPF65	St. Peters	121	67	82
20TLF98	Sackville	121	72	261
19TFA42	Dumbarton	120	56	30
19TFA40	Oak Bay	119	60	34
19TGA08	Sheffield	119	51	105
20TMF38	Northport	118	92	113
19TFA20	Little Ridge	118	64	37
20TMF60	Urbania	117	87	264
19TGA25	Hampstead	117	82	54
20TLH10	CFB Chatham	116	63	107
20TLE48	Kingston	115	107	134
20TPF98	St. Andrew's Channel	115	70	10 (!)
19TFA86	Traceyville	115	54	61

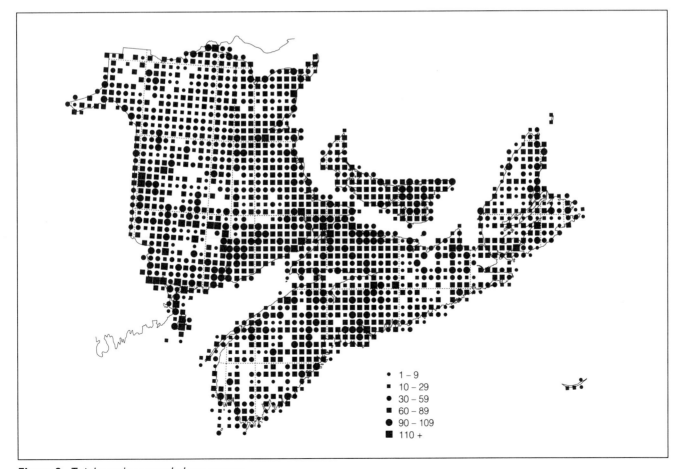

Figure 9. *Total species recorded per square.*

Owing to variation in effort (party-hours), no geographic pattern is immediately obvious, although more squares in southwest New Brunswick in blocks 19TFA and 19TGA were in the list than any other region. As the southwest New Brunswick squares listed all had less than 50 party-hours coverage, that is, they were worked by non-resident observers, these would have had even more species recorded if as much time had been spent there as in some other squares shown. The wide geographic scatter of squares with high total species suggests that continued efforts would have found more species in many other squares, even in many that were treated as complete.

The species found most widely

The Atlas may be looked on as a set of check-lists, one for each square. Some people will want to know which species appeared on most lists. The most frequently reported species (Table 2) were mostly birds that occurred everywhere and were abundant, but none had the same rank among the top species in each of the provincial lists. These widespread birds were all relatively common, but the habitats they frequented, and their relative ease of detection and of identification, also affected their rank.

Comparison of the number of squares in which a species was found (frequency of detection) with its population estimate suggests that familiarity of a species sometimes influences its detection as much as or more than its overall abundance. Species such as the Flicker, Crow, and Cedar Waxwing are conspicuous and easily recognized, but they were much less numerous than many species that were reported in fewer squares.

The "headline" records

Unique records included species confirmed as breeding in the Maritimes in the wild for the first time during the Atlas period:

Glossy Ibis (first breeding attempt in Canada)
Chukar—introduced (not mapped)
Wild Turkey—introduced (not mapped)
Solitary Sandpiper
Wilson's Phalarope
Blue-gray Gnatcatcher
Bohemian Waxwing
Pine Warbler
American Tree Sparrow
House Finch

However, successful nests were found only for two of these species. Most of these records were far from the

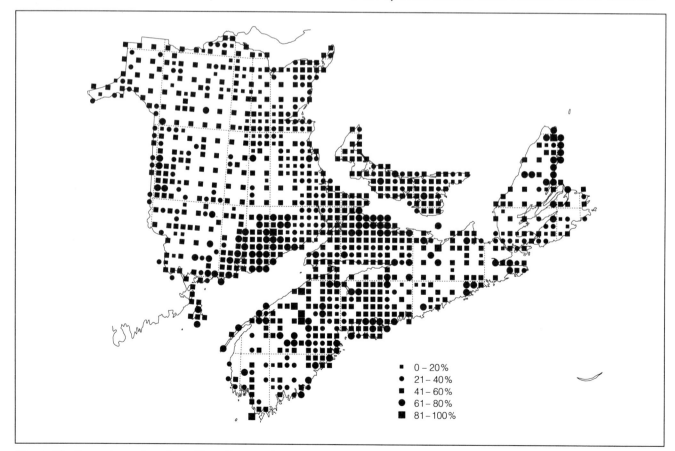

Figure 10. *Percentage of confirmed breeding species per square.*

Table 2.
The 25 species reported most frequently in the Maritimes as a whole and in each province separately, with the number of squares in which each was found. The sequence is for the Maritimes, and the species' rank in each province is also shown.

Species	Number of squares[1] with species found			
	Maritimes	**N.B. (rank)**	**N.S. (rank)**	**P.E.I. (rank)**
American Robin	1348	691 (1)	561 (1)	96 (2T[2])
White-throated Sparrow	1264	662 (2)	508 (5)	94 (5T)
American Redstart	1230	632 (3)	509 (4)	89 (14)
Common Yellowthroat	1212	616 (4)	500 (7)	96 (2T)
Song Sparrow	1194	563 (17)	534 (2)	97 (1)
Tree Swallow	1175	596 (6)	497 (9)	82 (24T)
Northern Flicker	1162	584 (10)	488 (11)	90 (10T)
Barn Swallow	1159	571 (13T)	498 (8)	90 (10T)
Dark-eyed Junco	1149	568 (15)	507 (6)	
Yellow-rumped Warbler	1140	571 (13T)	486 (12)	83 (21T)
Common Raven	1137	576 (12)	478 (13)	83 (21T)
Red-eyed Vireo	1132	586 (9)	459 (19)	87 (16)
Magnolia Warbler	1127	577 (11)	477 (14T)	
Black-capped Chickadee	1124	559 (19)	477 (14T)	88 (15)
American Crow	1115		511 (3)	95 (4)
Common Grackle	1111	562 (18)	456 (20)	93 (7T)
Cedar Waxwing	1107	601 (5)		85 (20)
Swainson's Thrush	1094	592 (7)		82 (24T)
Purple Finch	1091	590 (8)	437 (24)	
Hermit Thrush	1083	548 (22)	477 (14T)	
Ruby-crowned Kinglet	1067	530 (23)	468 (18)	
Blue Jay	1065	526 (24)	453 (21)	86 (17T)
American Goldfinch	1047		489 (10)	82 (24T)
Ovenbird	1035	558 (20)		
Northern Parula Warbler	1022	523 (25)		
Chipping Sparrow		565 (16)		86 (17T)
Rose-breasted Grosbeak		556 (21)		83 (21)
Black-and-white Warbler			473 (17)	
Yellow Warbler			449 (22)	94 (5T)
European Starling			442 (23)	90 (10T)
Golden-crowned Kinglet			436 (25)	
Savannah Sparrow				93 (7T)
Red-winged Blackbird				91 (9)
Bobolink				90 (10T)
Belted Kingfisher				86 (17T)
Rock Dove				82 (24T)
Bank Swallow				82 (24T)

[1] Total squares in the Maritimes 1682; 1539 squares had at least one record, but only 1208 squares had records of 30 or more species, hence the clustering of species near the latter level.
[2] T = Tied for rank shown

known breeding ranges of the species and were probably exceptional. Only Pine Warbler and House Finch were expectable extensions northeastward from their previously mapped ranges in Maine.

Species known to have bred previously but not confirmed as breeding, although recorded, during the Atlas period were:

Least Bittern
Redhead
Laughing Gull
Red-headed Woodpecker
House Wren
Loggerhead Shrike
Field Sparrow

Previously reported breeders not recorded at all in breeding season and habitat during the Atlas period were:

Horned Grebe
Red-necked Grebe
Northern Gannet
Bufflehead
Willow Ptarmigan (introduced)
Northern Bobwhite (introduced)
Passenger Pigeon (extinct)

Species accepted as at least possibly breeding but reported in only one square in the Maritimes were:

Snowy Egret
Glossy Ibis
Chukar (introduced)
Laughing Gull
Yellow-billed Cuckoo
Tufted Titmouse
Blue-gray Gnatcatcher
American Tree Sparrow

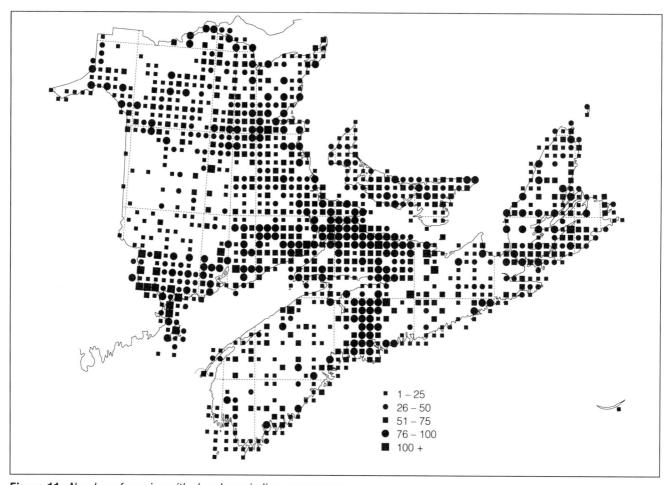

Figure 11. *Number of species with abundance indices per square.*

V

Species accounts: their structure and use, with comments on the Atlas data they present

General information

The accounts following, one for each species found in 1986–90 and accepted as breeding, summarize the distributional and population information gathered in the Maritimes Breeding Bird Atlas project. The presentation includes a map and a written account for each bird species determined to have bred in the Maritimes. Species detected during the Atlas, with only five or fewer records of probable or possible breeding, plus a few isolated records of confirmed breeding far outside the species' previously known range, were considered "peripheral"; these received shorter accounts and smaller maps, in a separate section following the breeding species. Species for which breeding had been confirmed in the past but which were not detected in 1986–90, species for which all submitted records were rejected as indicating non-breeding birds, and introduced game birds not yet clearly established here, were noted in a concluding list, without maps.

The maps follow the conventions now usual in breeding bird atlasses. Outline maps of the Maritimes—showing the external boundaries with Quebec and Maine, the grid of 100 x 100 km blocks, and the 60° W and 66° W zone lines—provide the setting on which the Atlas records are superimposed. In each 10 x 10 km square where a breeding species was detected, a symbol is placed, with different symbols representing possible, probable, or confirmed breeding (defined in Appendix B). The inset map presents the same data by 20 x 20 km groupings, or "20-squares," to help in visualizing overall ranges; see discussion of this presentation below. The page for each species (except peripheral species) also includes a line drawing of the bird, provincial population estimates, and a diagram summarizing the breeding season of the species, based on data in MNRS through 1989; dates were generally similar in all provinces, so all areas were combined in the spans of dates for eggs and for flightless young in the Maritimes.

The texts summarize general information on the species, with more detailed data arising from this project. Each account includes the worldwide range, based on the *A.O.U. Check-list* (1983) and Voous' *Atlas of European birds* (1960), and a general statement of the range within the Maritimes. Details of distribution may be seen on the maps, so are not treated in the text, except for species detected in fewer than 6 squares. Recent breeding ranges were based on Squires (1976) for New Brunswick, Tufts (1986) for Nova Scotia, Godfrey (1954) and Anon. (1988) for Prince Edward Island, and Adamus (1987) for Maine; maps from the forthcoming Quebec atlas of breeding birds (J. Gauthier and Y. Aubry, in prep.) were examined for certain species. These sources are to be understood as consulted, but were not cited explicitly in most species accounts. Details of habitats, nesting, or foods of the species were summarized in cases where these threw light on the present distribution pattern. The types of evidence most often leading to detection and confirmation of breeding (see Appendices B and E) were noted mostly when these were distinctive. As breeding of most water birds was confirmed by sighting of flightless young (FL), and of most small songbirds by sighting of either adults carrying food (AY) or of newly fledged young (FL), only departures from these general patterns were mentioned regularly, to minimize repetition. The inferred changes in numbers of the species in the Maritimes since the start of European settlement (cf. Chapter III) and the species' prospects for the future were also summarized. Trend data since 1966 were based on the Breeding Bird Survey (Robbins et al., 1986; Erskine, 1978, unpubl.; Erskine et al., 1990).

For additional information on a species, the numbers assigned to pertinent sources in the References are listed at the end of each account. General sources applying to many or most species are included in the Reference list but are not repeated for each species, e.g., Palmer (1962, 1976, 1988); Godfrey (1986); Tufts (1986); Squires (1976); Christie (1979); and Bent (Life Histories series). Many unpublished theses (Ph.D., M.Sc., B.Sc.), based on field studies of birds in the Maritimes, were included in the Reference list, as were specific published articles.

To save space, short forms (capitalized) of the "official" species names were used in many species accounts, wherever no ambiguity arose. For example, "the Catbird" in the account headed "Gray Catbird" obviously indicates this species, and in the account headed "Red-winged Blackbird" reference to "Redwing" means the blackbird and not a European thrush.

How well do the mapped data represent species ranges?

The main maps show the actual squares in which each species was found. The maps show many gaps, some being real, others meaning only that coverage was incomplete or lacking. Comparison with the maps showing party-hours, total species, and species confirmed per square (Figs. 8, 9, 10) will help the reader to recognize which gaps are artifacts of incomplete coverage.

The inset map for each species summarizes the records by 20 x 20 km unit, or "20-square," each such unit comprising one priority square (completed) and three other squares (varying numbers sampled and completed). This presentation uses the nearly complete coverage of priority squares to simulate a Maritimes atlas project with all units of a 20-square grid completed, allowing easier interpretation of the data for common and widely distributed species by largely ignoring the uneven coverage of secondary and tertiary squares. The "20-square maps" do not show all the detail available in areas where many or most secondary and tertiary squares had been worked. Also, by extrapolation of the distribution in the sampled squares across the entire "20-square," they may suggest, for scarce and unevenly distributed species (over one-third of all species), wider occurrence than really exists.

Some striking discontinuities on the main maps were produced by the uneven coverage. For example, the abrupt drop in frequency of detection of most species east of longitude 63° W in Nova Scotia was solely because Regions 13, 14, and 18 to the west were more completely surveyed than Regions 19 and 20 to the east. Such apparent "boundaries" are spurious, and must be allowed for in using the main maps. On the other hand, the straight NW-SE boundary in southeast New Brunswick on maps of open-country species, for example, Savannah Sparrow, is a real one, where farmland gives way to continuous forest.

Keep in mind that records mapped as "possible breeding" are just that, possible, and they must be interpreted in the context of other records for that species. If "possible" records were widely separated from other records, they probably represented wandering individuals rather than breeding birds. If they were in squares adjoining or interspersed among squares where breeding was confirmed, they were likely to represent breeding. A similar caution applies to isolated records of "probable breeding," some of which (P,T) involved sightings only, with no distinctive breeding behaviour reported; for example, because many ducks form pairs before migrating north, "P" records for ducks during May often were no more "probable" to be breeding locally than records of lone drakes mapped as "H."

Coverage

Although Atlas field work was deliberately dispersed across the 1682 sampling units or squares in the Maritimes, the sampling effort by volunteers was not uniform, and species were seldom if ever detected in direct proportion to their occurrence and numbers.

Here we discuss the varying detection of different species in various areas.

As outlined in Appendix B, the field effort was directed first to a uniformly spaced 25 per cent sample, totalling 416 priority squares, to which were added 34 special squares known to contain unique habitats or rare species. Complete coverage of all 450 priority and special squares was the only explicitly stated coverage goal that applied across all regions, and this was essentially achieved. Only one of these squares was not visited during the Atlas period; that square was in the target area of the CFB Gagetown artillery range. About 96 per cent (433) received complete coverage, and 439 had useful species lists. Varying proportions of the secondary and tertiary squares were sampled in different regions; overall, about one-quarter of the other squares were completed, and about one-third had useful species lists. Some regions had all or nearly all squares completed, whereas 12 of the 23 regions had less than 40 per cent of their squares at that level.

The instructions for Atlas field work were drawn up on the assumption that observers everywhere would be striving for complete coverage of a square. Observers were urged to visit all habitats in a square, as they usually could not visit all areas. The instructions set out criteria for assessing completeness of coverage, in terms of numbers of species detected and confirmed as breeding relative to the number predicted to occur there. The predictions varied widely, and some were much too high; many were reduced after two or three years, but some unrealistic high figures persisted until the end of the project. Some very effective observers completed a square in 5 party-hours on a single visit, but some others spent 50 or more party-hours each year in their square without completing it in five years. Most squares for which completion was seriously attempted received 10–30 hours field work, as they were worked, usually during a few visits, by observers who lived elsewhere.

Rapid "square-bashing" was often the only way to ensure at least minimal coverage of squares remote from observers' homes, but it placed emphasis on meeting arbitrary completion criteria more than on ensuring that all habitats in a square were sampled. The "square-bashers" were experienced observers, who were unlikely to miss scarce species if the appropriate habitats were readily accessible and visited. The chances of missing less common birds were greater with only a single visit or a few visits than they would have been with repeated coverage. People who did a "fine-tooth comb" job on their squares over several years placed more emphasis on sampling all areas and less on merely meeting coverage criteria; they were less likely to miss species through not having visited an appropriate habitat, but some less expert observers may not have detected or recognized all inconspicuous or "difficult" species, even after prolonged coverage. In general, the squares in which more time was spent had more species detected, but "diminishing returns" applied everywhere, and the better observers usually ended work in a square after 20–40 party-hours. Many squares, especially secondary or tertiary squares, were visited only once, for 1–5 hours, without expectation of completing the coverage; others showed only incidental records; 143 squares (8 per cent of the total) had no data at all.

We thus did not achieve ideal coverage for all species. In several regions, few of even the priority squares received much more than 30 hours field work. The varying amount of time spent in different squares affected the mapped distribution of each species, but not all species were influenced to the same extent.

Varying detectability of species

No species was detected in every square where it bred. Some species are intrinsically more conspicuous than others, e.g.:

More obvious birds	Less obvious birds
large, colourful, noisy	small, dull, secretive
flocking, colonial-nesting	solitary-nesting
in open habitats	in forests
using dry land areas	using marshes or islands
active by day	active by night
active all summer	active for shorter period
present every year	irregularly present

Common species, other things being equal, were missed less often than scarce species. Although Bay-breasted Warblers are more numerous in the Maritimes than are Bobolinks, they were not detected in as many squares during Atlas field work. Ease of detection also affected the apparent distribution of some species more than others.

The following widely distributed species were expected (not always correctly—see italicized notes) to be seriously under-represented in the Atlas data, most often because of the listed difficulties in their detection:

Leach's Storm-Petrel	nocturnal, burrowing, on offshore islands
Sharp-shinned Hawk	elusive, in forests
Northern Goshawk	elusive, in forests
American Woodcock	nocturnal, in woodlands
Black-billed Cuckoo	irruptive (no major incursion during Atlas period)
all owls	nocturnal, most in forests
Whip-poor-will	nocturnal, in forests
Yellow-bellied Flycatcher	call easily confused with Least Flycatcher
Brown Creeper	inconspicuous, weak song, in forests
Golden-crowned Kinglet	inconspicuous, in forests *(but found widely nonetheless)*
Warbling Vireo	song sometimes confused with Purple Finch
Philadelphia Vireo	song easily confused with Red-eyed Vireo
Cape May Warbler	inconspicuous, weak song, in forests
Blackburnian Warbler	weak song, in forests *(but found widely nonetheless)*
Palm Warbler	habitat (bogs) not always visited
Bay-breasted Warbler	inconspicuous, in forests
Sharp-tailed Sparrow	scarce habitat (saltmarsh)
Red Crossbill	irruptive, breeding outside usual spring/summer period (no incursion during Atlas period)

A species missing from the list above can not be assumed to have been detected in proportion to its numbers. It merely was less likely to have been seriously under-represented. An example is the White-winged Crossbill, of which an irruption flooded the central Maritimes in 1988, although the species was scarce or absent in the other years of the Atlas period. Species with limited distribution in the Maritimes may have been unfamiliar to some observers, and thus more often missed even where they did occur.

In using any species map, one should compare it first with the overlay map of coverage/effort, so as to distinguish where a patchy distribution only mirrored incomplete coverage, instead of arising from difficult detection or genuine scarcity.

There is no way to be sure that a species was represented in proportion to its numbers and distribution, unless we knew the real population size and breeding range. However, if we knew those already, there would be little point in undertaking an atlas project. We can only compare (subjectively) the representation of one species with those of other species. A species that was represented as well as or better than others with apparently similar detectability was likely to be represented acceptably. Conversely, a species less well represented than most others of similar detectability probably would have benefited by more intensive coverage. In some cases, our maps are not at all helpful in portraying the breeding range of a species, e.g., Long-eared Owl. Nevertheless, judgments of representativeness of data for a species are more or less subjective, and they are all comparative rather than absolute assessments.

Common Loon   Huart à collier

Gavia immer

The Common Loon breeds across the boreal forests of North America and into the tundra on Baffin Island, Greenland, and Iceland. It nests on the ground by shores of lakes larger than 40 ha in area, mainly where plant growth is sparse. All known breeding in the Maritimes is on fresh water, but loons, mostly subadults, regularly summer and moult along the coasts in all three provinces. Loons bred recently on some artificial marshes impounded for waterfowl in the Maritimes; this seems not to have been reported elsewhere and reflects adequate fish prey in such sites. Loons were less frequent in areas underlain by sedimentary rocks (eastern New Brunswick, northern Nova Scotia, Prince Edward Island), where suitable lakes are scarce or absent; most use of impoundments was in sedimentary areas. Loons are large, conspicuous, noisy birds, which were detected easily. Many reports of probable breeding (P,T) and most of possible breeding (H) were likely of non-breeding birds. Most confirmed breeding reported to the Atlas involved sightings of flightless young (FL, in 185 squares), with nests report-ed in only 9 per cent (47) of 506 squares with Common Loons.

Probably loons occupy most of their historic range within the Maritimes, except where lake shorelines have been too densely settled. Most lakes with loons contain only one or two pairs. Regular use of motorboats in summer does not always deter breeding loons, if the birds are not harassed. Although some are shot wantonly, more are prevented from breeding successfully through harassment or thoughtless visitation. Their numbers are certainly below primaeval levels, we estimate by one-third to one-half, but the species is still common and widespread in our region. Early writers stated that loons bred in Prince Edward Island, but there have been no acceptable records since 1900, and few freshwater lakes of suitable size exist there. The species is vulnerable to disturbance, and even more to water-level fluctuation, but the BBS indicated a significant increase here in recent decades. The effects of acid rain on fish production in lakes and rivers of southwest Nova Scotia, but not as yet in southwest New Brunswick, pose a threat if recent trends in acidification continue.

Refs. 76, 142

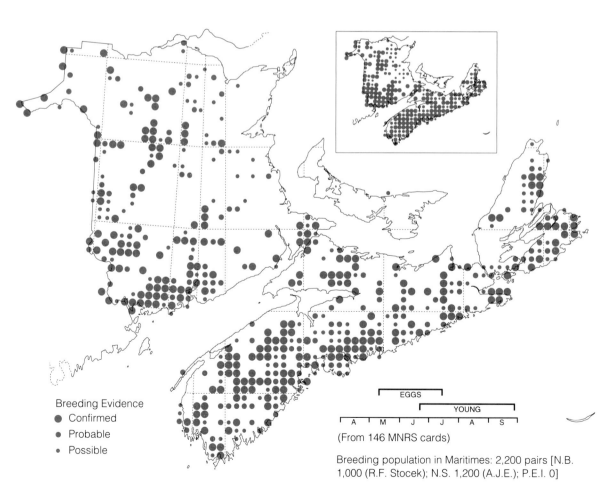

Breeding Evidence
- Confirmed
- Probable
- Possible

EGGS
YOUNG

A M J J A S

(From 146 MNRS cards)

Breeding population in Maritimes: 2,200 pairs [N.B. 1,000 (R.F. Stocek); N.S. 1,200 (A.J.E.); P.E.I. 0]

Pied-billed Grebe
Grèbe à bec bigarré

Podilymbus podiceps

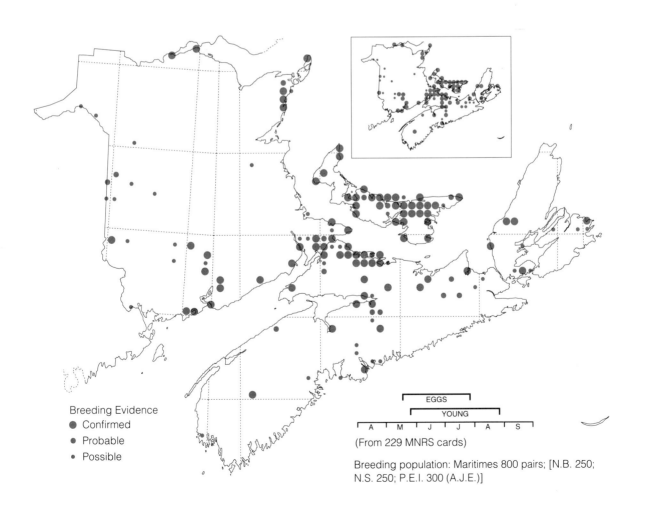

This small grebe breeds in North and South America, from southern Canada to Argentina and Chile. It is the only grebe breeding widely in eastern Canada. In the Maritimes, as elsewhere, its range was discontinuous. Pied-billed Grebes were generally restricted to the more fertile wetlands, especially in the New Brunswick–Nova Scotia border region and in eastern Prince Edward Island, with scattered records in other locations on flood-plains or underlain by sedimentary rocks. Many records from recently impounded marshes may reflect use of the first flush, after impoundment, of invertebrate life and the small fish that feed on it, and use of such areas may decline later. Breeding densities at the Amherst Point Migratory Bird Sanctuary, N.S., in 1982-83 were among the highest recorded anywhere for this species. The calls of Pied-billed Grebes in spring and summer are loud and far-carrying, so the birds should have been easily detected where present. Their absence from some suitable farm ponds and sewage lagoons, e.g., in Kings Co., N.S., may be a result of the large numbers of gulls in that area. Fledged young (FL) were by far the most frequent breeding evidence obtained (in 48 per cent of 141 squares with grebes), with nests reported in 11 squares.

Some marshes that may have harboured this species were lost to drainage, and others to urbanization, in the past. Pied-billed Grebes were shot during the subsistence period, when any bird large enough to eat was hunted. However, they were probably never so abundant or desirable as to have been sought-after deliberately by hunters. Losses to drainage and hunting in the past have been made up by the creation of new marshes in the past 25 years, and Pied-billed Grebes are now as common as or more common than in any former time. Efforts directed to marsh conservation for waterfowl will continue to benefit grebes as well. The impact of acid rain here will be less on the fertile and well-buffered marshes used by this species than in other wetlands in the Maritimes. The future for these birds in our area looks reasonably secure.

Ref. 48

Breeding Evidence
● Confirmed
● Probable
● Possible

EGGS

YOUNG

A M J J A S

(From 229 MNRS cards)

Breeding population: Maritimes 800 pairs; [N.B. 250; N.S. 250; P.E.I. 300 (A.J.E.)]

25

Leach's Storm-Petrel
Pétrel cul-blanc

Oceanodroma leucorhoa

"Look after the ... birds called Petrels; they are very peculiar in their way of life ... " (G. White, 1788). Our only member of the storm-petrel family is peculiar in many ways, and few atlassers have ever seen one, but it is quite numerous here. Leach's Storm-Petrels breed locally on offshore islands in the northern oceans, the Maritimes population being a southern extension from its stronghold in Newfoundland waters. These birds breed on islands with no terrestrial predators, as they are vulnerable while inside and when entering and leaving their underground burrows. Storm-petrels feed far out on the oceans, even in breeding season, when each bird incubates for up to five days while the other forages. They return to land only at night, to avoid attacks by gulls and other predators, so their detection requires overnight visits to the islands, when their calling is conspicuous. Undoubtedly they breed on more islands than the few visited during the Atlas, but they were confirmed as breeding in 24 of the 27 squares

where found, usually with nests (NY,NE,ON). Although no early estimates of storm-petrel numbers existed, the dogs, cats and rats that accompanied human settlement of our offshore islands wreaked havoc among the nesting birds here, and some colonies were extirpated. With abandonment of most human settlements on our outer islands, the introduced predators also disappeared, and the storm-petrels have increased. We have no estimates of total numbers. In New Brunswick, there are guesses of 1,000 pairs on Kent Island and 100 pairs on Machias Seal Island; in Nova Scotia, extrapolations from sample plots suggested at least 50,000 pairs on Bon Portage Island and 2,000 pairs on Pearl Island (in 1975; since reduced by gull predation). All known colonies are around the cold-water coasts of the Maritimes, and it is unlikely that the species breeds in the southern Gulf of St. Lawrence. If the "greenhouse effect" eventually warms the seas around the Maritimes, this may limit the numbers of breeding storm-petrels here, but only a large change could be detected.

Refs. 61, 72, 73, 81, 88, 157

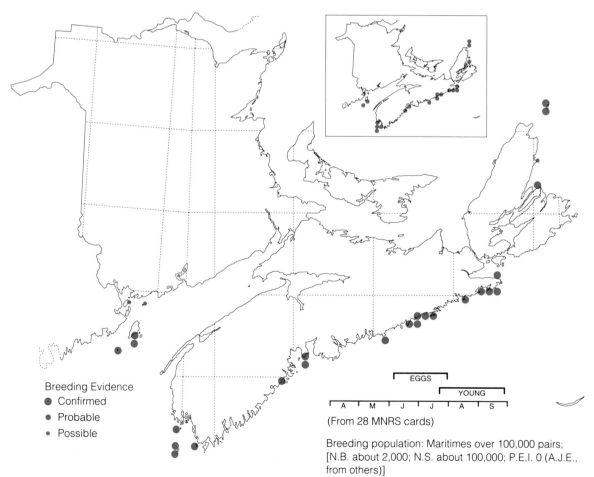

Breeding Evidence
- ● Confirmed
- ● Probable
- ● Possible

EGGS
YOUNG

| A | M | J | J | A | S |

(From 28 MNRS cards)

Breeding population: Maritimes over 100,000 pairs; [N.B. about 2,000; N.S. about 100,000; P.E.I. 0 (A.J.E., from others)]

Great Cormorant
Grand Cormoran

Phalacrocorax carbo

This large cormorant is widespread in temperate and tropical regions. It breeds from New Zealand and Australia to South Africa, and from China to Norway and the British Isles. In North America it breeds only around the Gulf of St. Lawrence and in eastern Nova Scotia; it may be the only bird species of which most (70 per cent) of the American population breeds in the Maritimes. Although Great Cormorants often breed inland in Eurasia, they are strictly coastal breeders here, nesting on sea-cliffs or low rocky islands by seas with plentiful stocks of fishes. Confirmed breeding records in the Atlas were from known colonies, in 8 squares in Prince Edward Island and 29 squares in Nova Scotia. Sightings (H,T) of these birds where no other evidence existed of breeding, including a few inland sites, were considered to involve non-breeders. A few Great Cormorants may breed on islands off southwestern New Brunswick, probably amidst colonies of Double-crested Cormorants, as with more southern records of this species in Nova Scotia. Breeding of Great Cormorants in New Brunswick

occurred in the past, and they were found breeding in Maine in the 1980s after an absence of 100 years.

Great Cormorants were more general in the northeast before Europeans settled here and declared war on rival fish-eating species. By the mid-1800s this bird was gone from New England, and by 1900 it was thought extirpated from North America, including the Maritimes. However, a few colonies persisted, mainly on Anticosti Island, Que., and gradually more sites were re-occupied, or recognized as including this species, confusion with Double-crested Cormorants having muddled the picture throughout. Cormorants were not protected by the Migratory Birds Convention of 1916, and persecution of these birds persists to this day, even though the fishes they eat are of little, if any, commercial value. They have increased gradually, as persecution was not continuous, and successful reproduction in one year out of three can maintain their numbers. Given the prejudice against black birds of all kinds, and increasing human exploitation of fish resources, further increases of cormorant numbers are unlikely, but the species is not in any jeopardy at present.

Refs. 36, 68, 86, 105, 130

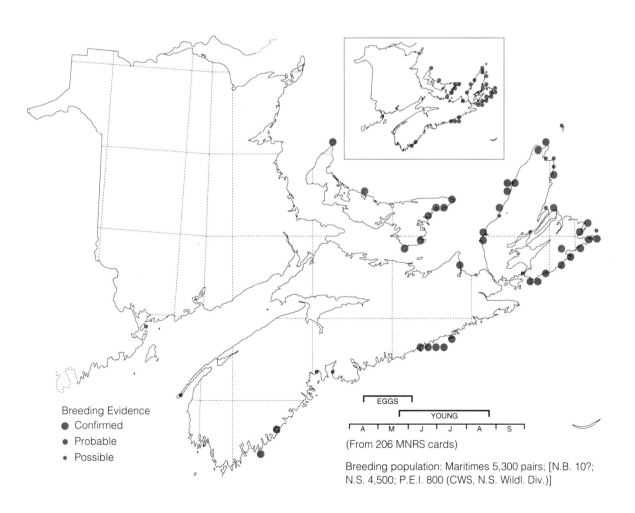

Breeding Evidence
● Confirmed
● Probable
● Possible

EGGS

YOUNG

A M J J A S

(From 206 MNRS cards)

Breeding population: Maritimes 5,300 pairs; [N.B. 10?; N.S. 4,500; P.E.I. 800 (CWS, N.S. Wildl. Div.)]

Double-crested Cormorant
Cormoran à aigrettes

Phalacrocorax auritus

This familiar species breeds locally across
Canada and the U.S.A. It is the only large cormorant in America that breeds by both fresh
and salt waters. In the Maritimes, it was mainly a
coastal breeder, but some inland breeding was noted in
southwest Nova Scotia during the Atlas, with a few
reports earlier, near McAdam and Grand Lake in New
Brunswick. The nests are on sea-cliffs or in trees, but,
on islands where all trees have been killed by the birds'
droppings, cormorants often breed on the ground after
the trees have fallen. A colony on old wharf pilings in
Pictou Harbour, N.S., about 50 m from a main highway,
is familiar to many people, but the largest colonies
(5,000 nests at Ram Island, P.E.I., 2,100 nests at
Manawagonish Island, N.B.) can only be viewed from
boats. Nest-sites that are safe from terrestrial predators
and within commuting range of an adequate food-sup-
ply, in this case fish, are the main factors limiting
breeding distribution of Double-crested Cormorants in
the Maritimes. Nests/colonies (NY,ON), noted in 93 of
189 squares with the species, were the usual evidence
of breeding. These cormorants often frequent
lakes and rivers in spring, and adults with
newly flying young also move to inland
waters in mid-summer. Many sightings
(H,T) inland were probably of non-breeders.

Cormorants, like other fish-eating
birds, were long persecuted as predators
competing with humans. This prejudice persists despite
many studies showing that cormorants eat few fish of
interest to people. By 1900, they had been reduced to
remnant numbers in the Maritimes, but they have
increased steadily since then, even though not protected under the Migratory Birds Convention of 1916.
Recent surveys, by provincial and federal officials, indicated at least 25 colonies in New Brunswick, 80 in Nova
Scotia, 10 in Prince Edward Island. They may still be
increasing, but pressure from fishing interests to reduce
this alleged "competition" also has increased in the last
decade. Shooting of cormorants, whether illegal or
authorized, is frequent, and further increases in cormorant numbers in the Maritimes appear unlikely. At
present population levels, such persecution poses no
threat to the species' continued presence here.

Refs. 68, 86, 105, 130

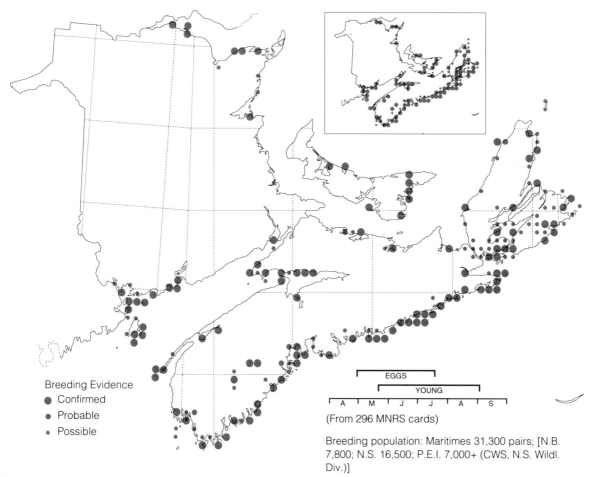

Breeding Evidence
- Confirmed
- Probable
- Possible

EGGS
YOUNG

A M J J A S

(From 296 MNRS cards)

Breeding population: Maritimes 31,300 pairs; [N.B.
7,800; N.S. 16,500; P.E.I. 7,000+ (CWS, N.S. Wildl.
Div.)]

American Bittern
Butor d'Amérique

Botaurus lentiginosus

This bittern breeds across Canada in boreal and temperate regions, and south to California, Texas, and Maryland; similar but larger species replace it in other continents. In the Maritimes it occurred widely in most regions, but was scarce in the high country of northwest and north-central New Brunswick, and on the Atlantic slope and Cape Breton Island in Nova Scotia, where marshes are few and relatively infertile. Bitterns were most often found in or near fresh water marshes, including very small wet patches in fields and settled areas, but they breed locally in salt marshes. Bitterns most often nest on heaps of vegetation in the marsh, only occasionally in hay fields, so their nests were seldom found (in 11 squares only); most reports of confirmed breeding were of young, out of the nest but still flightless (FL), and these were found in only 32 squares (9 per cent of 338 with bitterns). The far-carrying call or "thunder-pump," described as "ka-choong-ah!," was easily detected (contributing to the 272 squares with H or T) and bitterns were often sighted flying over marshes, so the species should not have been much under-represented, despite the difficulty of access to these habitats.

Although draining and filling-in of fresh water marshes in the past reduced the habitat of bitterns in the Maritimes, the recent development of impounded marshes for waterfowl management (by DUC) has partly offset the earlier losses. Indiscriminate shooting of bitterns and other marsh birds during duck hunting season, mainly on opening day, continues as a minor drain on the species. Overall, its numbers decreased since European settlement began, but should have stabilized in recent decades; we have no hard data on trends. Our population estimates are complicated by having to average relatively dense populations in extensive marshlands with other birds thinly scattered over minor wet areas. Wetland conservation efforts directed at waterfowl will continue to benefit bitterns and should ensure their survival here.

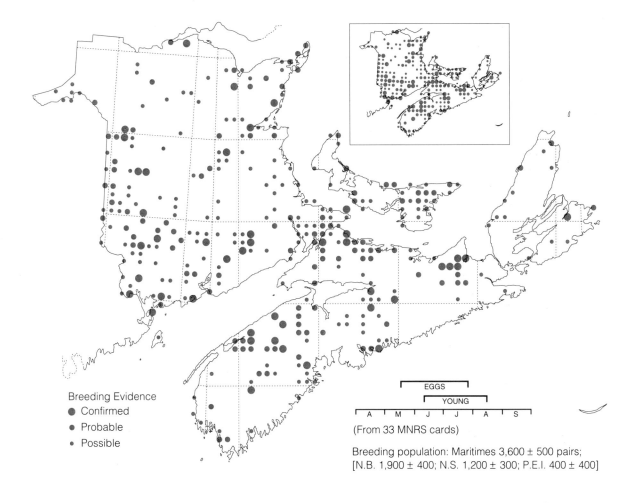

Breeding Evidence
- Confirmed
- Probable
- Possible

EGGS
YOUNG

A M J J A S

(From 33 MNRS cards)

Breeding population: Maritimes 3,600 ± 500 pairs;
[N.B. 1,900 ± 400; N.S. 1,200 ± 300; P.E.I. 400 ± 400]

Great Blue Heron
Grand Héron

Ardea herodias

There are large, adaptable herons (incorrectly called "cranes" by many people) on most continents. This is the North American species, which breeds across southern Canada, except Newfoundland and the British Columbia interior, and south through the U.S.A. to Mexico and the West Indies. They are colonial birds that often fly a long way from their nesting areas to feed, and small inland colonies are seldom detected. Colonies also shift location from time to time, as the trees supporting nests are gradually killed by the herons' droppings and fall down. The mapped distribution in the Maritimes undoubtedly missed many inland-nesting birds, but it probably represents the population fairly well, as most herons nest near the coasts. The scarcity of fishes in the often acidic waters inland limits nesting away from the sea. Breeding was confirmed in 105 squares, several with more than one colony. The more numerous single sightings (H, 288 squares) certainly involved many foraging birds that nested in other squares, and many others were excluded on that basis before records were

sent to the Atlas. Great Blue Herons often fly 20–30 km to forage, so inland records farther than this from confirmed breeding sites may indicate unknown colonies in those areas, especially in New Brunswick.

Although many former colony areas have been logged since Europeans came to the Maritimes, land-use practices in general have not limited numbers of these birds, which often nest in low patchy woodland. More influential, but mostly in the past, was the killing of these and other fish-eating birds as predators perceived as threatening human interests. This practice has been illegal since 1917, and, except very locally, heron numbers here are probably limited mainly by food supply. Present numbers are thought to be stable, and the Maritimes Heron Inventory in 1980–81 noted about 50 colonies in New Brunswick, about 80 in Nova Scotia, and about 10 in Prince Edward Island. Projected trends for the future seem unlikely to pose more serious threats than in the past, although increases in aquaculture and other private fish-rearing operations will lead people to again see herons as competitors.

Refs. 93, 122

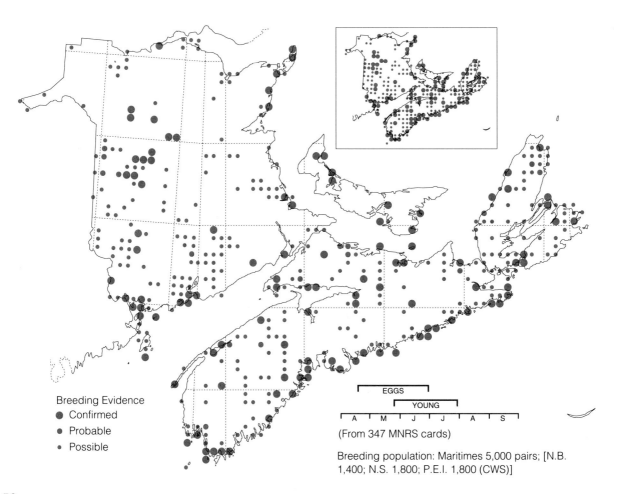

Breeding Evidence
● Confirmed
● Probable
● Possible

EGGS
YOUNG

| A | M | J | J | A | S |

(From 347 MNRS cards)

Breeding population: Maritimes 5,000 pairs; [N.B. 1,400; N.S. 1,800; P.E.I. 1,800 (CWS)]

Green-backed Heron
Héron vert

Butorides striatus

Recent merging of related small herons has given this species a wide range also in South America and the Old World tropics. In North America, its range is split by the treeless plains, as breeding is in wooded swamps and floodplain forests. One form occurs from southwest British Columbia to Utah, Texas, and Panama, and another breeds from Minnesota and New England to the Gulf Coast. In Maine, breeding by Green-backed Herons was proved as far northeast as Bangor, with only scattered sightings beyond, north and east to the New Brunswick border. The Maritimes Atlas produced a few sightings each year (14 squares in all), with breeding confirmed in 6 squares in New Brunswick, from Edmundston south to St. Stephen and Moncton. Most records were along the St. John River Valley in far western New Brunswick, and some other sightings may also have represented breeding birds, although vagrants appear annually, especially in May and August, in Nova Scotia as well as New Brunswick. The species is considered accidental in Prince Edward Island, whence the Atlas received only one record. Recently fledged young (in 5 squares) were the usual evidence of breeding, but one nest was also found.

Previous evidence of breeding in the Maritimes was limited to one clutch collected, apparently before 1900, near Washademoak, N.B. Earlier reports termed the species "rare" or at most "uncommon," although breeding was suspected over most of the area in which Green-backed Herons were documented in this Atlas. A small and well-scattered breeding population evidently persists inconspicuously in western New Brunswick, with probably only sporadic breeding elsewhere in the province. Global warming may improve conditions for Green-backed Herons in our region, and their dispersed population is less vulnerable to localized adverse influences such as tree-cutting or pollution than other herons. The persistence of breeding here over the past century, despite such low numbers, is encouraging.

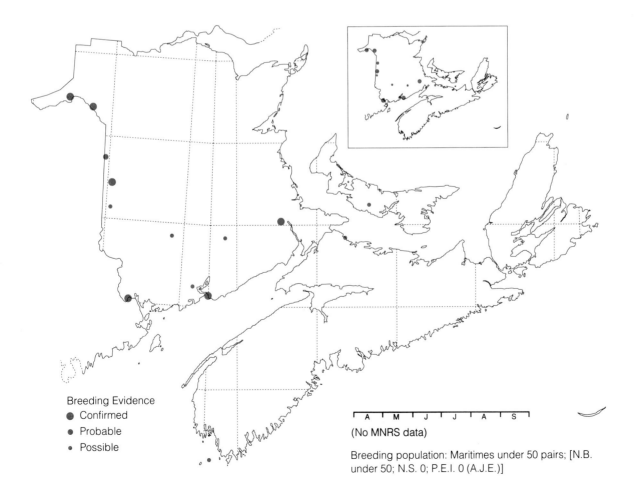

Breeding Evidence
- Confirmed
- Probable
- Possible

A M J J A S
(No MNRS data)

Breeding population: Maritimes under 50 pairs; [N.B. under 50; N.S. 0; P.E.I. 0 (A.J.E.)]

Black-crowned Night-Heron
Bihoreau à couronne noire

Nycticorax nycticorax

This night-heron is widely distributed in tropical and temperate regions of the Old and New Worlds. Its breeding range everywhere is fragmented, owing to (former) persecution and to destruction of wetland habitats. In North America it scarcely enters the boreal region. Its few colonies in the Maritimes represent easterly outliers of a more continuous range along the Atlantic coast and in the St. Lawrence Valley. Except near Edmundston, all known breeding here is coastal, usually on islands and often with Great Blue Herons, the only common colonial heron here. Most colonies are thus in spruce or fir, the common trees in coastal areas. The largest numbers are in the northeast, where the Inkerman, N.B., colony was estimated at 700 nests in 1981 (but less during the Atlas period). Most colonies include fewer than 40 nests. Night-herons often feed at considerable distances from the colony, and most records of possible breeding (H, in 33 of 48 squares with night-herons) were proba-

bly foraging birds from colonies in nearby squares. The few sight records of adults in summer far from known colonies suggest that no important concentrations were missed.

Collapse of nest-trees, killed by the droppings of these and other herons, forces most colonies to change location every few decades (as may have happened at Inkerman recently).

Ecological changes in the Maritimes since white settlement give few clues to earlier occurrence of night-herons here. This species, like most other herons, was hunted in its main range occasionally for food and more often for plumage in the past, but it was seldom a preferred species for either purpose. It was already known to breed in New Brunswick in the late 1800s, and all three main breeding areas in that province were known long before the Atlas period. Breeding in Nova Scotia was not detected until 1977 and is thought unlikely to have occurred earlier, so some expansion may be occurring. The future is uncertain for any species at the limits of its range. Global warming (the greenhouse effect) may improve conditions for this southern species in future.

Ref. 123

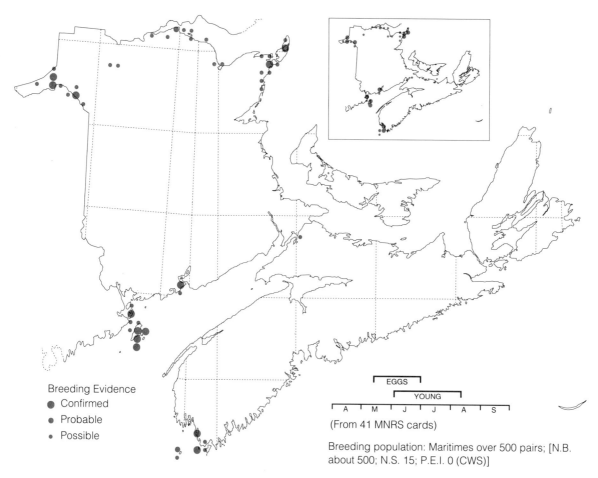

Breeding Evidence
- Confirmed
- Probable
- Possible

EGGS
YOUNG
A M J J A S

(From 41 MNRS cards)

Breeding population: Maritimes over 500 pairs; [N.B. about 500; N.S. 15; P.E.I. 0 (CWS)]

Canada Goose
Bernache du Canada

Branta canadensis

This species breeds all across North America, from the low arctic tundra south far into the United States, and it was introduced and thrived in many areas beyond its original range, as in Europe and New Zealand. In the Maritimes, it formerly bred on Miscou and Grand Manan islands in New Brunswick, but was extirpated there by 1900. Some recent records from coastal islands may represent re-occupation by wild birds of former breeding sites, from which breeding geese were excluded by human settlement (and predation) during the 19th century. Most breeding now is by descendents from introduced geese or escaped captives. Some aviculturists illegally allow their geese to fly free, and some records in the Atlas were doubtless of such feral birds; it is impossible to define which birds are truly "wild," and we have included all records of breeding except those on or adjacent to known aviculture stations. Among 104 squares with geese, 42 had records of broods of flightless or newly flying young (FL) with parent geese, and 19 others had nests (ON,NE,NY) found.

Canada Geese are conspicuous birds, which usually breed in small groups in open areas where predators can be seen at a distance while the geese forage on grasses or sedges on land. Nesting is often on islands (in lakes, marshes, or coastal bays), for safety. Present breeding numbers are evidently higher than at any time since 1900, and they may approach primaeval levels. Probably they breed in more places, owing to protection from disturbance and reduction of mammalian predators, than before European settlement. Their numbers will probably increase, as few except in Prince Edward Island are in places where they may cause problems in agricultural or recreational areas, as often occurred with introduced geese elsewhere. Far more geese (> 100,000) pass through the Maritimes in spring and fall than breed here, and several thousands winter in southern Nova Scotia.

Ref. 91

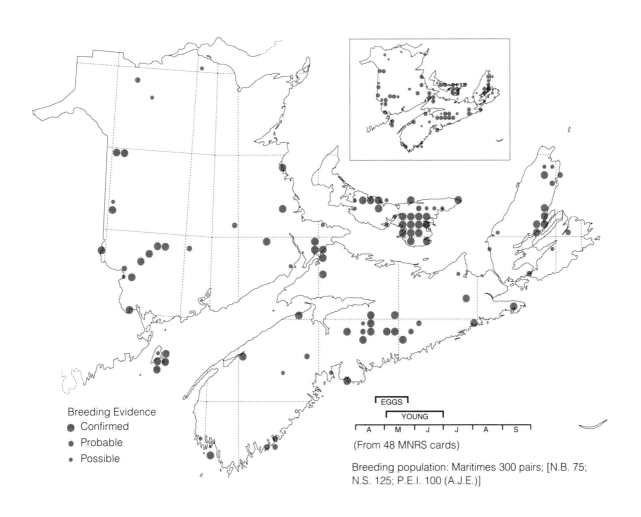

Breeding Evidence
- Confirmed
- Probable
- Possible

EGGS
YOUNG
A M J J A S

(From 48 MNRS cards)

Breeding population: Maritimes 300 pairs; [N.B. 75; N.S. 125; P.E.I. 100 (A.J.E.)]

Wood Duck
Canard branchu

Aix sponsa

This is the only North American species among the "perching ducks," which now are thought to be closely related to other dabbling ducks. These brightly coloured birds breed mainly in the U.S.A., in the far west from California into southern British Columbia and Alberta, and east of the prairies from Texas and Cuba up to southern Manitoba and the Maritimes. In our area, the Wood Duck's original range was restricted to flood-plain forests of a few rivers, especially the St. John River, where large trees with suitable cavities were situated close to fertile waters. The erection of nest-boxes, by DUC and private citizens, provided nesting sites in other areas and induced breeding on many small natural or man-made wetlands across our region, though still in only a few valley sites in northern New Brunswick. Nest-box use, as in other peripheral parts of the Wood Duck's range, is quite irregular, some boxes being used immediately whereas others remain vacant for years. Most records of confirmed breeding were of broods (FL, in 109, 37 per cent, of 290 squares with Wood Ducks); nests (NE,ON), even in boxes, were noted in only 21 squares.

Presumably Wood Ducks inhabited our river flood-plain forests in primaeval times, as at present. Despite some cutting, those forests were less affected by subsequent settlement by Europeans than were upland areas, so woodland nest-sites remained available. However, this species was decimated by unrestricted hunting in the U.S.A. and Canada in the 18th and 19th centuries and was reduced in the Maritimes also. It was given complete protection under the Migratory Birds Convention from 1916 until about 1940, and it has regained former abundance through much of its range. The Wood Duck here remains, as a cavity-nester in former times had to be, one of our scarcer ducks. It has been increased appreciably by nest-box programs, especially in Nova Scotia and Prince Edward Island where it rarely bred in the past. With efforts being intensified under the North American Waterfowl Management Plan, and the continuing interest in encouraging this colourful species, we may expect its persistence here in future. Global warming, if continued, may assist the expansion of southern species such as the Wood Duck.

Ref. 121

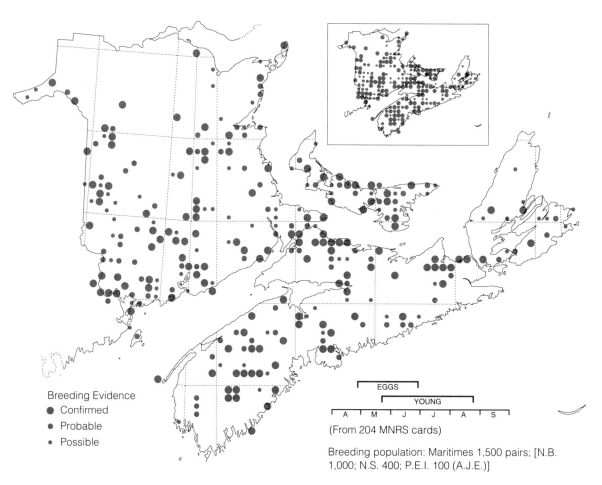

Breeding Evidence
- Confirmed
- Probable
- Possible

EGGS
YOUNG

A M J J A S

(From 204 MNRS cards)

Breeding population: Maritimes 1,500 pairs; [N.B. 1,000; N.S. 400; P.E.I. 100 (A.J.E.)]

Green-winged Teal
Sarcelle à ailes vertes

Anas crecca

Our smallest duck, this species breeds across Canada and the U.S.A. from near the arctic treeline south to California, Nebraska, and New Jersey, as well as all across Eurasia, in the boreal and temperate zones. Green-winged Teals reach highest density in the prairies, but, unlike many dabbling ducks, they are not restricted to open, fertile marshes. They breed also in forested regions and occasionally in brackish marshes near the coasts. Their range in the Maritimes seemed to be concentrated in the major wetland regions: Prince Edward Island, northern Nova Scotia, the New Brunswick–Nova Scotia border area, and the lower St. John River Valley. Atlas field work detected few teals elsewhere, but aerial surveys by government agencies confirmed that these ducks were distributed in all regions at low densities. More intensive surveys of wetland habitats, as in the Antigonish region, would have found Green-winged Teals more widely distributed, as this species is much less conspicuous than most ducks.

Broods (FL) were the most common evidence of confirmed breeding, noted in 152 squares (43 per cent of the 353 squares with these birds), whereas nests were found in only 4 squares.

Green-winged Teal habitat in the Maritimes suffered in the past through draining and dyking of marshes, and their numbers also were affected by unregulated hunting. To this day, this species ranks second after the Black Duck in annual duck hunting kill here despite its small size, because so many of these birds from Labrador and Ungava stop here during the migrations. Although major breeding concentrations were found here in regions where artificial waterfowl impoundments have been created, this species seems not to take advantage of these areas as much as other dabbling ducks. Its numbers have not increased in recent years, and probably have declined, but it is still one of our more numerous breeding waterfowl. The Green-winged Teal will remain a common duck in the Maritimes, with increased conservation efforts under the North American Waterfowl Management Plan, but it is little studied and poorly understood everywhere, notwithstanding the large numbers taken by hunters.

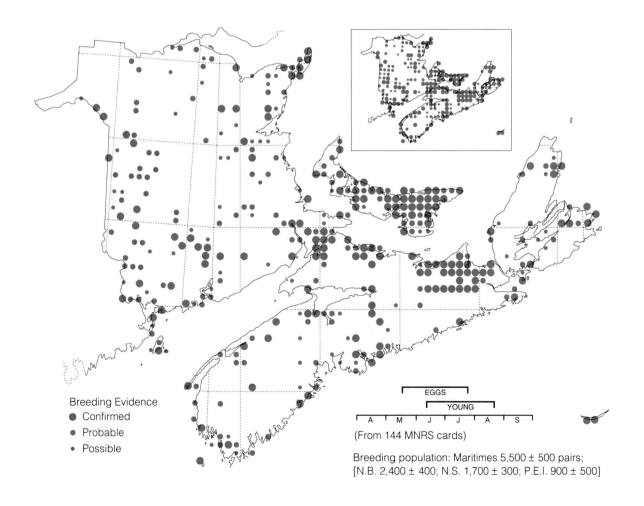

Breeding Evidence
- Confirmed
- Probable
- Possible

EGGS
YOUNG
A M J J A S
(From 144 MNRS cards)

Breeding population: Maritimes 5,500 ± 500 pairs;
[N.B. 2,400 ± 400; N.S. 1,700 ± 300; P.E.I. 900 ± 500]

American Black Duck
Canard noir

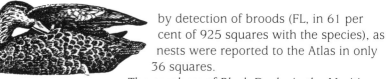

Anas rubripes

The Black Duck is the eastern North American representative of the widespread Mallard species-complex. It breeds from Manitoba eastward to the Atlantic Ocean, from the arctic tree-line south into the United States. Some birds winter as far north as open water persists inland and in tidal areas. In western and southern parts of this range, it has been largely replaced by the typical Mallard-plumaged form, through inter-breeding with wild and released birds during the last 40 years. In the Maritimes, Black Ducks bred almost everywhere except in regions with steep terrain, where suitable wetland habitat is lacking. Often the nest, on the ground under or among dense vegetation, is a kilometre or more from the marsh or other wetland where the brood will be reared. Broods appear in fresh marshes, ponds, river backwaters, and salt marshes, as well as in man-made wetlands such as sewage lagoons, wild-rice plantations, and Ducks Unlimited Canada impoundments. Breeding was almost always confirmed by detection of broods (FL, in 61 per cent of 925 squares with the species), as nests were reported to the Atlas in only 36 squares.

The numbers of Black Ducks in the Maritimes have decreased from former times, by all accounts, through the combined effects of formerly unrestricted hunting down the flyway, of habitat loss in breeding and wintering areas, and of pollution, including lead poisoning. Replacement by Mallards has not yet had a measurable effect on Black Duck numbers in the Maritimes, although there are suggestions of such replacement in areas where free-flying Mallards are fed through the winter. Estimates of Black Duck numbers from the Atlas compared reasonably well with earlier extrapolations from two independent data sets. With major management efforts devoted to ensuring its continued availability as a game bird, the adaptable Black Duck is likely to remain one of our common ducks, unless and except where it is replaced by the still more adaptable Mallard, an event of which the scale and timing cannot yet be foreseen.

Ref. 41, 44, 67, 70, 134

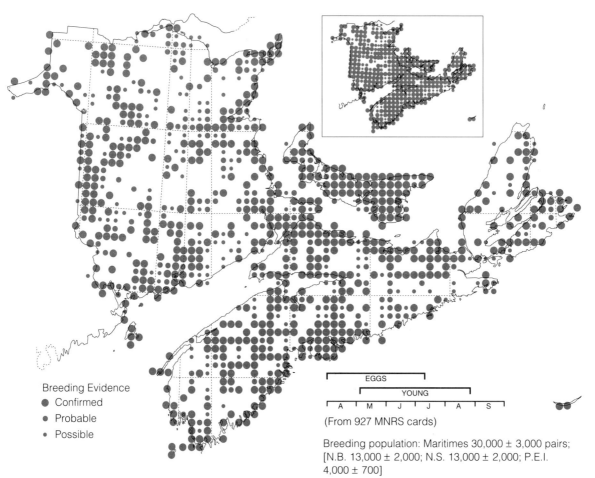

Breeding Evidence
● Confirmed
● Probable
● Possible

EGGS
YOUNG

A M J J A S

(From 927 MNRS cards)

Breeding population: Maritimes 30,000 ± 3,000 pairs; [N.B. 13,000 ± 2,000; N.S. 13,000 ± 2,000; P.E.I. 4,000 ± 700]

Mallard   Canard colvert

Anas platyrhynchos

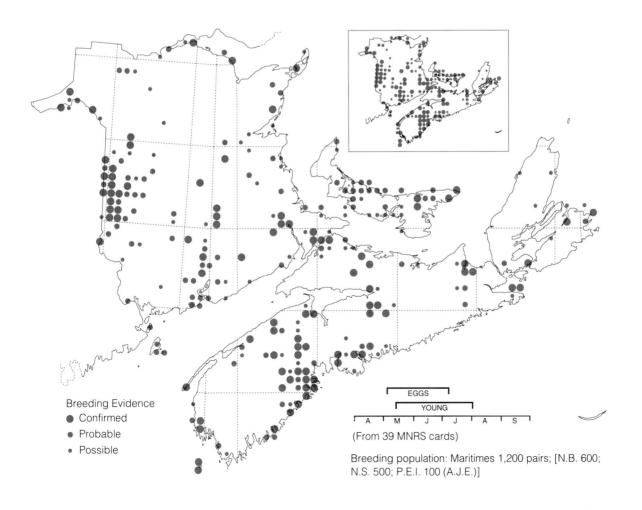

The Mallard is the most common, most adaptable, and most widely distributed duck on Earth, breeding in all wetland habitats of temperate and boreal zones of the Northern Hemisphere. Various ducks in which both sexes resemble female Mallards, including the American Black Duck, are closely related and interbreed freely with this species. Records of Mallards paired with Black Ducks were mapped as Mallards. Most forms of domestic ducks were derived from Mallards, and many still resemble the wild form. Hand-reared Mallards were widely released in New Brunswick and Nova Scotia, mostly in the 1950s and 1960s, and free-flying birds from city parks also have established sedentary stocks, which depend to varying degrees on local feeding in winter. It was usually impossible to determine whether Mallards seen in the Atlas period were naturally wild or derived from domestic stock, including releases; we excluded birds breeding in urban areas around known release sites, but many others of similar origin must have been missed. The map shows mostly where domestic or released stocks have contributed to local breeding. Mallards in the western St. John River Valley had been thought to be derived from birds that moved in from farther west, but the range in Maine and Quebec is not continuous with this area. Most records of confirmed breeding (36 per cent of 263 squares with Mallards) were of flightless young (FL); nests were reported in only 12 squares.

Before 1920, the Mallard was considered a rare fall vagrant in the Maritimes. Its increase here was aided by deliberate releases, but it also followed the dramatic increase in southern Ontario and Quebec since 1950. The situation here now raises concern that Mallards might replace Black Ducks as the common breeding duck in the Maritimes; this process required only 15–20 years in eastern Ontario, starting with Mallards no more frequent there in 1955 than at present in the Maritimes. The major difference is that agricultural lands and urban wintering areas, of which Mallards seem more tolerant, made up a much larger proportion of the area in Ontario than here, and replacement may be restricted to these habitats. The situation seems likely to change in the near future, but the extent of change cannot be foreseen.

Ref. 14

Breeding Evidence
- Confirmed
- Probable
- Possible

EGGS
YOUNG

| A | M | J | J | A | S |

(From 39 MNRS cards)

Breeding population: Maritimes 1,200 pairs; [N.B. 600; N.S. 500; P.E.I. 100 (A.J.E.)]

Northern Pintail
Canard pilet

Anas acuta

The Pintail is widespread around the northern hemisphere, breeding from the low arctic tundra south to California, central Europe, and Manchuria. Its breeding in eastern America south of the tundra is very localized, extending south to Massachusetts but with no breeding at all in Maine in 1978–83. Everywhere, it frequents open habitats, such as prairies, tundra, extensive bogs, salt marshes, and coastal grasslands. Thus, in the Maritimes it was found mainly in the extensive marshes of the New Brunswick–Nova Scotia border region and of central Prince Edward Island, and on coastal grasslands and sewage lagoons of northeast New Brunswick. It is one of the larger and more distinctive dabbling ducks, so is likely to have been detected in most areas where it occurs, although it is not a common bird here. Most records were either of broods (FL) or of paired adults (P) (in 41 and 22 squares, respectively, of 92 with the species). Only two nests were found here in 1986–90.

Pintails seem to have settled in the Maritimes rather recently, with first recorded breeding around 1938, in the New Brunswick–Nova Scotia border area. Although recolonization after earlier extirpation is an alternative explanation, it seems unlikely that these striking birds could have been missed completely had they been breeding here in the 1800s, when they were known only as migrants. Before then, it is plausible that there was insufficient open marsh habitat, here and generally through eastern North America, to attract this species. After 1940, Pintail numbers in the Maritimes increased rapidly, and they spread to Prince Edward Island, but the increase ceased soon after 1960. Numbers now are certainly lower, by 50 per cent or more, both in the border region and in Prince Edward Island, than 25 years ago. Estimates from the Atlas data are somewhat lower than those from other sources. There are no obvious explanations for the recent decline, but increasingly intensive use of open fields near wetlands may inhibit nesting. Pintails have not been conspicuous in adopting newly impounded areas, the most frequent form of habitat manipulation undertaken to enhance duck numbers in the Maritimes. We cannot expect that their numbers will increase much in future.
Ref. 7

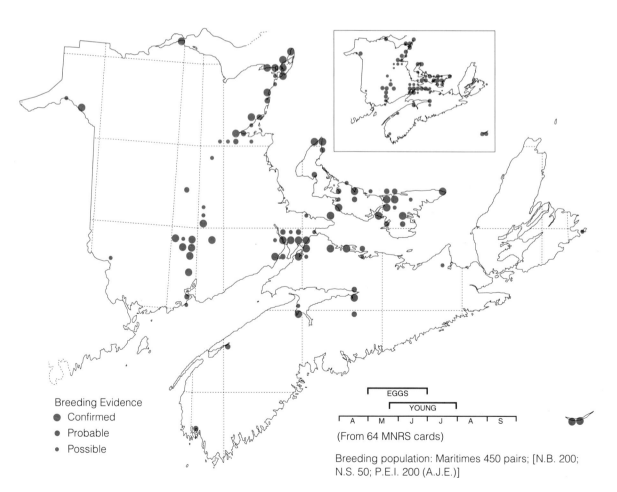

Breeding Evidence
● Confirmed
● Probable
● Possible

EGGS
YOUNG
A M J J A S
(From 64 MNRS cards)

Breeding population: Maritimes 450 pairs; [N.B. 200; N.S. 50; P.E.I. 200 (A.J.E.)]

Blue-winged Teal
Sarcelle à ailes bleues

Anas discors

This small duck breeds across the U.S.A. and Canada, more widely in the west where it reaches southern Alaska and Mackenzie, whereas it reaches its northeast limits in the Gaspé Peninsula and southwest Newfoundland. The Blue-winged Teal is a bird of open, fertile marshes, including brackish areas around estuaries; scarcity of such habitats is reflected in its discontinuous range away from the coasts in the Maritimes, except in Prince Edward Island. It was virtually lacking from many upland regions of both New Brunswick and Nova Scotia. Outside Prince Edward Island, Blue-winged Teals were widespread only in the coastal regions of northeast New Brunswick and northern Nova Scotia, the New Brunswick–Nova Scotia border area, and the lower and middle parts of the St. John River Valley. Earlier assessments of the species as generally "common" seem to have been optimistic. As with other ducks, breeding by this species was confirmed almost entirely by seeing broods (in 158, 48 per cent, of 332 squares with these birds), and nests were seen in only 15 squares.

Blue-winged Teals presumably were scarcer here in the distant past, before Europeans set about clearing forest lands and dyking marshes in the Maritimes. Similarly, they may have suffered from unregulated hunting in the more recent past. With development of artificially impounded marshes, mostly in open, relatively fertile areas, during the last 25 years, the species has probably increased. Its numbers may now be the highest ever, though the Atlas estimates are somewhat less than the numbers estimated a decade ago from earlier CWS data. Most Blue-winged Teals leave the Maritimes before the duck hunting season starts, which gives them an advantage over most other locally breeding species. Increasing efforts in waterfowl management and higher summer temperatures through global warming are likely to allow this species' numbers to remain high in the future.

Ref. 41

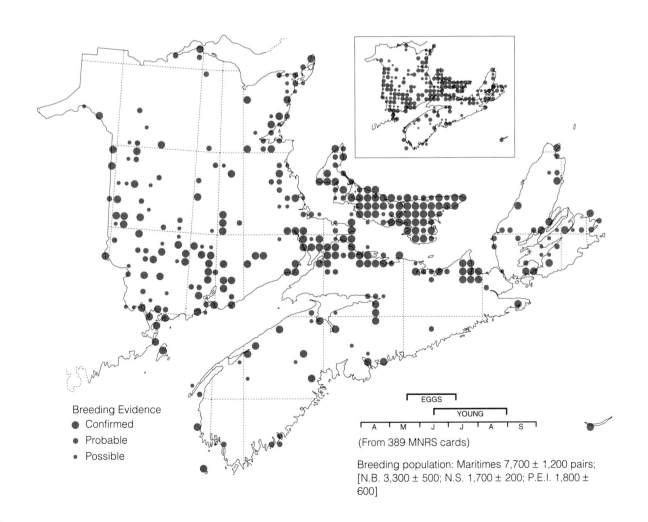

Breeding Evidence
● Confirmed
● Probable
● Possible

(From 389 MNRS cards)

Breeding population: Maritimes 7,700 ± 1,200 pairs; [N.B. 3,300 ± 500; N.S. 1,700 ± 200; P.E.I. 1,800 ± 600]

Northern Shoveler
Canard souchet

Anas clypeata

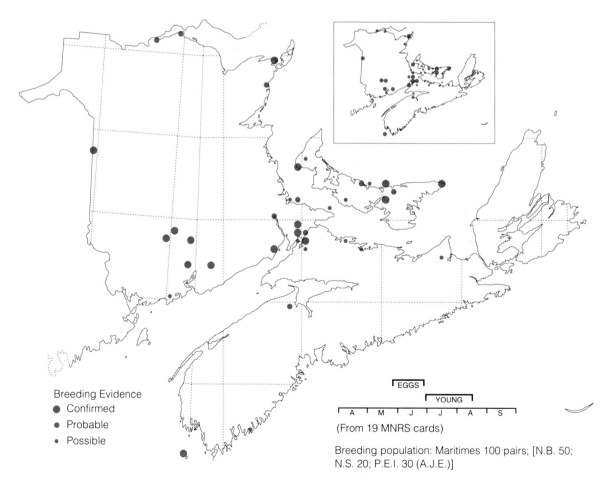

The Northern Shoveler breeds across Eurasia from eastern Siberia to Ireland, and in western North America, with several related species of similar form in the southern hemisphere. In North America it is most common in the prairies, but it also occurs in suitable marshy ponds from Alaska to California and through the inter-mountain plateaux. Breeding in the east is much less general, locally in southern parts of Ontario and Quebec and in the Maritimes. Shovelers here bred mostly in the New Brunswick–Nova Scotia border marshes, in Prince Edward Island, and in the lower St. John River Valley, (the main fertile wetlands in this region) with only a few records elsewhere, often in sewage lagoons. Although most records were near the coasts, and these birds often were found on brackish pools in salt marsh during migration, breeding was almost entirely by fresh water.

As with other ducks, breeding usually was confirmed by sightings of broods (in 13 of 37 squares with Shovelers).

Shovelers were not known to have bred in eastern North America before 1948, when a nest was found near Sackville, N.B., so our population is of very recent origin. Breeding became somewhat more frequent with the artificial impoundment of shallow marshes by Ducks Unlimited Canada, from about 1968, but the Shoveler is still an uncommon bird in our region. The present breeding population is probably still too small to ensure the species' continued survival here under any serious pressure from hunting or environmental threats. At this time, no such pressures have been identified, and continued development of suitable marshes, by DUC and others, may help to ensure its breeding here in the future.

Breeding Evidence
- ● Confirmed
- ● Probable
- • Possible

EGGS
YOUNG

A   M   J   J   A   S

(From 19 MNRS cards)

Breeding population: Maritimes 100 pairs; [N.B. 50; N.S. 20; P.E.I. 30 (A.J.E.)]

Gadwall
Canard chipeau

Anas strepera

The Gadwall is an inland duck, breeding mainly in the drier regions of western North America and western Eurasia, but with widely scattered groups far beyond that range, for example, in Iceland, formerly on Anticosti Island, Que., and on islands in the Pacific Ocean. In recent decades it appeared in eastern North America, where small numbers now breed locally but regularly in southern parts of Ontario and Quebec and in the U.S.A. from Wisconsin to North Carolina to Massachusetts. The Maritime picture agrees with the vagrant or pioneering aspects of Gadwall distribution, as it included inland marshes in the New Brunswick–Nova Scotia border area, brackish estuarine marshes in Prince Edward Island, sewage lagoons in northeast New Brunswick, and offshore islands (Grand Manan Island, N.B., and [in 1981] Bon Portage Island, N.S.). Only the first two seem to be regular breeding areas for Gadwalls, which are more restricted than sev-

eral other birds of fertile wetlands that breed mainly in these regions, e.g., Pied-billed Grebe, Northern Shoveler, Sora. Breeding was confirmed in 11 of 26 squares with Gadwalls, the broods being easily seen in open marshes; one nest also was found.

Except for one report in 1861, the Gadwall was not known to have bred in eastern America until 1939. Since then, scattered breeding groups have developed, including some introduced birds, in many states and provinces. Breeding in the Maritimes was not suspected before the first record in 1974 and has been regular only since 1978. Although a few reports of breeding came from areas where Gadwalls have not since become established, numbers seem to have stabilized. It is too early to conclude that the species will persist here indefinitely, as the present numbers are too few to be self-sustaining in the absence of immigration. The pattern of spread and increase in eastern North America in the recent past gives hope for their continued presence here.

Ref. 24, 97

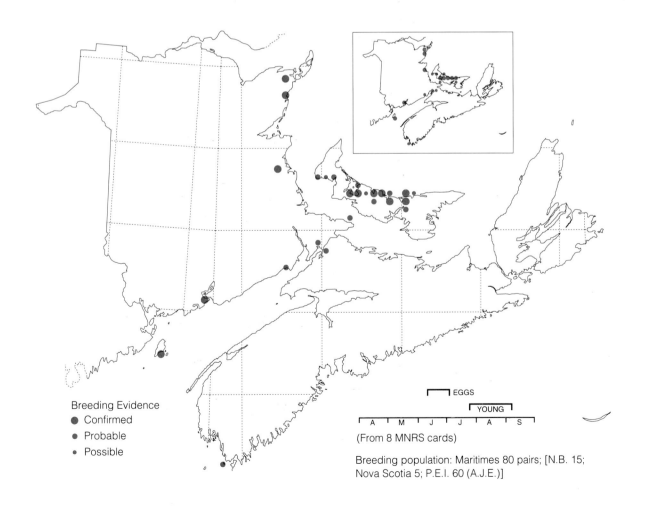

Breeding Evidence
● Confirmed
● Probable
● Possible

EGGS
YOUNG

| A | M | J | J | A | S |

(From 8 MNRS cards)

Breeding population: Maritimes 80 pairs; [N.B. 15; Nova Scotia 5; P.E.I. 60 (A.J.E.)]

American Wigeon
Canard siffleur d'Amérique

Anas americana

This wigeon, or "Baldpate," is a largely western American duck, breeding from Alaska to Hudson Bay, and from the tundra edge south to the central U.S.A. In the east, breeding is very discontinuous, mainly in agricultural areas, where it frequents impounded marshes and sewage ponds that simulate the habitats used farther west. Breeding in the Maritimes was not detected until 1957. Atlas workers found it concentrated in the more extensive marshes along the St. John River, around the estuaries and barrier-beach lagoons of northeast New Brunswick, near the New Brunswick–Nova Scotia border, and in eastern Prince Edward Island, with scattered pairs elsewhere near the coasts, except the outer Atlantic coast of Nova Scotia. The virtual absence of the species from the forested interiors of New Brunswick and Nova Scotia agrees with the pattern elsewhere, as the American Wigeon is generally much more common in open habitats than in the boreal forests. As with other ducks,

sightings of broods accompanied by females (FL) were the most frequent confirmation of breeding (44 per cent, 75 of 169 squares with wigeons) in the Atlas study, and nests were found in 7 squares.

The colonization of the Maritimes by American Wigeons in recent decades seems likely to have been new, rather than a re-occupation of former eastern range from which the species had been eliminated by hunting during the 19th century. Given the wigeon's preference for open marsh habitats, but not including salt marshes to any great extent here, this is unlikely to have been a common breeder here in former times, when the country was more forested. Thus, its present breeding population may be the highest ever, although it is still not a common species. Continuing development of impounded marshes by Ducks Unlimited Canada, for waterfowl, wild rice, and muskrats, should help to expand the breeding here of these "prairie ducks."

Ref. 7

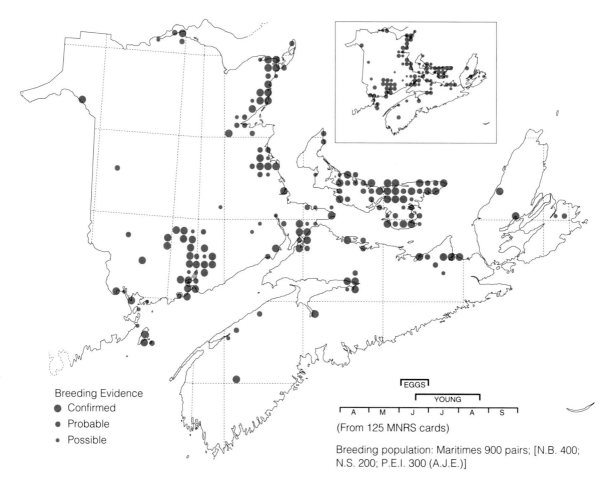

Breeding Evidence
● Confirmed
● Probable
● Possible

EGGS
YOUNG
A M J J A S

(From 125 MNRS cards)

Breeding population: Maritimes 900 pairs; [N.B. 400; N.S. 200; P.E.I. 300 (A.J.E.)]

Ring-necked Duck
Morillon à collier

Aythya collaris

This species breeds across the forested regions of Canada, from Alaska to Newfoundland, and south into the northern tier of the United States. In the Maritimes, it is the most generally distributed inland diving duck. It is most characteristic of lakes, ponds, and stillwaters, where sedge or shrub marshes border much of the shorelines. The recent development of impounded marshes, often dominated by cattail beds but with deeper water than is typical of most natural cattail marsh here, has also favoured Ring-necked Ducks. The species' frequency in the New Brunswick–Nova Scotia border region and in eastern Prince Edward Island reflects its acceptance of artificial marshes there. Its nests are in the marsh, in clumps of sedge, sweet gale, or leatherleaf. Broods of flightless young (FL) provided the most frequent confirmation of breeding (in 41 per cent of 447 squares with the species); only 15 nests were found during Atlas work. Ring-necks are late breeders, with few broods appearing before late June, and many are flightless through August.

The Ring-necked Duck was very scarce in the Maritimes and elsewhere in the northeast for several decades before the 1930s, when a rapid increase in numbers began. A popular explanation had this as a western species that had expanded its range eastward at that time. However, evidence of its occurrence in small numbers in widely scattered localities across the northeast between 1850 and 1930 suggested an alternative explanation: that the former eastern breeding population had been decimated by unrestricted hunting earlier, with the recent increase representing re-occupation of former range after hunting pressure was reduced. This reconstruction remains unproved, but seems plausible. The proliferation of suitable habitat in DUC ponds has contributed to further increases, and the species may well be more common now than in even the distant past. Population estimates from Atlas work were of similar order to those from other sources. Hunting pressure on the species in the Maritimes now is light, as most Ring-necked Ducks migrate south before the duck hunting season begins here. The future for these birds seems assured.

Ref. 98

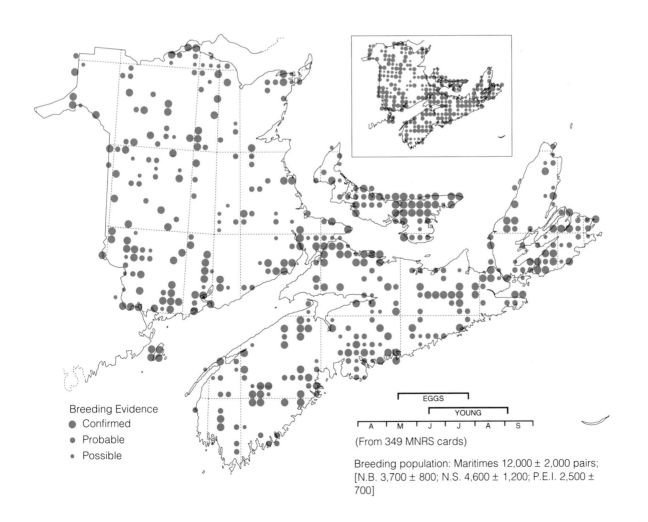

Breeding Evidence
- Confirmed
- Probable
- Possible

EGGS

YOUNG

A M J J A S

(From 349 MNRS cards)

Breeding population: Maritimes 12,000 ± 2,000 pairs; [N.B. 3,700 ± 800; N.S. 4,600 ± 1,200; P.E.I. 2,500 ± 700]

Greater Scaup
Grand Morillon

Aythya marila

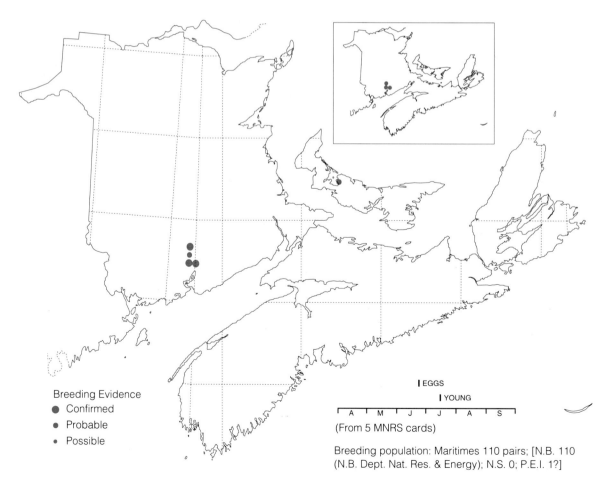

The main range of the Greater Scaup, both in Eurasia and North America, is in the subarctic transition between forest and tundra, with Lesser Scaup being the more expectable scaup species in southern Canada. Where cold waters occur farther south, in large lakes and in southward-flowing currents, there are isolated pockets of breeding Greater Scaups, on islands in Lake Winnipeg, Man., and on coastal barrens near Cape Race, Nfld. Southern breeding in Eurasia has been thought to represent relicts of a formerly wider distribution. Breeding by this species in the Maritimes does not fit any patterns found elsewhere. Undetailed reports from Tabusintac, N.B., in 1917 and Cape Breton Highlands in 1968 preceded breeding at Indian River, P.E.I. (brood, 1980; nest, brood later, 1981). Lingering singles and pairs occur irregularly, in spring migration stopover areas in the Maritimes and elsewhere in southern Canada, through mid-June or later without any breeding evidence emerging, and most of these summering birds may be ignored as non-breeders. The only confirmed breeding reported to the Atlas was in a migration area in the St. John River around Grassy Island, N.B., where nests were found annually from 1986. Broods were seen later up- and down-river from the main nesting site. This breeding far from other stocks of scaups points to ecological flexibility and readiness to occupy new areas unexpected in this typically subarctic bird.

Although the Maritimes had a subarctic climate and habitat at various times during the glacial epoch, there is no suggestion of such drastic cooling during the Little Ice Age of 1350–1850. The recent breeding of Greater Scaups in the Maritimes can hardly be re-occupation of traditional breeding areas. At Grassy Island, estimates increased annually from 1986 (21 nests) to 1990, which may suggest that a tradition for breeding in such atypical situations is developing. If so, many other areas in the Maritimes might be occupied by Greater Scaups in the future. Equally, this surprising addition to New Brunswick's breeding avifauna may disappear in a few years. We can only wait and see.

Ref. 96

Breeding Evidence
- Confirmed
- Probable
- Possible

EGGS

YOUNG

A M J J A S

(From 5 MNRS cards)

Breeding population: Maritimes 110 pairs; [N.B. 110 (N.B. Dept. Nat. Res. & Energy); N.S. 0; P.E.I. 1?]

Common Eider
Eider à duvet

Somateria mollissima

Eiders are among the most northerly breeding ducks, and their down, collected from nests, is famous for its warmth. They breed along the arctic marine coastlines of North America, Europe, and far east Asia, and secondarily southward in boreal waters to Maine, Nova Scotia, Eire, and Denmark. Nesting islands free of predatory mammals, with shoal waters around reefs and rocky islets to which the ducklings may be led after hatching, are scarce or lacking along coasts underlain by sedimentary rocks; this explains the absence of breeding eiders in the southern Gulf of St. Lawrence, including Prince Edward Island. They feed on mussels and other shellfish, including sea urchins, which are characteristic of rocky coasts.

Eiders winter around the southwestern Maritimes and return to breeding areas here during April. After breeding, the males move away during June to moult, in company with non-breeding subadult birds, often along coasts where few if any breed. Moulters and non-breeders account for most summer records of eiders away from previously known breeding areas. Atlas data were obtained in nearly all areas with previous breed-ing records, with broods of flightless young (FL, in 56 squares) the most frequent evidence. Relatively few nesting islands were visited during Atlas work, and nests (in 29 of 105 squares with the species) made up only 33 per cent of confirmed breeding records.

Breeding eiders in the Maritimes, abundant at first European contact, were reduced to remnant levels by unrestricted hunting during the 19th century. With protection under the Migratory Birds Convention since 1916, they have recovered much of their former numbers. Their status in the Maritimes was reviewed in the early 1980s, and extrapolations based on the Atlas records are compatible with the earlier figures. Unstandardized observations suggest a continuing increase, particularly in southwest Nova Scotia. Potential threats to the Maritime stocks include oil spills near colonies, such as the *Kurdistan* spill off eastern Nova Scotia in 1979; predation of newly fledged young by Great Black-backed Gulls, when herring were scarce in the Bay of Fundy in 1985–86; and increased hunting of sea ducks when Black Duck hunting was restricted in recent years. As with other seabirds, eider populations are well-adapted to low breeding success, but they recover slowly from losses of adult birds.

Refs. 55, 92, 106, 125, 132

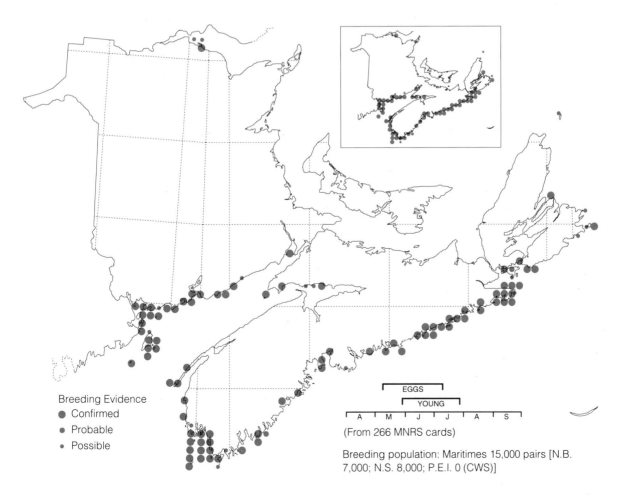

Breeding Evidence
- Confirmed
- Probable
- Possible

EGGS
YOUNG

A M J J A S

(From 266 MNRS cards)

Breeding population: Maritimes 15,000 pairs [N.B. 7,000; N.S. 8,000; P.E.I. 0 (CWS)]

Common Goldeneye
Garrot à œil d'or

Bucephala clangula

The Common Goldeneye's breeding range is largely restricted to the boreal forest regions of North America and Eurasia. It breeds across Canada from northern British Columbia to southern Labrador, but barely into extreme northern U.S.A. A tree-cavity nester, it breeds only in wooded areas, sometimes in nest-boxes. In the Maritimes Atlas, breeding goldeneyes were found mainly in the northwest highlands and the Miramichi and St. John River valleys of New Brunswick, and in the valleys of northern Cape Breton Island. No breeding records were obtained from the Nova Scotia mainland or Prince Edward Island, where floodplain forests are few and of small extent, and none were known there previously. Records of confirmed breeding were mostly of broods (FL, in 71 squares, 42 per cent of 167 squares with the species). Nests were reported in only 8 squares.

The goldeneye's nesting habits always limited its numbers and range to areas near water and with trees large enough to contain suitable cavities. In the Maritimes, the mature forests of the pre-settlement era included more large trees than are now available. Subsequent logging, with large trees being especially selected, and unrestricted hunting of all ducks likely reduced goldeneye numbers below their original level, until the early part of this century. The Common Goldeneye was still considered one of the most abundant breeding ducks of the St. John River marshes in the 1930s–1950s, so the earlier reduction may not have been large, or some recovery may have occurred by then. Other recent population estimates were based largely on data from the 1950s and do not provide a useful picture of the current situation, which indicates a continuing decline, at least in the St. John River Valley. Continued cutting in the forests of northern New Brunswick also can hardly benefit the species. Until recently, the Common Goldeneye ranked third in the Maritimes duck harvest (after Black Duck and Green-winged Teal), but it has slipped gradually farther behind since 1975. Even though much of the kill is of birds from more northern areas, there may be grounds for concern over the status of this duck in our area.

Refs. 23, 121

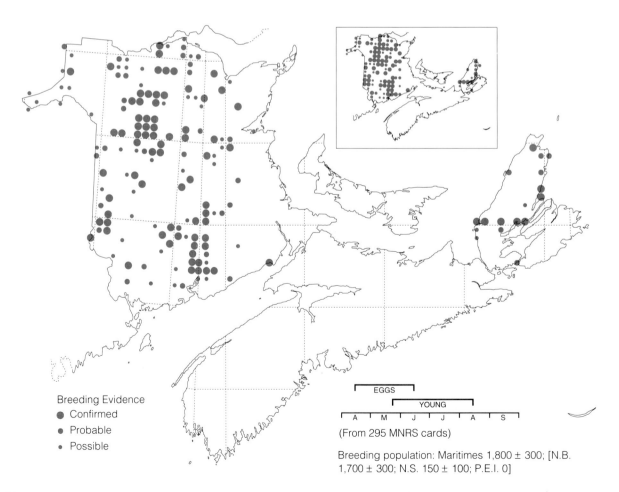

Breeding Evidence
- Confirmed
- Probable
- Possible

EGGS

YOUNG

A M J J A S

(From 295 MNRS cards)

Breeding population: Maritimes 1,800 ± 300; [N.B. 1,700 ± 300; N.S. 150 ± 100; P.E.I. 0]

Hooded Merganser
Bec-scie couronné

Lophodytes cucullatus

Our smallest merganser is a strictly North American species, breeding in southern Canada and across the United States, south to Oregon, Iowa, and Tennessee. It is absent in the treeless prairies, as its nests originally were all in hollow trees and large woodpecker holes. Widespread erection of nest boxes, usually for Wood Ducks, has also helped Hooded Mergansers throughout their range. In the east, the range corresponds with the northern half of the eastern broad-leafed forest biome, which reaches its northeast limit in western New Brunswick. More eastern breeding records in the Maritimes resulted largely from Wood Duck nest-box programs, especially at artificial ponds created by Ducks Unlimited Canada. Most reports of confirmed breeding by Hooded Mergansers were of broods (FL, in 58 of 158 squares with these birds), with nests in only 7 squares.

As this species depended originally on cavities in large trees for nest sites, clearing and cutting in the primaeval mixed forests of New Brunswick somewhat reduced its breeding opportunities here during the past two centuries. This bird is more often found by lakes and less by river flood-plains than is the Common Goldeneye, which uses similar nest-sites.

Unrestricted hunting of all waterfowl prior to 1916 also may have reduced its numbers. Development of impoundments and nest-box erection in the last 25 years contributed to the scatter of records beyond the former range in western New Brunswick, but are unlikely to have increased the total Maritimes population greatly. While impoundments and nest-boxes continue to proliferate, and if global warming develops and persists long enough to influence our forests, Hooded Merganser numbers may well increase somewhat in future. There are no obvious factors working against their continuing as breeding birds here.

Ref. 121

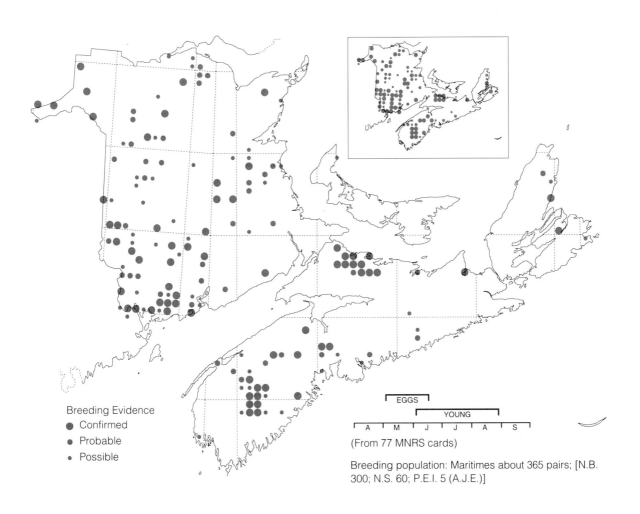

Breeding Evidence
- Confirmed
- Probable
- Possible

EGGS

YOUNG

A M J J A S

(From 77 MNRS cards)

Breeding population: Maritimes about 365 pairs; [N.B. 300; N.S. 60; P.E.I. 5 (A.J.E.)]

Common Merganser
Grand Bec-scie

Mergus merganser

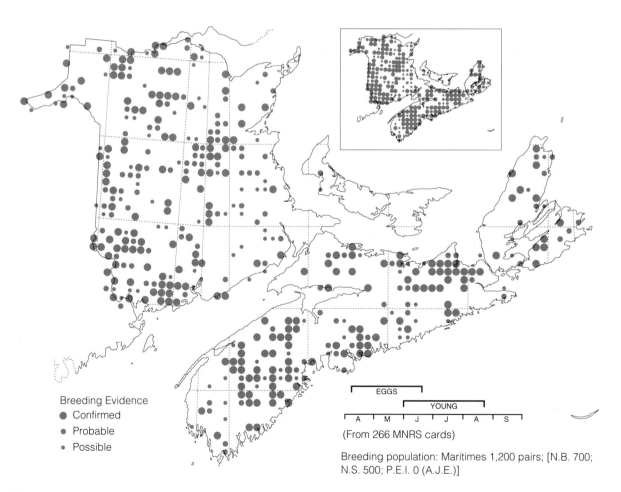

This species breeds across North America, Europe, and Asia in the boreal regions, with southward extensions or isolated stocks in mountain areas. It feeds almost entirely on fishes, which it obtains in lakes, rivers, and streams with relatively clear waters. It nests most often in hollow trees and stubs, but also in nest-boxes, buildings, holes in cliffs, or under brush or low branches on the ground, near the shores of suitable waters. The young are led to water after hatching, and broods often move gradually downstream to tidal waters before attaining flight; on rivers, some broods recorded in the Atlas may have been in squares other than those where the birds nested. Few nests (only 13, in 418 squares with this species) were found during Atlas field-work, and most records were of broods (FL, 195 squares) or pairs (P, 109 squares).

These mergansers were formerly more common, as were salmon and trout, the young of which often are the main foods of mergansers breeding along rivers. Persecution of mergansers as predators of young sport fishes was often the abuse of a scape-goat, to avoid acknowledging excessive exploitation of the fish by humans, rather than a scientific management activity; some shooting of mergansers, now illegal except in hunting season, persists. Accumulation through food chains of DDT used against spruce bud-worm, combined with shooting, greatly reduced numbers of breeding mergansers in New Brunswick between 1950 and 1970; evidently they have not yet fully regained their former abundance there. Although Common Mergansers were stated to have bred in Prince Edward Island in the past (and this seems plausible), no definite records are known in the last 40 years, and none was obtained during Atlas field-work. The species is not rare and cannot be viewed as under serious threat. Although DDT use here is past history, some habitat has been lost through pollution of waters, and more through damming of rivers. Loss of trees suitable for nest-cavities imposes little restriction, as the birds tolerate alternative sites. Even direct persecution, alone, has never eliminated these birds for long from extensive areas. Their future in the Maritimes seems reasonably secure.

Refs. 35, 156

Breeding Evidence
- Confirmed
- Probable
- Possible

EGGS

YOUNG

A M J J A S

(From 266 MNRS cards)

Breeding population: Maritimes 1,200 pairs; [N.B. 700; N.S. 500; P.E.I. 0 (A.J.E.)]

Red-breasted Merganser
Bec-scie à poitrine rousse

Mergus serrator

This, the most northerly of the mergansers, approaches its southern limit in our area. Red-breasted Mergansers breed in the boreal to low arctic zones across North America and Eurasia, with only local or irregular breeding farther south, as in northern Maine. It is a ground-nester and so is not restricted to forested regions like the other mergansers. Its largest numbers in the Maritimes nest amid low brush and driftwood on coastal islands and sandbars, sometimes in loose groups with up to 50 nests on one island, usually in association with gulls or terns. These birds were peculiarly distributed here in summer, with confirmed breeding in eastern New Brunswick and western Prince Edward Island, on Cape Breton Island and in the Grand Manan archipelago and St. John River estuary but only scattered records elsewhere. Breeding along the south shore of Northumberland Strait had never been confirmed before 1990, and Atlas records in eastern Prince Edward Island were also inconclusive (mostly P). A few Red-breasted Mergansers nest inland, but after hatching they take their broods downriver to estuarine habitats. Breeding was confirmed, mostly by seeing broods (FL), in 68 squares (45 per cent of 151 with this species), and nests were found in 13 squares.

Habitat changes since the start of European settlement included, for many years, human habitations on coastal islands, mainly for fishing. Such dwellings, with accompanying predation by domestic animals and by humans, may have reduced this species' use of the Atlantic coastal zone in summer in the past, but recovery would have been expected by now. Similarly, unrestricted duck hunting and persecution in the name of protecting fishery interests are largely in the past, but this species remains rather scarce as a breeding bird compared to the many thousands that pass through in migrations to and from Labrador. Although global warming might create less favourable conditions for a northern species inland, the use by these mergansers of coastal habitats, which are less likely to become much warmer, will reduce impact on this species from this cause.

Ref. 159

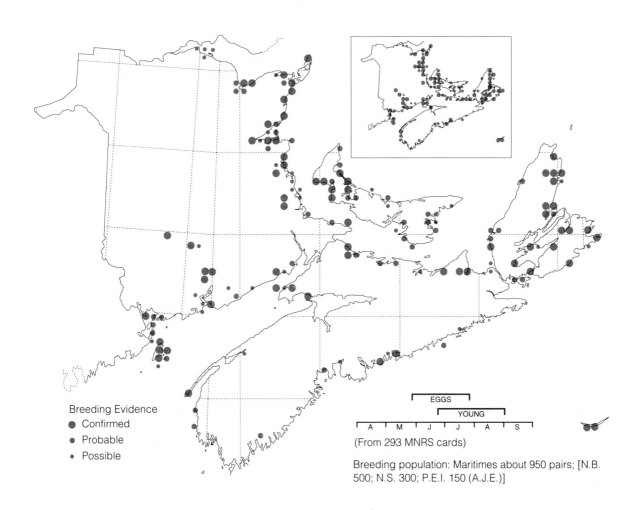

Breeding Evidence
● Confirmed
● Probable
● Possible

EGGS

YOUNG

| A | M | J | J | A | S |

(From 293 MNRS cards)

Breeding population: Maritimes about 950 pairs; [N.B. 500; N.S. 300; P.E.I. 150 (A.J.E.)]

Osprey   Balbuzard

Pandion haliaetus

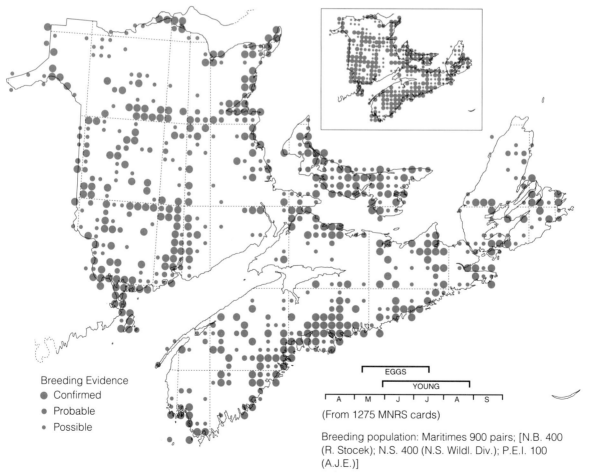

Ospreys are found all around the Northern Hemisphere; many are also found farther south, including on the coasts of Australasia. They breed throughout Canada and the U.S.A. from Alaska and southern Labrador into Central America. High-density breeding is mainly in coastal areas with extensive shallow waters, but they also nest inland near lakes with abundant fish. In the Maritimes their distribution was uneven, with obvious gaps in highland areas (north and southeast New Brunswick, and in the Cobequid Hills and Cape Breton Highlands, N.S.), and also in lowlands of east-central New Brunswick. Coastally, they were lacking around the upper Bay of Fundy, where muddy waters prevent visual detection of fish, and along other steep coasts. The Osprey is a large conspicuous bird, which feeds in the open over water, so its occurrence was very well represented. Breeding was confirmed in more squares (321, 49 per cent of 650 with the species) than for any other raptor, though American Kestrel was detected in more squares. Osprey nests (in 276 squares) were easily detected, even in the hinterlands; use of power-line poles is increasing, and artificial nest-platforms are readily accepted, so suitable nest-trees may be dwindling.

Probably Ospreys were always common in the Maritimes. The widespread shooting of "hawks," which decimated many large species before they were given legal protection in the 1940s to 1960s, affected Ospreys less than most. The accumulation of DDT metabolites, which reduced reproduction of Ospreys in the eastern U.S.A. in the 1960s, may have caused some reduction. DDT use against spruce bud-worm in New Brunswick continued from 1952 to 1967, and it still persists in South American wintering areas. This species was designated in New Brunswick as "endangered" in 1974, partly in reaction to alarms farther south as no population data were available here before then. Increasing numbers of known nests since then suggest that numbers were lower before the first extensive surveys in 1974–75 and may still be growing. Unless commercial use expands to include the coarse fish usually taken by Ospreys, no conflict exists between this species and human interests, and its presence is widely appreciated. We will continue to enjoy seeing these striking birds along our coasts in summer for years to come.

Refs. 6, 10, 47, 120

Breeding Evidence
- Confirmed
- Probable
- Possible

EGGS

YOUNG

A   M   J   J   A   S

(From 1275 MNRS cards)

Breeding population: Maritimes 900 pairs; [N.B. 400 (R. Stocek); N.S. 400 (N.S. Wildl. Div.); P.E.I. 100 (A.J.E.)]

Bald Eagle
Pygargue à tête blanche

Haliaeetus leucocephalus

The Bald Eagle, national symbol of the United States, breeds all across Canada as well as in the U.S.A. (including Alaska), but its range does not extend into the arctic tundra. It feeds largely on fish, as well as carrion of many species. Its nests, mostly close to water, are placed in large trees but often not above the forest canopy, and nest-territories, some of which include several nests, are used year after year. The breeding distribution of Bald Eagles thus is limited by availability of suitable nest trees within reasonable distance of waters with adequate fish supply. As the birds are large, scarce, and conspicuous, most nest sites in settled areas are well known, and many records (47 per cent of 328 squares with eagles) during the Maritimes Atlas study were of occupied nests (ON,NE,NY). The concentration of Bald Eagles breeding on Cape Breton Island is the largest in eastern North America.

Bald Eagles remained common in the Maritimes through the settlement era into this century. In many areas they were reduced or disappeared between 1945 and 1973 as a result of DDT accumulation through food chains. DDT use in forest and agricultural spraying in New Brunswick until 1967 probably contributed to losses here, including the virtual disappearance of the Bald Eagles from Florida which summered in New Brunswick until the 1950s. Eagles in the Maritimes had already been reduced by senseless persecution, and by 1970 the species bred here at only 20–30 sites away from Cape Breton Island. Since DDT use was banned in North America, eagles have re-occupied many traditional areas in central Nova Scotia and southwest New Brunswick, and also in Prince Edward Island where the species had been extirpated before 1945. Some apparently adult-plumaged eagles, seen in squares without known nests, may not yet be sexually mature, as these birds do not breed until four or five years old. Not all possible areas are occupied, but the large rivers of northern New Brunswick mostly lack suitable fish, and large further increases are not anticipated. The species was viewed as endangered as in the U.S.A. and was so designated in New Brunswick in 1976. This no longer seems appropriate, and reduction to "regionally endangered" has been recommended. COSEWIC does not consider the Bald Eagle as vulnerable nationally, and its numbers in our area are stable.

Ref. 56, 143

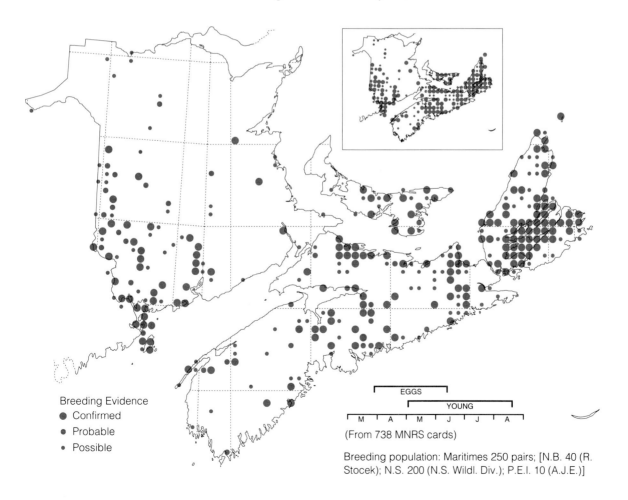

Breeding Evidence
- Confirmed
- Probable
- Possible

EGGS
YOUNG

M   A   M   J   J   A

(From 738 MNRS cards)

Breeding population: Maritimes 250 pairs; [N.B. 40 (R. Stocek); N.S. 200 (N.S. Wildl. Div.); P.E.I. 10 (A.J.E.)]

Northern Harrier
Busard Saint-Martin

Circus cyaneus

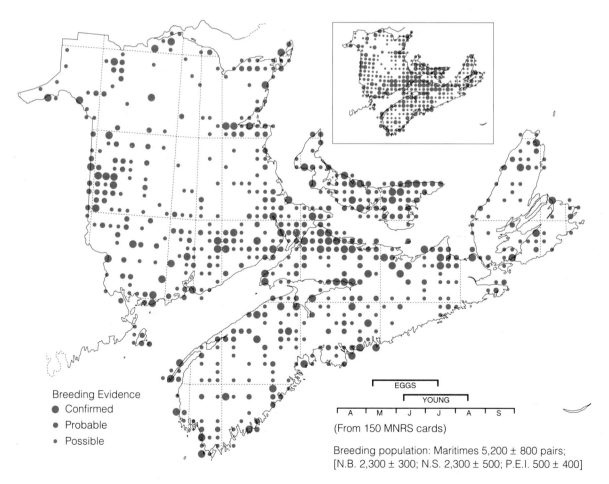

This harrier, formerly known as the Marsh Hawk, is widespread across North America as well as in Eurasia. Closely related harriers occur in South America and Africa. All of these birds breed in open habitats, especially marshes and damp meadows, in temperate and boreal regions. Their distribution is thus discontinuous. Here, harriers were most general in the agricultural areas, the St. John River Valley, southeast New Brunswick and adjacent parts of Nova Scotia, and in Prince Edward Island; they were also found scattered elsewhere except in thickly forested upland areas of northwest and central New Brunswick, southwest interior Nova Scotia, and the Atlantic slope of Nova Scotia. As these birds frequent open areas, where they cruise low over the fields, they are easily detected, but breeding is less easily confirmed than in many "hawks." Nests (NY,NE,ON) were found in only 32 (5 per cent) of 639 squares with the species, most records being single (H, 339 squares) or repeated (T, 104 squares) sightings of hunting birds. Although small mammals, especially meadow voles, are the most regu-

lar prey here, as elsewhere, birds up to the size of ducks (often injured or sick individuals) and frogs are taken when available.

Harriers, being visible and slow-moving, probably suffered to some degree from the indiscriminate shooting of all "hawks" that persisted throughout the Maritimes until recent decades. It is unlikely that wetland drainage before the last quarter-century was sufficient to have restricted available breeding habitat appreciably. Harrier numbers were well maintained, and in years of vole outbreaks they were and are locally common; in 1980, R. Simmons found 35 nests in an area of 300 ha in the New Brunswick–Nova Scotia border area, several times the density there in more typical years. More intensive draining of moist lowland fields, and reversion to forest of other marginal farmland, may reduce the traditional areas available to harriers for foraging in future, but breeding habitat is unlikely to be greatly restricted. Small numbers hunt over the large open areas of forest clear-cuts and young plantations. Harriers should be able to persist here in numbers approaching present levels.

Ref. 136

Breeding Evidence
● Confirmed
● Probable
· Possible

EGGS
YOUNG

A M J J A S

(From 150 MNRS cards)

Breeding population: Maritimes 5,200 ± 800 pairs; [N.B. 2,300 ± 300; N.S. 2,300 ± 500; P.E.I. 500 ± 400]

Sharp-shinned Hawk
Épervier brun

Accipiter striatus

Small hawks of the accipiter group occur all over the Earth, the Sharpshin occupying North America from the tundra edge south to Mexico and the Greater Antilles. Some of our birds are year-round residents of the breeding range, but others, especially immatures, migrate south in fall, the survivors returning in April. In breeding season they frequent conifer and mixed forests, their nests usually being in spruce trees. In contrast to their visible activity around winter bird-feeding stations, Sharp-shinned Hawks are very inconspicuous in summer, and the Atlas data undoubtedly under-represent their status in the Maritimes. They were found in all regions and probably occur generally throughout the area. Only 17 per cent (67 of 402 squares with this species; 15 with nests) had any confirmed breeding evidence, and the majority (276 squares, 69 per cent) had single sightings only (H). Their presence in summer is the only evidence that can be obtained without intensive study; a Ph.D. research

project in Fundy National Park, N.B., came up with only 12 nests in seven years.

Before the 1960s, most raptors were persecuted as rival predators considered harmful to human interests, but the small and elusive Sharpshin suffered less than larger species. Any losses from this cause were balanced by improved hunting opportunities in the woodlands opened up by human settlement, and also through access to introduced House Sparrows and Starlings in urban and suburban areas during the winters. Related accipiter species in Great Britain and the U.S.A. suffered lower reproductive success through bioaccumulation of chlorinated pesticides in the 1960s and 1970s; although pesticide residue levels also increased in Sharp-shinned Hawks, there is no evidence of reduced populations here. With a poorly represented species such as this, our population estimates are minimal. Sharp-shinned Hawks should be around in the Maritimes as long as there are small birds for them to eat, as trends in land use or in the overall environment, including global warming and acid precipitation, seem unlikely to affect them.

Refs. 46, 101

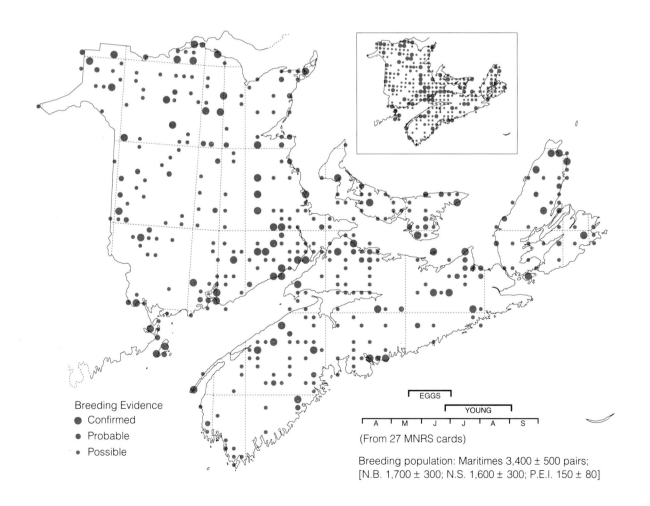

Breeding Evidence
● Confirmed
● Probable
· Possible

EGGS
YOUNG

A M J J A S

(From 27 MNRS cards)

Breeding population: Maritimes 3,400 ± 500 pairs;
[N.B. 1,700 ± 300; N.S. 1,600 ± 300; P.E.I. 150 ± 80]

Cooper's Hawk
Épervier de Cooper

Accipiter cooperii

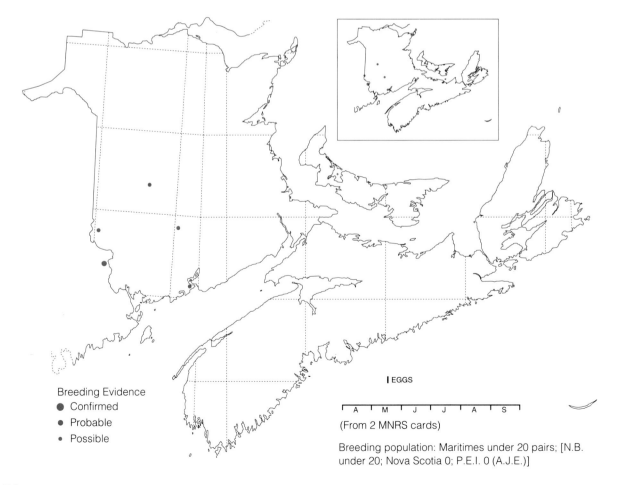

This is a bird of broad-leafed and mixed woodlands, often hunting along wood-edges in settled areas, from northern Mexico across the U.S.A. to southern Canada (British Columbia to Quebec). Being intermediate in size, it is easily confused with related species (Sharp-shinned Hawk, Goshawk), and many records in peripheral areas are suspect, as both other species extend farther north. Confirmed breeding in the Maine atlas was only in the extreme south and west, but other sightings were widely scattered east to the New Brunswick border. Former breeding in the Maritimes was proved by eggs collected in 1889 in New Brunswick and 1906 in Nova Scotia, but only a few reports have emerged more recently. The Atlas received one record of a group of recently fledged young (FL, 1987). The latter birds could have been hatched in Maine (within 1 km), so this record was mapped as probable breeding only. Four other sightings (H) were also considered plausible. Breeding is uncommon and seldom detected, even in southwest New Brunswick, and Cooper's Hawks may not breed here every year.

The egg collections suggest that the species may have been somewhat more regular a century ago in the Maritimes. Shooting of all "hawks," viewed as predators, up into the 1960s could have reduced their already low numbers. Cooper's Hawks, across their main range in the U.S.A., were reduced more than most other raptors in the 1950s and 1960s by accumulation of chlorinated hydrocarbon residues. Nevertheless, this species has persisted here at the fringe of its range, though in precariously low numbers. If Cooper's Hawks could survive here through the past century, we may predict that only drastically adverse conditions would eliminate them completely from our fauna. Global warming, if it comes to pass, might even allow this southern species to become more frequent here in future.

Breeding Evidence
- ● Confirmed
- ● Probable
- • Possible

EGGS

A M J J A S
(From 2 MNRS cards)

Breeding population: Maritimes under 20 pairs; [N.B. under 20; Nova Scotia 0; P.E.I. 0 (A.J.E.)]

Northern Goshawk
Autour des palombes

Accipiter gentilis

The Goshawk breeds throughout the northern forest zones of Eurasia and North America, south in the western mountains to north Mexico but only to Maryland in the east. Although it is more general in the boreal forest region, likely because less often disturbed there, it is also widespread in more temperate habitats. Goshawks are permanent residents, although some move south in fall most years. They were found scattered throughout the forests of the Maritimes, but their low density led to under-representation, especially in areas covered only on brief visits. Except where nests (in 34 squares) were found, the species was most often seen fleetingly, with only single sightings (H) in two-thirds of 214 squares with Goshawk records. The extreme scarcity of records in two thoroughly worked areas, around the lower St. John River and in northwest Nova Scotia, both with extensive road networks and many small farms and woodlots, suggests that some gaps in the mapped range are real, perhaps resulting from sustained persecution in the past.

Hawks and owls were generally more common in the Maritimes before Europeans came here. The actions of the Goshawk, a formidable predator of free-ranging poultry, as well as of hares and grouse, in rural areas, may have set off the persecution by humans of all these birds, which persisted well into the 20th century. However, the Goshawk, because it is less visible and faster-moving, was less vulnerable to indiscriminate shooting than most other raptors, and it remained relatively common in some major farming areas, such as the Annapolis-Cornwallis Valley and the middle part of the St. John River Valley. Continued large-scale forest cutting may hinder increases in the future, even now that these birds are protected by law, but Goshawks seem likely to persist at present low levels indefinitely.

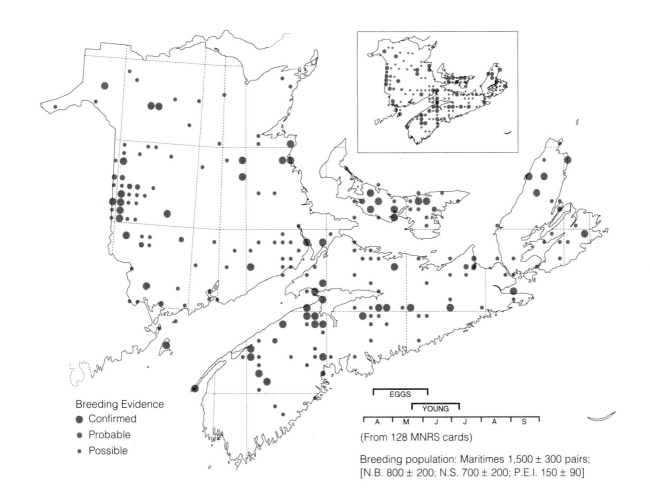

Breeding Evidence
- Confirmed
- Probable
- Possible

EGGS
YOUNG
A M J J A S

(From 128 MNRS cards)

Breeding population: Maritimes 1,500 ± 300 pairs;
[N.B. 800 ± 200; N.S. 700 ± 200; P.E.I. 150 ± 90]

Red-shouldered Hawk
Buse à épaulettes

Buteo lineatus

The Red-shouldered Hawk is a more southern species than our other Buteo hawks, with its main range in the eastern temperate forest of North America. It breeds from the Gulf Coast states and northern Mexico to California, Nebraska, Wisconsin, southern Ontario and Quebec, and Maine, mainly in lowland broad-leafed forests. In the Maritimes Atlas project, about 30 records were received, mostly from southwest New Brunswick, north to near Plaster Rock and east to Moncton. Several records were rejected, owing to inadequate or unconvincing documentation, as this bird is often confused with other Buteo hawks. Several earlier reports of breeding also had proved unsatisfactory, and the only accepted breeding record before the Atlas was of eggs collected in 1896 at Washademoak Lake, N.B. The Atlas accepted one record of confirmed breeding, near Hampstead, N.B., (FL). The species is not known to have bred in Nova Scotia or Prince Edward Island; of the five sightings received from Nova Scotia, the one mapped was an adult in mid-June. The others evidently were of stray birds that had overshot the usual range.

The primaeval landscape of the Maritimes included more mature broad-leafed forest, the typical habitat of Red-shouldered Hawks, than now exists; the climatic cooling of the Little Ice Age (1350–1850) would have made the area less attractive to these southern birds. The meagre historic record, with a single collection of eggs, suggests that this species was never more than a marginal breeder here, as it is now. The New Brunswick range mapped is continuous with that in Maine, so we conclude, as did previous compilers, that a few breed in New Brunswick.

Red-shouldered Hawks have declined throughout their range, but they are not obviously scarcer now in New Brunswick than in the past century. They easily could disappear without anyone here being aware of their withdrawal until long afterwards, as a population this sparse cannot be monitored effectively without efforts even more extensive and intensive than the Atlas project.

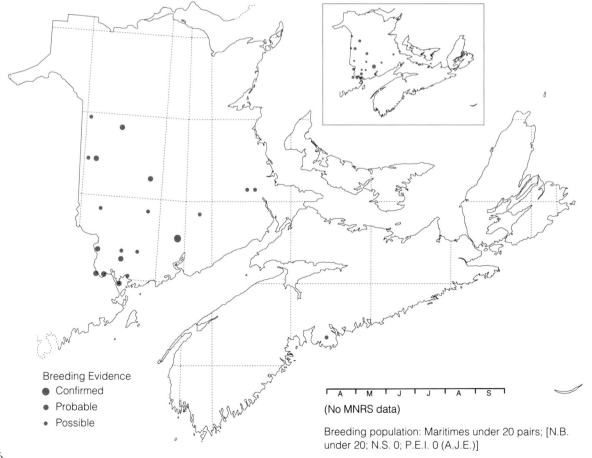

Breeding Evidence
● Confirmed
● Probable
● Possible

A M J J A S
(No MNRS data)

Breeding population: Maritimes under 20 pairs; [N.B. under 20; N.S. 0; P.E.I. 0 (A.J.E.)]

Broad-winged Hawk
Petite Buse

Buteo platypterus

This species breeds throughout the eastern broad-leafed forest from Florida and Texas north into southern Canada, where it extends west through the parklands to Alberta. It was found widely in New Brunswick, except in the extreme northwest, but in Nova Scotia it was regular only in the Northumberland Strait lowlands; elsewhere it was scarce and local, virtually absent on Cape Breton Island and the southeast mainland. Three birds sighted during Atlas work were the first ever recorded in Prince Edward Island; the Broad-winged Hawk may be starting to occupy the island's regenerating forests. This species is seldom seen in summer away from closed stands of broad-leafed or mixed woods, where it feeds on small mammals, frogs, and large insects. Its call is neither very loud nor very striking, and the species is easily missed. Most records (in 60 per cent of 412 squares) were of single sightings (H) only; confirmed breeding records were more often of nests (NY,ON, in 30 squares) than of fledged young (FL, 21 squares) or adults with food (AY, 11 squares).

In 1600, the forests of the Maritimes included more Broad-winged Hawk habitat than they do at present. Subsequent clearing of mature hardwood and mixed forests restricted their nesting opportunities, and indiscriminate shooting of raptorial birds reduced their numbers. The nearly complete removal of mature forests in Prince Edward Island by 1900 probably explains the species' absence there in the past. The increase in Broad-winged Hawk records in Nova Scotia during the last 40 years parallels the growing interest in birdwatching, but it may point to a real increase; there is no indication of such a change in New Brunswick in that period. Forest management emphasis on increasing the conifer component of Maritimes forests leaves the impression that Broad-winged Hawks will be lucky to maintain their present numbers in future.

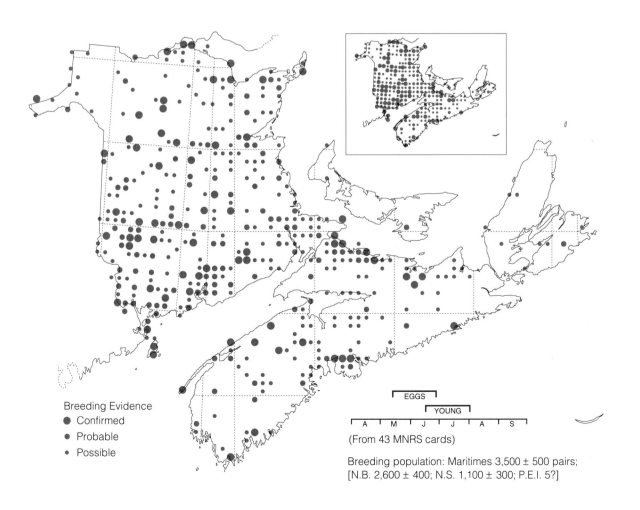

Breeding Evidence
- Confirmed
- Probable
- Possible

EGGS

YOUNG

A M J J A S

(From 43 MNRS cards)

Breeding population: Maritimes 3,500 ± 500 pairs;
[N.B. 2,600 ± 400; N.S. 1,100 ± 300; P.E.I. 5?]

Red-tailed Hawk
Buse à queue rousse

Buteo jamaicensis

This is the largest and most widespread of our Buteo hawks, breeding throughout North America up to the limits of continuous forest. Closely related forms inhabit South America and most of Eurasia. Red-tailed Hawks mostly nest in woodlands; although cliffs may be used where trees are scarce, especially in the prairies, this has not been reported in our area. However, they often forage in open areas including forest cut-overs, unlike the strictly woodland-dwelling Broad-winged Hawk. In the Maritimes, Red-tails were termed uncommon to rare in the past, but they were found widely during the Atlas period, especially in Nova Scotia. Breeding was confirmed in 100 squares (18 per cent of 547 squares with these birds; similar to most other hawks), with nests seen (NY,NE,ON, in 42 squares, 8 per cent) almost as often as newly flying young (FL, in 48 squares). Although Red-tails were often detected, usually calling in flight over woodlands, they were seldom noted twice in the same area (T, in only 58 squares, vs. 314 with single records, H); perhaps they range widely while breeding here.

The opening-up of the Maritimes forests since the start of the settlement period created more Red-tailed Hawk habitat than existed previously here. However, this species probably suffered more than most others from the persecution directed at all birds of prey, for supposedly competing with people for game species, up to 1940 or later. The scarcer status attributed to this bird in earlier summaries probably reflected populations reduced by indiscriminate shooting, from which recovery still may not be complete. There are no useful data on trends in numbers. Given the great adaptability of this species across its broad range, predicted changes in land use and climate in the Maritimes seem unlikely to hinder the Red-tailed Hawk from maintaining or exceeding its present numbers.

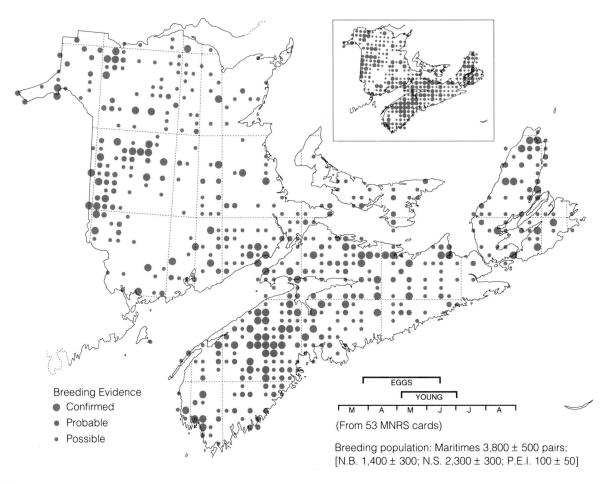

Breeding Evidence
- Confirmed
- Probable
- Possible

EGGS
YOUNG

M A M J J A

(From 53 MNRS cards)

Breeding population: Maritimes 3,800 ± 500 pairs;
[N.B. 1,400 ± 300; N.S. 2,300 ± 300; P.E.I. 100 ± 50]

American Kestrel
Crécerelle d'Amérique

Falco sparverius

This species is the western hemisphere member of a world-wide group of small falcons of similar colour and habits. It breeds throughout North and South America, wherever open areas for foraging occur near trees, cliffs, or buildings containing cavities suitable for nesting. In the Maritimes, it breeds almost everywhere, except in the few continuously forested areas and along the Atlantic slope of Nova Scotia. Possibly, suitable nest-cavities are scarce in the generally smaller trees in the latter region, and grasshoppers, one of the main foods, may also be scarce in the coastal barrens and poor farmland there. The Kestrel eats small mammals and small birds when these are easily available, but it seldom relies on these for food, unlike the similarly sized Merlin and Sharp-shinned Hawk. It is most frequent here in the better farming areas and in cut-over forest lands, where it is conspicuous and often noisy, hence easily detected, unlike many other predatory birds. Nests, especially in tree cavities, are often found, but the contents are less easily ascertained; 100 (44 per cent) of 228 squares with breeding confirmed involved nests found (mostly ON), and recently fledged young (FL, in 111 squares) were only slightly more frequent than nests. Breeding was confirmed in 31 per cent of the 743 squares with Kestrels, a higher percentage than for other diurnal raptors except the much larger Bald Eagle and Osprey.

The clearing, for settlement and later for forestry, of many parts of the forested Maritimes since 1600 expanded greatly the opportunities for Kestrels here. Persecution by Europeans of hawks and owls, viewed as rival predators on game birds, extended even to this harmless species, but probably reduced it less than larger raptors. It was earlier considered to be the most common hawk in Nova Scotia and to have increased noticeably in recent years in New Brunswick. More recently, there is an impression of decreased numbers, at least in farmlands, with no plausible explanation, but it remains one of our common raptors. Our "crystal ball" has revealed no obvious causes to anticipate future increase or decline of Kestrels in the Maritimes.

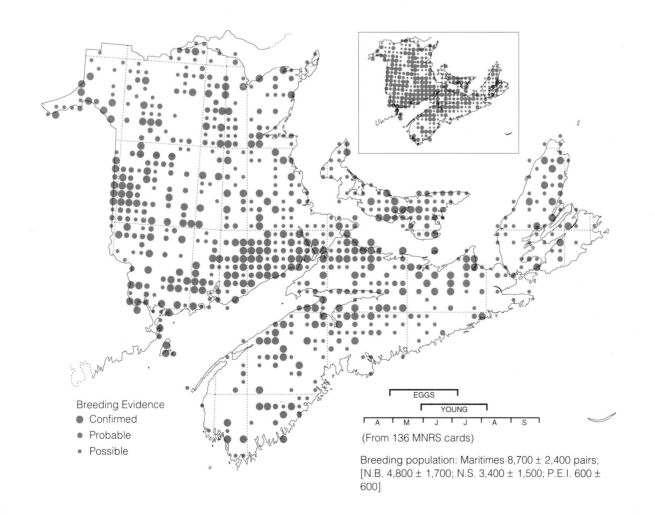

Breeding Evidence
- Confirmed
- Probable
- Possible

EGGS
YOUNG
A M J J A S

(From 136 MNRS cards)

Breeding population: Maritimes 8,700 ± 2,400 pairs;
[N.B. 4,800 ± 1,700; N.S. 3,400 ± 1,500; P.E.I. 600 ± 600]

Merlin
Faucon émerillon

Falco columbarius

This little falcon breeds all around the Northern Hemisphere in boreal and north temperate regions. It nests on the ground in open lands north of or altitudinally above tree-line, but in forested areas such as the Maritimes it often uses old crow nests in trees. The Merlin feeds on small birds and is most often seen, in the Maritimes, harrying migrating shore-bird flocks or coastal concentrations of migrant land birds during the autumn months. Its status as a breeding bird here seems to have changed recently, perhaps shortly before the Atlas period. Through 1985, New Brunswick sources showed no nest found in that province, and fewer than a dozen nests in Nova Scotia and two in Prince Edward Island had been reported. The Atlas map shows that Merlins bred, albeit rather sparsely, throughout the Maritimes in 1986–90. Confirmed breeding, mostly involving adults carrying food (AY) or family groups of fledged young (FL), was

noted in 18 of the 23 Atlas regions (52 of 240 squares with Merlins), with nests found in 16 squares. Several sight records (H) during August were excluded, as Merlins, possibly of more northern origin, are seen here pursuing migrant shorebirds from mid-July onward.

Merlins are not notably wary birds, and their numbers in the Maritimes probably were reduced during the years when any raptorial bird here was a target for gunners. They remained common on the island of Newfoundland throughout. The recent evidence of widespread breeding here might represent a recovery towards former larger numbers, but the fact that the egg-collectors up through the early 1900s, who found nests of the now much rarer Cooper's and Red-shouldered Hawks, failed to detect this species argues against it. The Merlin is not an obvious bird while breeding, except in the immediate vicinity of its nest, and our records undoubtedly under-represent it. As long as stocks of breeding song birds persist in the Maritimes, small numbers of these attractive predators are likely to continue to nest here.

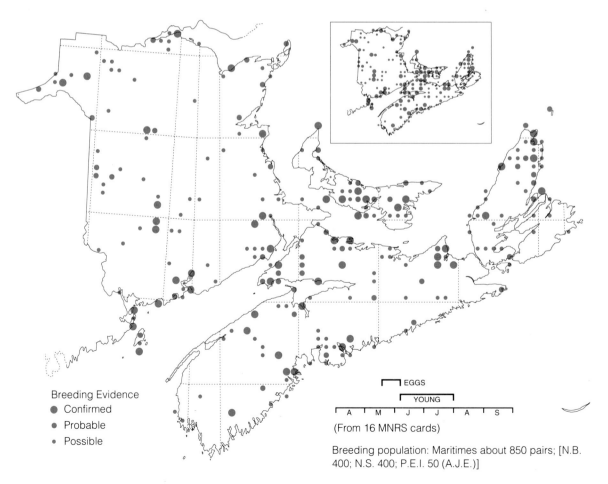

Breeding Evidence
- Confirmed
- Probable
- Possible

EGGS
YOUNG
A M J J A S

(From 16 MNRS cards)

Breeding population: Maritimes about 850 pairs; [N.B. 400; N.S. 400; P.E.I. 50 (A.J.E.)]

Peregrine Falcon
Faucon pèlerin

Falco peregrinus

This falcon breeds, or bred, almost everywhere on Earth except in the extremes of hot and cold deserts and tropical rain-forests. Before World War II, Peregrines bred throughout North America, but by 1960 they had disappeared entirely from the east. The cause was reproductive failure resulting from DDT and other pesticides accumulated through their prey species. Present breeding birds in the Maritimes resulted from deliberate releases of captive-hatched young birds at and near traditional nesting cliffs. Breeding to date has never occurred at the points of release, but at suitable sites not far away, in one case under a high-level road bridge over Saint John Harbour. The records accepted involved actual nestings, or pairs occupying cliff ledges, and are too few to allow generalization. Concern over this endangered species (designated by COSEWIC) ensured that all reports of suspected nesting were checked by authorized people.

Peregrine Falcons were never common breeding birds, in the Maritimes or elsewhere in forested areas of North America. Here they relied for food mainly on small- to medium-sized forest birds such as Flickers, Blue Jays, and Robins until late July when sandpipers, migrating from the arctic, arrived on nearby beaches. Even before the pesticide years, Peregrine numbers had been reduced by human persecution of all predatory birds. Fewer than a dozen nest-sites were known to have been occupied in the Maritimes through 1960, and potential numbers here were estimated at fewer than 20 pairs. Peregrines disappeared here after 1960. Re-introductions began in 1982 and will continue until at least five pairs are established and breeding. The first recent nesting occurred in 1989. To ensure their continued success, the breeding locations, except the well-publicized site at Saint John are obscured in this map. The future of the species here is still precarious, but governments are committed to re-establishment of Peregrines in the Maritimes.

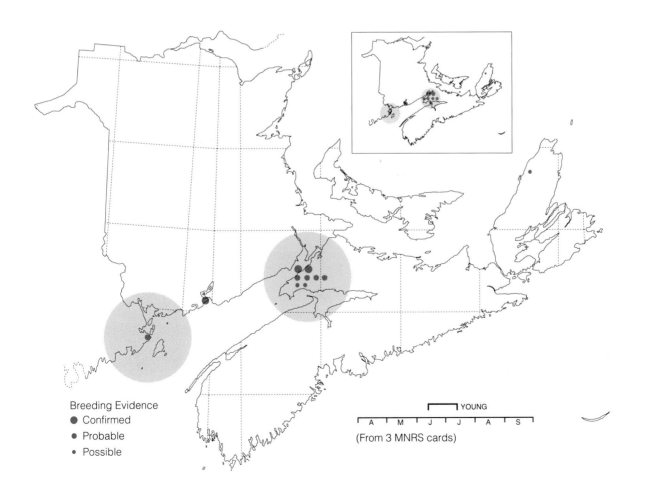

Breeding Evidence
● Confirmed
● Probable
• Possible

YOUNG

A M J J A S

(From 3 MNRS cards)

Gray Partridge
Perdrix grise

Perdix perdix

The Gray Partridge is native to eastern Europe and western Asia, from whence it was introduced widely in North America as a game bird. The populations established in the Maritimes are far separated from the main range in America, which extends from southwest Quebec and New York to the Rocky Mountain foothills of Alberta and Colorado, with others in the Great Basin from British Columbia to California. Throughout their range, these are birds of open lands, which in the Maritimes means mainly farmland. Deep snow cover limits their winter survival, but areas with strong winds that blow the ground clear, as in Prince Edward Island and the prairies, may retain their stocks while others disappear. Crusting of snow in damp coastal areas may be implicated in disappearance of some Maritimes stocks, as "Huns" feed entirely

on the ground, on waste grain and weed seeds, and in spring and summer on insects. In their open habitats, partridges are easily detected while foraging, and most records (29 of 51 squares with the species) were of family groups (FL).

Gray Partridges were first released in each of the Maritime Provinces in 1926–27. Populations peaked around 1940, when hunting was first permitted, and no substantial numbers were released after 1960. The New Brunswick birds were scarce by 1960, with last records near Moncton and Shediac about 1981, and the Nova Scotia birds also declined to remnants of their former abundance. Only in Prince Edward Island have partridges done well. Even if the "greenhouse effect" accentuates the trend to reduced snow cover in winter, Gray Partridges may disappear gradually in Nova Scotia, though they should persist in Prince Edward Island.

Ref. 74

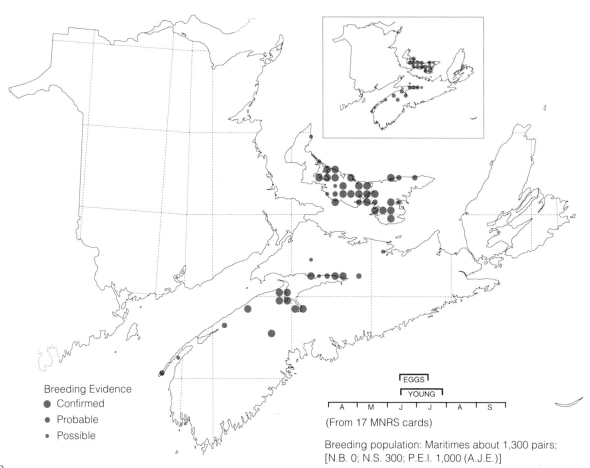

Breeding Evidence
● Confirmed
● Probable
● Possible

EGGS
YOUNG

A M J J A S

(From 17 MNRS cards)

Breeding population: Maritimes about 1,300 pairs;
[N.B. 0; N.S. 300; P.E.I. 1,000 (A.J.E.)]

Ring-necked Pheasant
Faisan de chasse

Phasianus colchicus

This bird of the Eurasian steppes was introduced long ago to western Europe, and thence to many parts of the Earth's temperate zones, as a game species. Its occurrence in Canada thus is erratic, depending on where it was introduced, whether the habitat is suitable in winter as well as summer, and whether the birds are fed artificially in winter. In the Maritimes, as elsewhere, it was found as a permanent resident in some farming regions, and particularly in the Annapolis Valley, the New Brunswick–Nova Scotia border and adjacent areas in both provinces, and Carleton Co., N.B. Other clusters seemed to be focused on suburban areas in Nova Scotia near Yarmouth, Bridgewater, Halifax, and Truro, probably because of more numerous winter feeding stations there. Except in southwestern Nova Scotia, Pheasants probably persist here only with the aid of supplemental feeding in winter. Hunting has been restricted for many years to put-and-take "pheasant preserves," and survivors from such continuing releases account for some of the widely scattered records in the Atlas. Breeding was usually confirmed by broods, sighted in 100 squares (43 per cent of 230 with these birds), and nests were noted in 10 squares (4 per cent).

Pheasants were first introduced to the Maritimes in the 1850s, but they did not become established until the 1920s or 1930s. As with some other exotic species released here, they increased greatly for a few years, until set back by hard winters, and the anticipated hunting opportunities proved short-lived except where maintained by ongoing releases. Introductions other than on preserves virtually ceased after 1965. The established population, independent of deliberate human support, is certainly small and is probably limited by snow cover on natural feeding areas in winter. At present, they are probably still outside their limits of tolerance for unaided occurrence in the Maritimes. If global warming comes about in future, a reduction in snow cover might well assist Pheasants to a less precarious residency here.

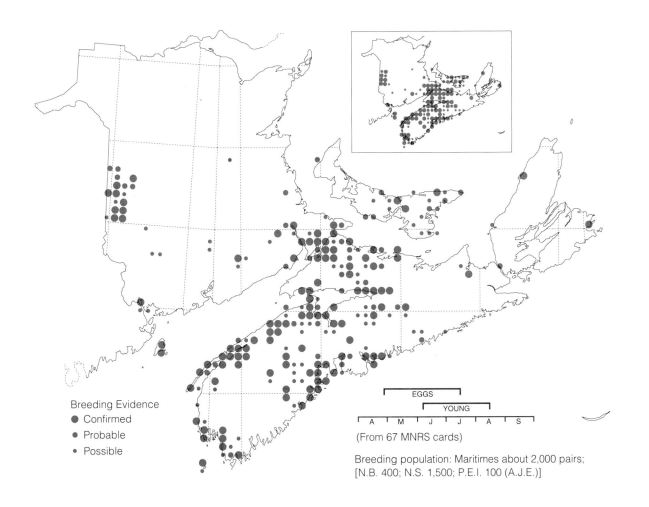

Breeding Evidence
- Confirmed
- Probable
- Possible

EGGS

YOUNG

A   M   J   J   A   S

(From 67 MNRS cards)

Breeding population: Maritimes about 2,000 pairs;
[N.B. 400; N.S. 1,500; P.E.I. 100 (A.J.E.)]

Spruce Grouse
Tétras du Canada

Dendragapus canadensis

The "fool hen" breeds through the boreal regions of Alaska and Canada, and south to Oregon in the western mountains. It is the grouse of spruce forests, where it prefers more open areas, including regenerating second-growth and forested bogs, and also jack pine stands. In the Maritimes, the Spruce Grouse was widespread in forested areas of Nova Scotia, but evidently scarcer in northwest and southeast New Brunswick than in other parts of that province; it was absent from hardwood-dominated areas, including the St. John River Valley, and from Prince Edward Island where there is no certain evidence that it ever occurred. Its reluctance to cross even narrow straits is shown by its failure also to reach Newfoundland until introduced in 1964, but its absence from the Grand Manan archipelago until recently could have followed extirpation of a former population. Its unfamiliarity to many birders reflects the urban habitat and habits of people more than a real

scarcity of the bird. Its tameness and ground-nesting habit give it little chance for survival near human settlement among roving dogs and cats, but in its appropriate remote habitats, the Spruce Grouse is easily seen and identified. Most records obtained in the Atlas study were of broods of partly grown young (FL, half of 236 squares with the species, nearly all the others being of lone adults — 91H, 13T). Three nests were reported.

Its numbers may have gained as much from opening-up and replacement by conifers of the formerly more continuous mixed forests of the Maritimes as they have lost to settlement and to predation by people and their animals. The future for the species in the Maritimes depends on the degree of disturbance of remote woodland areas caused by forest management; the management effort is focused on maintaining conifer habitat, much of it in early growth stages, which should favour the species. Global warming, by encouraging more temperate forest types, may be expected to have negative effects, in the long term.

Refs. 75, 124

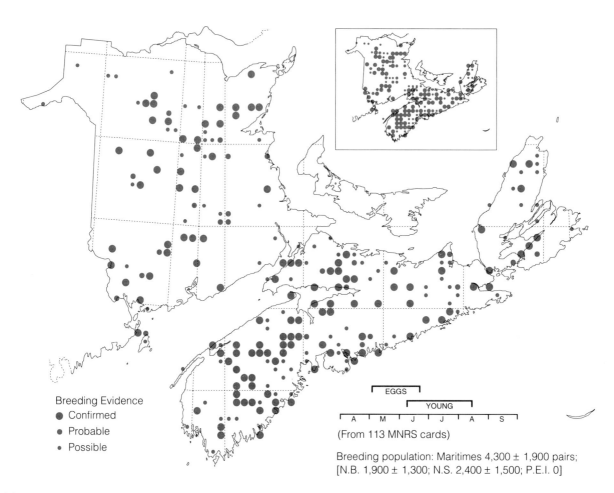

Breeding Evidence
● Confirmed
● Probable
● Possible

EGGS
YOUNG
A M J J A S
(From 113 MNRS cards)

Breeding population: Maritimes 4,300 ± 1,900 pairs;
[N.B. 1,900 ± 1,300; N.S. 2,400 ± 1,500; P.E.I. 0]

Ruffed Grouse
Gélinotte huppée

Bonasa umbellus

This species is a permanent resident and the most widespread grouse over most wooded parts of Canada and the U.S.A., wherever broad-leafed trees make up an appreciable proportion of the forest cover. Other grouse (e.g., Spruce, Blue) replace it where mainly conifer forests prevail, and still other species such as ptarmigans or Sharp-tailed Grouse occur in the open tundra and plains. Numbers of Ruffed Grouse vary widely but irregularly, over periods averaging about ten years. The grouse "cycle" seems not to be synchronous across our area, so these birds were scarce in some parts of the Maritimes but not in others during the Atlas period. Ruffed Grouse were noted in all regions, least generally in southeast New Brunswick, and not less regularly in regions with higher proportions of conifer forest (northwest New Brunswick, Atlantic slope of eastern Nova Scotia); they probably would have been found even more generally during a population "high." The species is easily detected, both from its "drumming" in spring and by its acrobatic displays when trying to distract people or other enemies from their broods. Most confirmations of breeding (in 566, 70 per cent of 809 squares with the species) were by seeing broods (FL, in 491 squares), as nests (NY,NE,ON) were found in only 53 squares (7 per cent).

The Ruffed Grouse was probably a common bird in the forests of the Maritimes before Europeans came here, subject to periodic variations in numbers as at present. Land-use and forest-cover changes in the past 400 years seem unlikely to have affected greatly the area's overall suitability for this species, even in Prince Edward Island where most forests were cleared in the past. Locally, forest succession in the past often led to solid conifer stands, which later were broken up by spruce budworm outbreaks, but such changes were superimposed on local variations without affecting the general abundance of Ruffed Grouse. No widespread trend data are available, as the BBS does not detect Ruffed Grouse in sufficient numbers to allow trend analysis. Like many grouse, these birds do not form pairs while breeding, but most records, of drumming males or feigning females, represent only one sex, so may be extrapolated in the same way as for other species. Projected trends suggest that Ruffed Grouse should remain plentiful here, with variations as in the past, in the coming decades.

Ref. 144

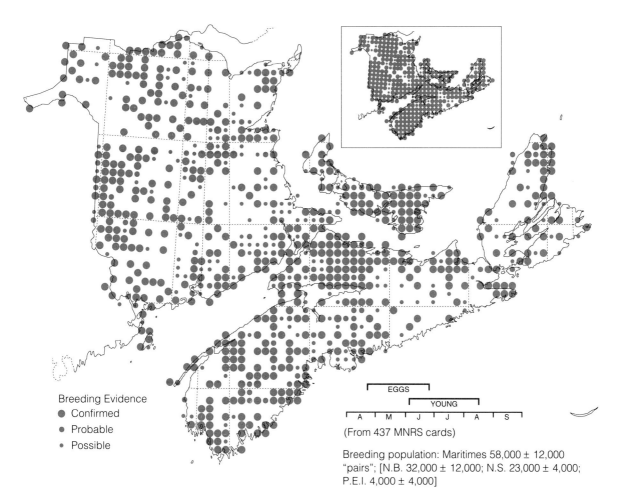

Breeding Evidence
- Confirmed
- Probable
- Possible

EGGS
YOUNG
A M J J A S

(From 437 MNRS cards)

Breeding population: Maritimes 58,000 ± 12,000 "pairs"; [N.B. 32,000 ± 12,000; N.S. 23,000 ± 4,000; P.E.I. 4,000 ± 4,000]

Virginia Rail
Râle de Virginie

Rallus limicola

Rails are secretive birds, and knowledge of their ranges must be built up from casual observations and responses to playback of tape-recorded calls. This species breeds across southern Canada, in suitable marsh habitats, from British Columbia to the Maritimes, and south through the U.S.A., Central America and South America. In our area, Virginia Rails are certainly more common and widespread than the few records show, as these birds are among the most poorly represented of any in our area. The concentration of reports in the three areas of relatively fertile marshes—lower St. John River, New Brunswick–Nova Scotia border, and east-central Prince Edward Island — reinforces the view that Virginia Rails are scarce or absent in many regions of the Maritimes. Except for 11 squares with sightings of flightless young birds (FL), most records (37 of 57) were of single (H) or repeated (T) detections, mostly by vocalizations.

Given the secretive nature and small size of Virginia Rails, we can infer that relatively few were shot for human food over the years since 1600, and that hunting had little effect on numbers and range of this species. Although breeding was not proved to occur until this century (1908 in New Brunswick, 1962 in Nova Scotia, 1970s? in Prince Edward Island), these birds were presumably present all along in their favoured marshes, and any decreases were mainly from loss of marsh habitat. With ongoing marsh creation and maintenance in the interests of waterfowl, rail habitat is more likely to increase than decrease in future, and Virginia Rails will continue to lurk, largely undetected, in our marshes in summer.

Ref. 103

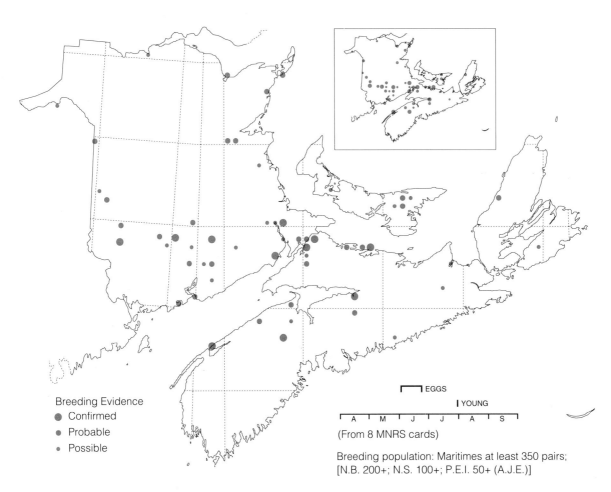

Breeding Evidence
- Confirmed
- Probable
- Possible

EGGS

YOUNG

A M J J A S

(From 8 MNRS cards)

Breeding population: Maritimes at least 350 pairs;
[N.B. 200+; N.S. 100+; P.E.I. 50+ (A.J.E.)]

Sora
Râle de Caroline

Porzana carolina

The most common of our rails, this species breeds across Canada and the United States, from southern Yukon to Baja California in the west, but only from Pennsylvania to the Maritimes in the east. In our Atlas area, it was scattered in many regions, but was most general in eastern Prince Edward Island and near the New Brunswick–Nova Scotia border. Both of these areas have many small, impounded, freshwater marshes on relatively fertile soils, and elsewhere in the Maritimes the Sora also occurred mainly in the more fertile marshes. Cattail and giant bur reed are the most usual vegetation in these habitats, and Sora nests are built in clumps of these aquatic plants. Like other rails, Soras skulk through the marsh and are seldom seen, but their "whinny" and "kee-eek" calls, once learned, betray their presence. The species was certainly under-represented in the Atlas records (only found in 193 squares), but probably less so than other rails. Fledged young (FL, in 45 squares) were the most frequent confirmation of breeding, as most records (135 squares) were of calling birds heard during single (H) or repeated (T) visits to an area. Three squares had nests, the only nests reported of any rail species, although searches of suitable marshes with a dog may turn up that many in a June morning.

Loss of former marshlands by drainage has been partly offset through the impoundment of wetlands for waterfowl management, mostly by DUC since 1965. Many of the new marshes are on dyked lands that were formerly salt marsh, which is not used by Soras, so the area of habitat available now may even approach that present when European settlement began. Soras may never have been much more numerous here than at present, as the widely scattered records away from the main concentrations may add up to perhaps one-third of the total. Ongoing marsh management for ducks will ensure that Sora breeding habitats in the Maritimes persist for the foreseeable future.

Ref. 103

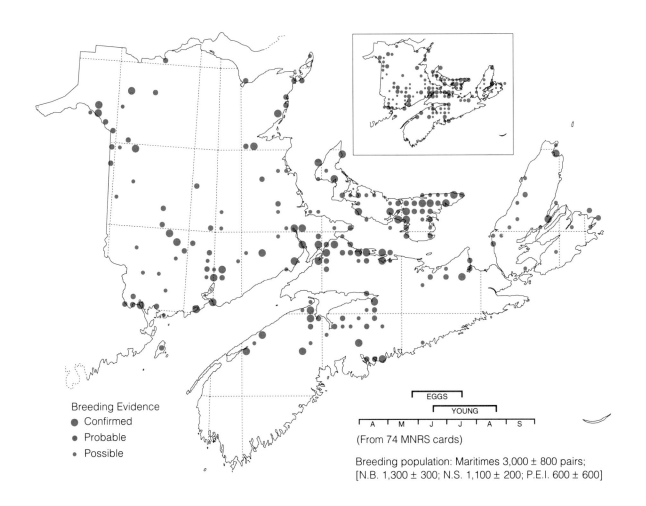

Breeding Evidence
- Confirmed
- Probable
- Possible

EGGS
YOUNG

A M J J A S

(From 74 MNRS cards)

Breeding population: Maritimes 3,000 ± 800 pairs;
[N.B. 1,300 ± 300; N.S. 1,100 ± 200; P.E.I. 600 ± 600]

Common Moorhen
Poule-d'eau

Gallinula chloropus

The Moorhen, a large aquatic rail, is distributed in temperate and tropical regions almost worldwide, with a replacement species in Australasia. In America, it breeds north to California, Nebraska, Michigan, and Massachusetts. It is resident year-round in much of its range, but migrates south from cooler areas, including southern Ontario, the only part of Canada where it is widespread. In Maine, breeding was confirmed in only 4 well-spaced squares in 1978–83, so the Maritimes are well beyond the main range. The regular occurrence here of Moorhens dates back to 1960, first at Red Head Marsh near Saint John. With the development since 1965 of impounded marshes by DUC, breeding was first reported at Amherst Point in 1976, and at other newly impounded areas in the New Brunswick–Nova Scotia border area since then. Atlas records also included confirmed breeding near Germantown and Hillsborough, N.B., and near Tatamagouche, N.S., with sightings of adults (H,P,T) in other marsh areas including Musquash, N.B., and Malagash and Windsor, N.S. Nests were found at Amherst Point in 1982–83, but all confirmed breeding in the Atlas was based on flightless young birds (FL, in 5 of 13 squares with Moorhens).

The Moorhen had not bred in the Maritimes and adjacent areas before the recent period, probably owing to the infertility of most marshes here. Although Moorhens use a wide variety of fertile wetlands in their main range, they seem to use only man-made habitats here, mostly during the first years after an area is impounded, when there is a flush of aquatic invertebrate life. Along with other species of similar habitats, such as Redhead and Black Tern, the Moorhen disappeared from Amherst Point within 10 years after its first breeding there. The total breeding population is very small, and the small local stocks could disappear from one year to the next without any cause being obvious. Ongoing efforts to create marsh habitat for waterfowl should help, but it is too early to forecast whether the Moorhen will persist as a breeder here. Global warming may assist it to do so.

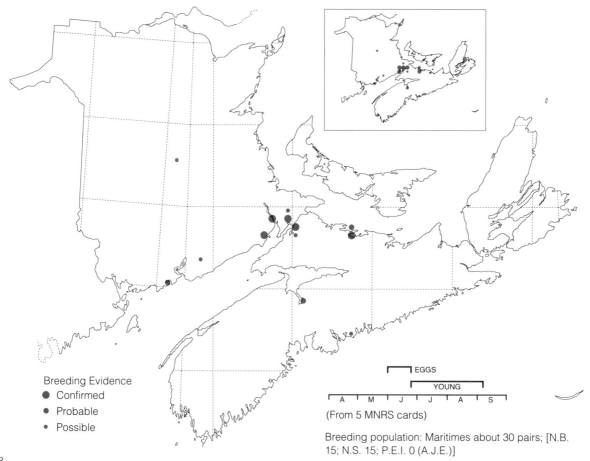

Breeding Evidence
● Confirmed
● Probable
• Possible

EGGS

YOUNG

A M J J A S

(From 5 MNRS cards)

Breeding population: Maritimes about 30 pairs; [N.B. 15; N.S. 15; P.E.I. 0 (A.J.E.)]

American Coot
Foulque d'Amérique

Fulica americana

There are coots of one or another species on most continents. This large aquatic rail breeds in marshes with open water through much of western Canada and the U.S.A., and into northern South America. East of the prairies, it is a scarce and local breeder, which has only recently begun to breed in the northeast, where it is a regular though scarce migrant in fall. Artificial impoundments for waterfowl and sewage lagoons are its chief haunts from Ontario to the Maritimes. In our area, the few breeding records (in 21 squares, 7 with broods) were mostly in the New Brunswick–Nova Scotia border region, with single records of confirmed breeding in eastern Prince Edward Island, northern Nova Scotia, and in the St. John Valley. As broods were reported in the past in areas of New Brunswick where Coots were not found later (e.g., Harvey Station in 1941; Williamstown Lake in 1944), some isolated sightings (H) in the Atlas may also have represented sporadic instances of breeding.

Only the more fertile marsh areas were used by this species, which was not reported as breeding in the Maritimes until 1941. It bred here quite irregularly until DUC impoundments appeared in the late 1960s, and it is still a rare species with us. Increased activity in habitat management for waterfowl should help the Coot to maintain its small population in our area.

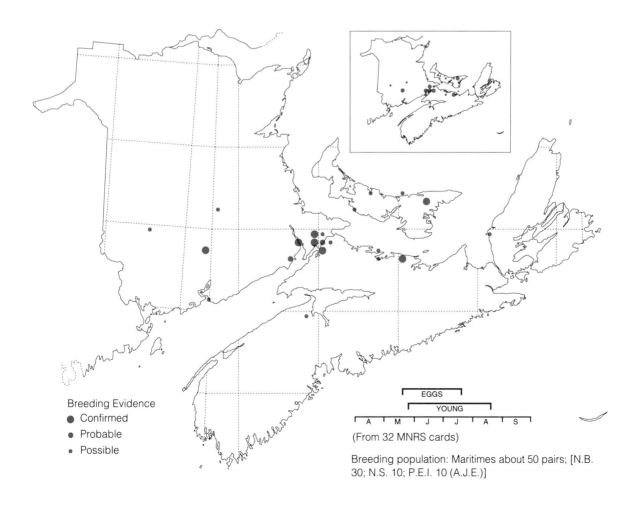

Breeding Evidence
● Confirmed
● Probable
● Possible

EGGS
YOUNG

A M J J A S

(From 32 MNRS cards)

Breeding population: Maritimes about 50 pairs; [N.B. 30; N.S. 10; P.E.I. 10 (A.J.E.)]

Semipalmated Plover
Pluvier semipalmé

Charadrius semipalmatus

This is the American member of a group of small, brown-backed, arctic-nesting plovers, distinct from the pale plovers of similar size that use sandy beaches in warmer areas. Semipalmated Plovers breed in the low arctic and subarctic regions of northern Canada and Alaska, extending south along coasts washed by cold sea waters to the Maritimes, James Bay, and the Queen Charlotte Islands. Ours is the southernmost breeding area for the species. Atlas records showed confirmed breeding continuing in some previously known areas: in Yarmouth Co., on Sable Island, and along Northumberland Strait in Nova Scotia, and at New London Bay, P.E.I., Isle Haute, N.S., and Maximeville, P.E.I., where breeding was unknown earlier. During the Atlas period previous breeding sites in the Grand Manan island group and at Waterside in New Brunswick had only pairs and courtship in spring, and the Englishtown, N.S., area was not used. Some reports of this species (not mapped) evidently reflected confusion with Killdeer, of which the young have only one breast-band in their first plumage, a point not mentioned in many field guides. Confirmed breeding was mostly based on broods (FL, in 6 squares), with nests in 2 squares.

The Maritimes breeding population of Semipalmated Plovers was first detected over a century ago, with eggs and young collected in the 1870s, so it is of long standing. The species may have found conditions more suitable here during the Little Ice Age (1300s–1800s), but it probably was never common at the southern extremity of its range. The Semipalmated Plover is somewhat less vulnerable than the Piping Plover, as the gravel beaches it uses are less in demand by people. Its numbers, however, are so small that local stocks can be lost easily from natural causes and without the loss being detected promptly. Monitoring their numbers may cause more disturbance than benefit, unless extreme caution is used.

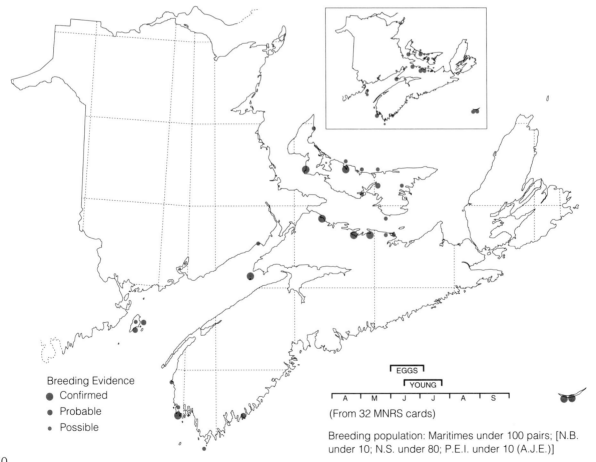

Breeding Evidence
● Confirmed
● Probable
· Possible

EGGS
YOUNG
A M J J A S
(From 32 MNRS cards)

Breeding population: Maritimes under 100 pairs; [N.B. under 10; N.S. under 80; P.E.I. under 10 (A.J.E.)]

Piping Plover
Pluvier siffleur

Charadrius melodus

This is one of several small plovers adapted to white sand beaches that, in various parts of the globe, have suffered from human uses of these sites. With its specialized habitat requirements, the Piping Plover always had a fragmented distribution, around shores of prairie lakes, formerly around the Great Lakes, and along the Atlantic seaboard from Newfoundland to North Carolina. Many local populations were lost in the past century, and the species is officially designated as "endangered" in both the United States and Canada. Piping Plovers breed in each of the Maritime Provinces, with the shores of the Gulf of St. Lawrence in New Brunswick and Prince Edward Island having more potential habitat but fewer inaccessible beaches than the much-fragmented outer coast of Nova Scotia. The birds nest and rear their young on open sandy shores and thus are easily detected by people alert to their presence. Most of the records received were of confirmed breeding, divided between nests (NE,ON) and broods of flightless young (FL), in 68 per cent and 19 per cent, respectively, of 57 squares

with the species.

This species must have declined in recent decades with the ever-increasing use of suitable beaches by summer visitors. Beaches now used intensively by people often lost part or all of their former breeding plover populations before these were documented. By comparison with areas of potential breeding habitat, the best guess is that one-half or more of the original Maritimes Piping Plovers have been eliminated, largely without destructive intent. Population estimates, from intensive surveys by CWS and co-operating agencies in 1987, showed that the Maritimes holds about one-third of the total population of Piping Plovers in Canada. The fact that one-quarter to one-half of each province's breeding birds are found in National Parks — Kouchibouguac, N.B., Kejimkujik Seaside Adjunct, N.S., and Prince Edward Island, (P.E.I.), complicates their future conservation. These areas have protective laws and staff to enforce them, but they also have very high demands for use of their beaches. Public pressure to conserve Piping Plovers is still needed, in addition to official designation as an endangered species.

Refs. 20, 21

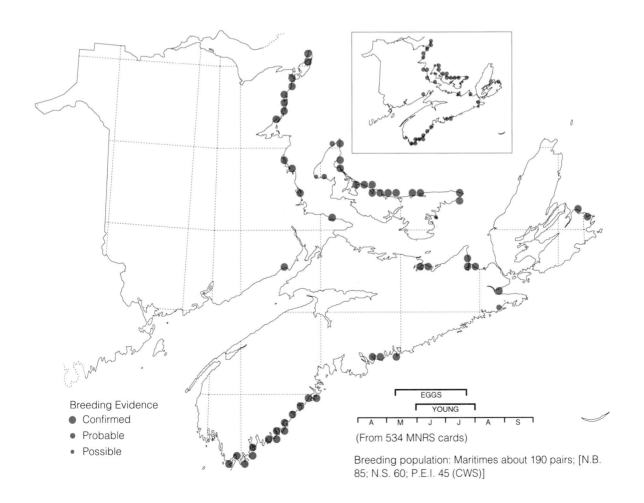

Breeding Evidence
● Confirmed
● Probable
• Possible

EGGS
YOUNG

A   M   J   J   A   S

(From 534 MNRS cards)

Breeding population: Maritimes about 190 pairs; [N.B. 85; N.S. 60; P.E.I. 45 (CWS)]

Killdeer   Pluvier kildir

Charadrius vociferus

This is the noisy plover of open lands throughout North America, an ecological counterpart to the unrelated Lapwing of Europe and Asia. Clearings amid the forests have allowed Killdeers to spread far beyond their original range, and they now breed locally north to Yukon Territory and Newfoundland. They became established in New Brunswick as breeding birds only in the 1940s, and in Nova Scotia and Prince Edward Island not until the 1960s. In the Atlas period, they were found almost throughout our area, though scarce in Nova Scotia on the Atlantic slope and on Cape Breton Island. Besides farmlands, they also bred in gravel pits, forest clear-cut areas, and open lands along the coasts. This was an easy species to detect and confirm as breeding, being large, noisy, and inhabiting open areas. Breeding was confirmed in 546 squares (71 per cent of 770 with the species), most commonly by seeing flightless young (FL, in 288 squares), or adults giving the "broken-wing act" (a distraction display, DD, in 132 squares), although nests (NE,ON) were noted in 122 squares (16 per cent).

Suitable habitat for Killdeers was scarce in the Maritimes, as elsewhere in the eastern forests, before European settlement. However, their late appearance here as breeding birds is surprising, as cleared land in the Maritimes reached its maximum area soon after 1900 and was decreasing before Killdeers first bred in New Brunswick. The still-sparse population in the discontinuous northern "corridor" of cleared land connecting source populations in eastern Quebec to those in northwest New Brunswick may have acted as a bottleneck to colonization here for several decades. The recent increase here, also documented by the BBS, corresponds in time with the widespread clear-cutting of forests since about 1960 more than with availability of agricultural lands. Present numbers may have reached a plateau, with most suitable areas occupied. Future land-use scenarios suggest no reduction in breeding habitat, and global warming seems unlikely to effect adverse changes for this species. Our Killdeers are likely to be with us for a long time.

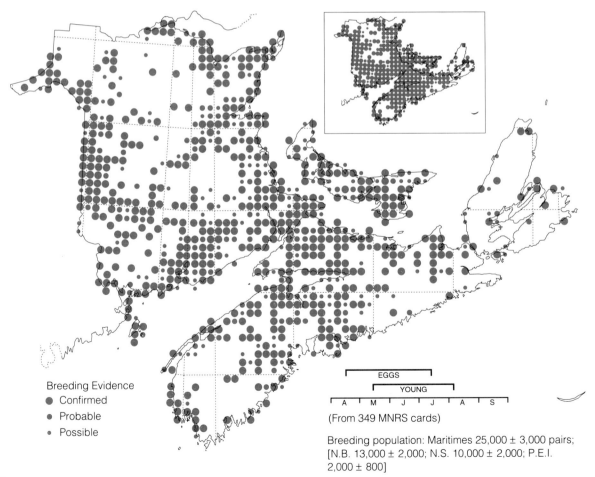

Breeding Evidence
- Confirmed
- Probable
- Possible

EGGS

YOUNG

A   M   J   J   A   S

(From 349 MNRS cards)

Breeding population: Maritimes 25,000 ± 3,000 pairs; [N.B. 13,000 ± 2,000; N.S. 10,000 ± 2,000; P.E.I. 2,000 ± 800]

Greater Yellowlegs
Grand Chevalier

Tringa melanoleuca

These large, noisy shorebirds are familiar sights in salt marshes and around ponds and rivers during migration, but their breeding habitat is very different. Greater Yellowlegs breed in wooded bogs and muskegs across the boreal forest from northern British Columbia and Mackenzie to Labrador, Newfoundland, and eastern Nova Scotia. Summering non-breeders on the coasts fill the gap in time between the last spring stragglers and the first fall migrants. Greater Yellowlegs are noisy and appear agitated at all seasons, therefore many sightings in habitats other than forest bogs, and especially records along the coasts, were considered not to have been breeding birds. Breeding evidence from Atlas field-work came largely from the Cape Breton Highlands (N.S.), with a few records from forested bogs on the cool Atlantic slope of east-central Nova Scotia. No records from New Brunswick or Prince Edward Island were accepted. Given the remote locations and difficulty of access to their preferred habitats, it is not surprising that most confirmed breeding was based on distraction displays (DD) and "strafing attacks," from which records of agitated birds (A) differed only in degree. Flightless young seen in 1956 and a nest with eggs found in 1974, both on the Cape Breton plateau, are the only conclusive records from earlier years.

Greater Yellowlegs, as breeding birds, apparently have not changed in status over the period of European settlement in the Maritimes. These birds were shot, for food or for sport, or simply from annoyance at their noisy advertisement of hunters lying in wait for other birds, but they were not obviously reduced in numbers before legal hunting ended in 1916. There is no impression of either increase or decrease over recent decades. Their present breeding numbers scarcely make up a viable population if subject to serious pressures. At present, we are not aware of obvious threats to their continuing existence as breeding birds here, but their long-term survival probably depends more on availability of winter habitats in Latin America and whether those areas are contaminated with toxic chemicals.

Ref. 90

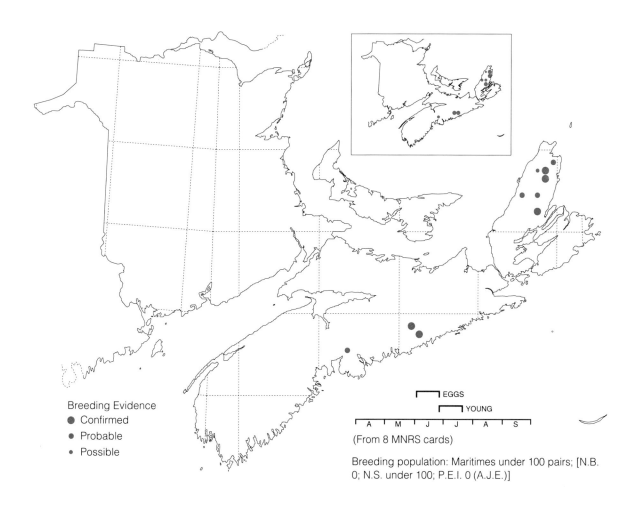

Breeding Evidence
- Confirmed
- Probable
- Possible

EGGS
YOUNG

A M J J A S

(From 8 MNRS cards)

Breeding population: Maritimes under 100 pairs; [N.B. 0; N.S. under 100; P.E.I. 0 (A.J.E.)]

Solitary Sandpiper
Chevalier solitaire

Tringa solitaria

This well-named species breeds in boreal forest regions from Alaska to Labrador, south barely into settled areas of Canada. Previous reports placed the southeast limits around Sault Ste. Marie and Temagami, Ont., Chibougamau, Que., and Churchill Falls, Labrador. Solitary Sandpipers are scarce until early June, in western New Brunswick, during spring migration and are not seen every year elsewhere in the Maritimes. They are more regular and more common, though hardly common, during the southward migration, with first sightings in mid- to late July. The scant six-week interval between the early and late summer sightings long suggested that breeding might occur less far away than proved by past records. The dispersed nesting habits, in remote forest pools and bogs, and their use (uncommon among shorebirds) of old nests of Robins and other tree-nesting species contributed to the difficulty

of studying this species at the edge of its range. During the Maritimes Atlas project, adults defending groups of newly flying young (FL) were found in New Brunswick near Cains River, 29 July 1988, near Hanford Brook, 17 July 1989, and near Juniper, 8 July 1989. Two other sightings also involved highly agitated adults (A) without young being seen. Whether or not these young birds were hatched in the precise squares where they were seen, they were evidently fledged in New Brunswick rather than north of the St. Lawrence estuary. We also mapped a few records (H, 3 in New Brunswick, 2 in Nova Scotia) of these birds in what was judged to have been possible breeding habitat before 20 July.

Solitary Sandpipers probably have bred sparsely in the northern Maritimes in the past, as birds have appeared away from breeding habitat by mid-July as early as records exist. Undoubtedly breeding here has been and remains scarce and widely dispersed. Future warming trends would work against this species' continuance here.

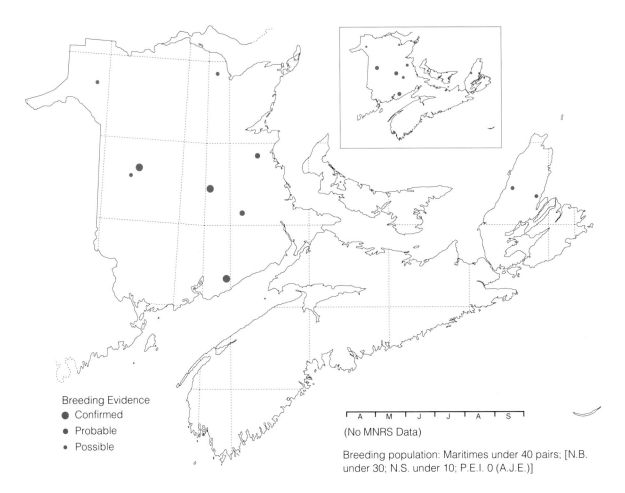

Breeding Evidence
- ● Confirmed
- ● Probable
- ∙ Possible

A M J J A S
(No MNRS Data)

Breeding population: Maritimes under 40 pairs; [N.B. under 30; N.S. under 10; P.E.I. 0 (A.J.E.)]

Willet
Chevalier semipalmé

Catoptrophorus semipalmatus

Willets breed in two widely separated regions: along the Atlantic coast from the Maritimes to Florida, and in the west from the Canadian prairies to Nebraska and California. In the Maritimes and other eastern areas, the birds feed mainly in vegetated salt marshes and nest in fields and other open areas nearby. Their distribution is strictly coastal, including barrier-beach ponds, tidal estuaries, and fringing salt marshes. Willets return to the Maritimes by early May and begin breeding at once. While they circle widely above the marsh and surrounding areas during courtship, their black-and-white wing-stripes and loud cries make them very conspicuous, so they are unlikely to have been missed where present, but on the ground they are much less evident. Most confirmed records in the Atlas (35 of 166 squares with Willets) were of young birds (FL) accompanying noisy adults, and nests were reported in 21 squares.

Eastern Willets were hard hit by market hunting during the 19th century, and only a small population in southwestern Nova Scotia survived north of Virginia by 1916. Dyking and draining of salt marshes reduced the available habitat here during the same period. With protection from hunting, the species spread greatly in the next 50 years, reaching southeast New Brunswick and Prince Edward Island by the late 1960s. It is not clear whether further spread northward in New Brunswick and in Prince Edward Island can be anticipated; climate and habitat appear suitable as far as the Baie des Chaleurs, but there has been little expansion recently. The Maritimes population has made a major advance since 1916, but it is not yet an abundant bird. It should be feasible to make an accurate census and to monitor the breeding populations of this conspicuous bird periodically, but this has not yet been attempted. With its restricted habitat requirements, proposals to alter the few remaining salt marshes (e.g., DUC impoundments) pose threats to the survival of Willets in the Maritimes, and the species may warrant such attention.

Refs. 31, 63

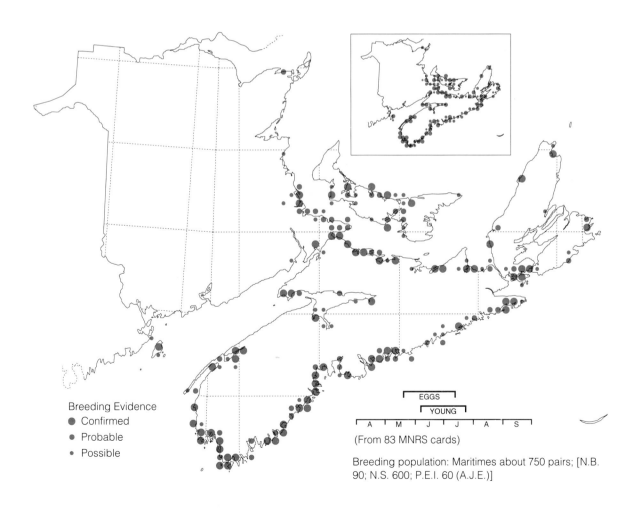

Breeding Evidence
- Confirmed
- Probable
- Possible

EGGS
YOUNG

A M J J A S

(From 83 MNRS cards)

Breeding population: Maritimes about 750 pairs; [N.B. 90; N.S. 600; P.E.I. 60 (A.J.E.)]

Spotted Sandpiper
Chevalier branlequeue

Actitis macularia

The most widespread of our breeding shore-birds, the Spotted Sandpiper occurs plentifully from Alaska across Canada south of the tundra, and in the northern U.S.A., more sparsely farther south, and scarcely reaches the Gulf of Mexico. The unspotted but otherwise similar Common Sandpiper of Eurasia is sometimes considered the same species. Their most familiar habitat in our area is open, gravelly shores of rivers, streams, and lakes, but they also breed around pools in other open habitats, near the sea-coasts and on islands, in gravel pits and quarries, and even in farmland. Spotted Sandpipers were found during the Atlas project throughout the Maritimes, more sparsely in forested regions, with the most obvious gaps in head-water areas between river systems; there were no obvious concentrations, as these birds do not cluster when breeding. Sightings of flightless young (FL) were the most common evidence confirming breeding (in 35 per cent of 903 squares with the species), but the nests, on the ground in open areas, were found more often (in 99 squares, 11 per cent) than for any shorebird except the Killdeer.

Although Spotted Sandpipers use open areas more than forest, availability of their water-edge habitats seems unlikely to have increased or decreased greatly as a result of human settlement in the Maritimes. This species, being small and solitary, was never hunted for food as were the larger shorebirds and those that occur in dense flocks. Its numbers now may differ little from those present when Europeans first came here. Its prospects for the future also are good, as there are obvious threats neither to its habitats or foods, nor to its direct survival. In Europe, acid rain has been considered a threat to habitats used by Common Sandpipers and Dippers, but our bird seems to be doing well even in the more acid-stressed areas of southwestern Nova Scotia.

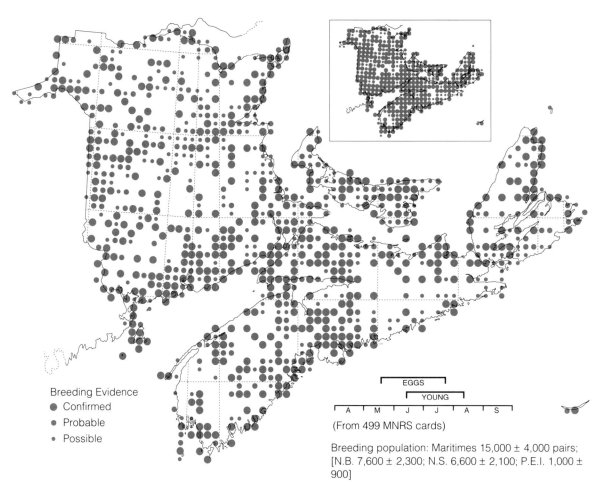

Breeding Evidence
- Confirmed
- Probable
- Possible

EGGS
YOUNG

A M J J A S

(From 499 MNRS cards)

Breeding population: Maritimes 15,000 ± 4,000 pairs; [N.B. 7,600 ± 2,300; N.S. 6,600 ± 2,100; P.E.I. 1,000 ± 900]

Upland Sandpiper
Maubèche des champs

Bartramia longicauda

This uncharacteristic "shorebird" frequents pastures and hayfields and similar open, grassy areas, and its range is fragmented almost everywhere. It breeds mainly in the prairies, from Alberta and Manitoba to Texas and Missouri; sparsely in mountain meadows and barrens from Alaska and the Yukon, south to eastern Oregon; and locally through the Great Lakes states and southern Ontario and Quebec to Maine. Recently a few have been proved or suspected to breed in all three Maritime Provinces, in the few extensive grasslands at least 20 ha in area. The 19 squares with this species show some concentration in southeastern New Brunswick, but are too scattered to be convincing. Confirmed breeding was based on sightings of young (FL, in 6 squares), with one nest (ON); a nest had been found earlier in southeast New Brunswick in 1974. Upland Sandpipers may have been under-represented,

despite their distinctive appearance and behaviour, as confirmation was obtained in one well-worked square only in 1990, 2–3 km distant from earlier unconfirmed sightings.

Upland Sandpipers probably were absent from the Maritimes through the early years of European settlement, owing to a scarcity of suitable grassland habitat. They probably bred here, locally and in very low numbers, long before breeding was first confirmed in New Brunswick in 1971 and Prince Edward Island in 1975; suitable habitat had been present, but unstudied, in those locations throughout this century. There is no evidence of change in its numbers in the Maritimes, though Upland Sandpipers have declined elsewhere in the east due to habitat loss and a tendency to earlier mowing of grasslands in recent decades. The Maritimes breeding population is precariously small. Its future persistence as a breeding species depends on local land-owners maintaining suitable habitat and farming practices, as Upland Sandpipers easily could be lost from our avifauna without anyone realizing this was about to happen.

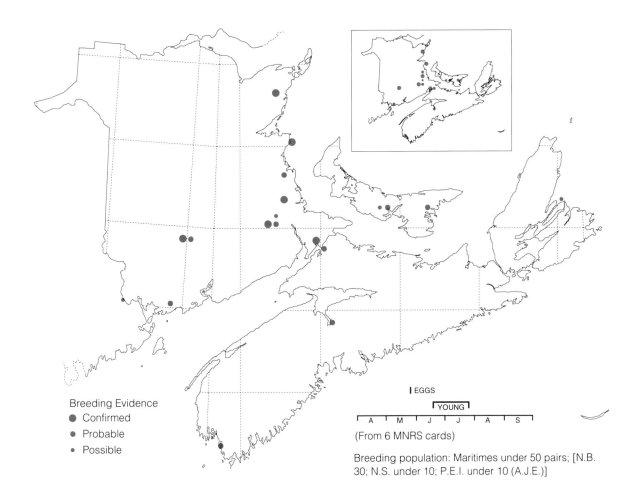

Breeding Evidence
● Confirmed
● Probable
● Possible

| EGGS

[YOUNG]

A M J J A S

(From 6 MNRS cards)

Breeding population: Maritimes under 50 pairs; [N.B. 30; N.S. under 10; P.E.I. under 10 (A.J.E.)]

Least Sandpiper
Bécasseau minuscule

Calidris minutilla

Least Sandpipers are the southernmost of the largely arctic-nesting "peep" sandpipers. They breed mainly in sedge meadows and bogs of the subarctic zone from Alaska to Newfoundland, but a few nest farther south along cool coasts, in the Magdalen Islands, Que., and Nova Scotia, and (once) in Massachusetts. Confirmed breeding records in the Atlas (all in Nova Scotia) were on Sable Island (FL,DD), whence came most earlier reports, and near Hartlen Point (FL). Other recent proved breeding locations in Nova Scotia included Cape Sable Island (1961), East Lawrencetown (1971), and Fourchu (1971). One mapped record, in a bog on the Cape Breton Highlands, more closely parallels the breeding habitat in the species' northern range than the coastal situations more often used here. Least Sandpipers occur commonly in the Maritimes as migrants, with scarcely a month (late May to late June) between regular spring and "fall" occurrence; most sightings other than of nests, broods,

or active courtship were inconclusive, or were not recognized by the observers as being of possible breeding significance. Thus, the species was presumably somewhat more widespread, in suitable areas, than the six squares in which it was found.

The habitats used for breeding by Least Sandpipers in the Maritimes have been little altered during historic times, and local breeding stocks of this species were little affected by the minimal exploitation of small migrant sandpipers before 1916, when legal hunting of most shorebirds ended. These birds are probably no more common or uncommon now than in recent centuries. The few known records, before and during the Atlas study, span most of the Atlantic coastline of Nova Scotia, albeit at low density, so the mainland breeding population may equal the earlier estimate of 75 pairs on Sable Island and seems likely to be stable. Global warming will tend to make these marginal areas less acceptable in future, but direct threats from human actions seem unlikely for these birds.

Ref. 102

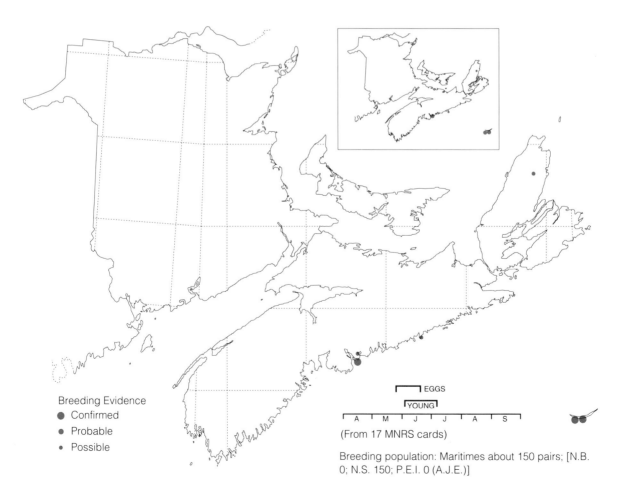

Breeding Evidence
- Confirmed
- Probable
- Possible

EGGS
YOUNG
A M J J A S
(From 17 MNRS cards)

Breeding population: Maritimes about 150 pairs; [N.B. 0; N.S. 150; P.E.I. 0 (A.J.E.)]

Common Snipe
Bécassine des marais

Gallinago gallinago

This Snipe breeds in temperate, boreal, and subarctic regions of Eurasia and North America; closely related forms occupy temperate and tropical regions in Africa and South America. Our Snipe is found in shallow marshes and bogs, where sedges or grasses are important cover species, often mingled with low shrubs. These birds are not often seen except in flight, as they seldom emerge on open shores and other wet areas to feed during breeding season. Snipes return to the Maritimes in mid-April, but song-flights ("winnowing") were the only breeding evidence noted before June. Most detections were during song-flights, which extend over wide areas, tending to result in over-representation. This may balance out its secretive nature in all other contexts, which points to under-estimates overall. Snipes were detected in most Atlas squares with adequate coverage but were scarce in the northwest highlands of New Brunswick and on the Atlantic slope of Nova Scotia where rocky terrain included little suitable habitat. Their nests are on the ground, usually in sedge clumps, but few nests were reported to the Atlas (26, in 694 squares where the species was found). Fledged young accompanying adults (FL) was the most frequent confirmation (70 squares), most records (76 per cent) being of single (H) or repeated (T) sightings.

There are no suggestions of historical changes in numbers over the centuries since Europeans came to the Maritimes, clearing in the forests having created damp fields as least as often as drainage eliminated shallow marshes. The BBS suggests numbers have been stable or slightly increasing since 1966. This is our second most widespread breeding shorebird, after Spotted Sandpiper, and it may well be the most abundant. Wetland conservation measures directed towards waterfowl will secure its habitat here for the future, perhaps balancing losses cause by intensification of use of agricultural lands. Land use in its wintering areas may hold the key to its future.

Ref. 145

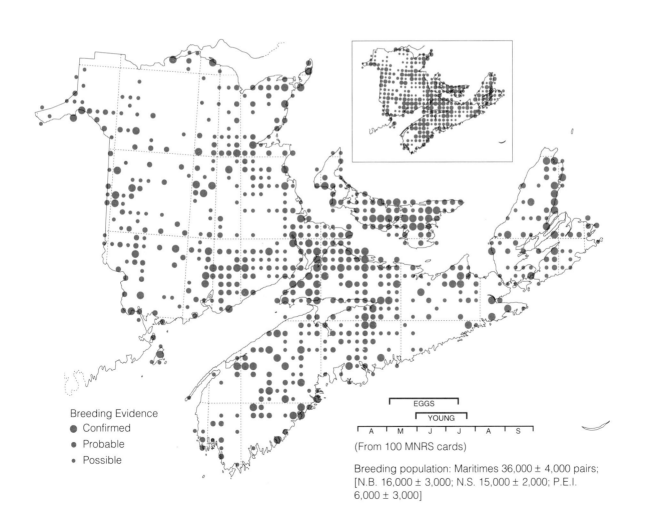

Breeding Evidence
- Confirmed
- Probable
- Possible

EGGS
YOUNG
A M J J A S

(From 100 MNRS cards)

Breeding population: Maritimes 36,000 ± 4,000 pairs; [N.B. 16,000 ± 3,000; N.S. 15,000 ± 2,000; P.E.I. 6,000 ± 3,000]

American Woodcock
Bécasse d'Amérique

Scolopax minor

This woodcock is a North American bird, replaced in Eurasia by related, but larger, species. It breeds from Louisiana and Texas north to southern Manitoba and the Gaspé Peninsula, so the Maritimes are near the northeastern limit. It is primarily a bird of broad-leafed forests and swamps, and its range peters out in the southern edges of the boreal conifer region. Woodcocks were found nearly everywhere in the Maritimes except in solid conifer forests, which may explain their relative scarcity in the northwest highlands of New Brunswick and near the Atlantic coasts of Nova Scotia. These birds are well-hidden while resting in dense tree or shrub cover by day, emerging at night to feed in swamp edges and other damp areas, so they were under-represented by the Atlas data. Most records, about half of 541 squares with the species, were of birds detected in their spring "song-flights." Breeding was confirmed most often (in 97 of 146 squares) by sightings of partly fledged young, which can fly before they are half-grown. Nests were found in 33 squares (6 per cent of those with Woodcocks).

Woodcocks were widespread in the hardwood forests of the Maritimes before European settlement. Clearing and cutting of the forests may have improved conditions, but the later reversion of cut-over areas to solid spruce or fir stands made large tracts unattractive to this species. The abandonment of marginal farmland, and its succession through old fields and shrubland to open woods, created a lot of good Woodcock habitat early in this century; much of this land is now forest again, and thus little used. Areas clear-cut by the forest industry are used by Woodcocks for a few years, but at lower densities than were old-field habitats. Thus, the numbers of this species in any area have gone up and down in response to availability of their habitat. As one of two erratic-flying ("sporting") shorebirds, Woodcocks are pursued by a few keen specialist hunters; despite liberal bag limits, the hunting kill, estimated at 12,000–15,000 annually in recent years, is thought unlikely to restrict numbers here. More intensive forest management in the future is likely to ensure that suitable habitats for Woodcocks will persist in the Maritimes, but numbers of these birds are likely to be fewer than in the early 1900s.

Ref. 27

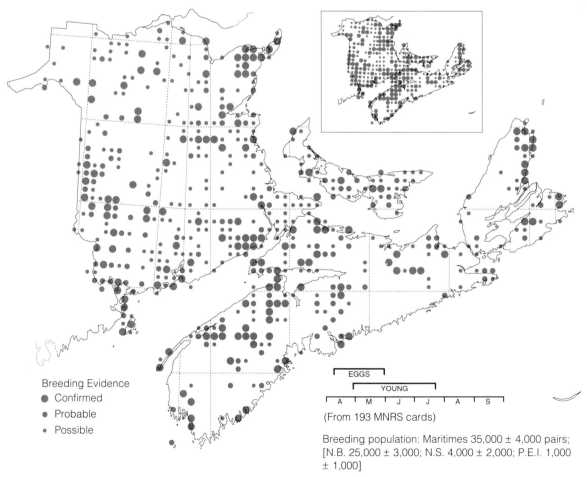

Breeding Evidence
- Confirmed
- Probable
- Possible

EGGS
YOUNG
A M J J A S

(From 193 MNRS cards)

Breeding population: Maritimes 35,000 ± 4,000 pairs; [N.B. 25,000 ± 3,000; N.S. 4,000 ± 2,000; P.E.I. 1,000 ± 1,000]

Wilson's Phalarope
Phalarope de Wilson

Phalaropus tricolor

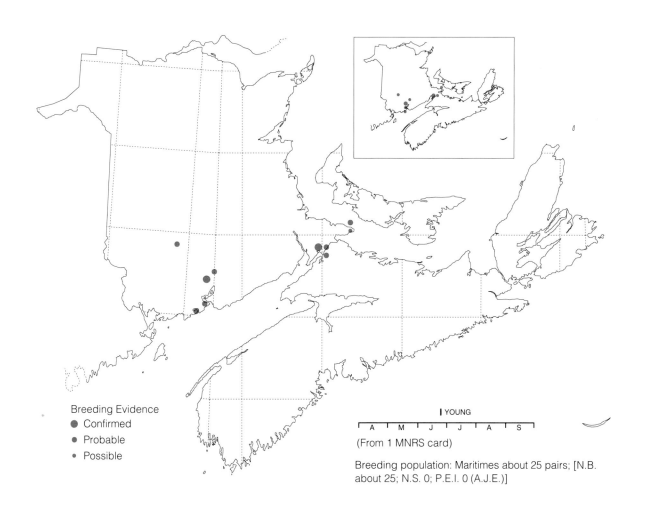

Phalaropes are specialized sandpipers that forage by swimming, and that winter at sea rather than on shores or marshes. The other two species nest in the arctic tundra, but Wilson's Phalaropes breed around fertile ponds and marshes from British Columbia to Manitoba and from California to Kansas. Only sparse and localized breeding is known farther east, mostly near the Great Lakes, where fertile sites have been used increasingly over the last century. Wilson's Phalaropes were first suspected of breeding in the Maritimes in 1978, and the first confirmation here was a brood on Grassy Island, N.B., in 1986. Other records in the Atlas were from waterfowl impoundments and other fertile marshes in the St. John River Valley and in the New Brunswick–Nova Scotia border area. Repeated sightings in the latter area since 1978, including groups of up to nine birds in mid-July to mid-August in 1988–90, indicated breeding nearby, although no nests or flightless young have been found. Single and paired birds from early May through June comprise the other evidence, with records in a total of 10 squares in 1986–90.

Wilson's Phalaropes are best known as birds of prairie sloughs and other fertile ponds in open areas. Their spread eastward, at least since World War II, coincided with the development of sewage lagoons and shallow water impoundments for waterfowl use that simulated the prairie slough habitat. Habitat management for waterfowl is still increasing and will continue to assist this species to become better established in our area.

Ref. 94

Breeding Evidence
- Confirmed
- Probable
- Possible

YOUNG

A   M   J   J   A   S

(From 1 MNRS card)

Breeding population: Maritimes about 25 pairs; [N.B. about 25; N.S. 0; P.E.I. 0 (A.J.E.)]

Ring-billed Gull
Goéland à bec cerclé

Larus delawarensis

Before the 1920s, this North American species was a breeding bird of the prairies, with only a few small colonies scattered east to the Atlantic. The Ring-billed Gull is now a successful follower of human settlement and has reached pest status around the Great Lakes and in the St. Lawrence valley. Like other large gulls, these birds are omnivorous, feeding in summer on insects and small mammals on agricultural fields, and on intertidal invertebrates. Fishes, and human garbage and sewage, are used at all seasons. Ring-billed Gull colonies in the Maritimes are mostly on islands, but some of the largest are on industrial waste areas around mainland factories at Dalhousie and Belledune, N.B.; to date, most are coastal. Only colonies (NE,NY,ON) were accepted as breeding evidence; birds in adult plumage often appear in summer inland, wherever feeding is easy. This is the gull most regularly encountered away from coasts, except at dumps, where the larger species predominate; it has learned to forage around fast-food outlets, picnic areas, shopping centre parking lots, and sports fields. Such sites are little used during the nesting season. In late summer, blueberries may be eaten, this being almost the only direct conflict with human interests here that involves this species, unlike the many problems with larger gulls.

Breeding by Ring-bills in the Maritimes was first detected in 1965, at Bathurst, N.B., perhaps an incursion from the long-known breeding areas on the north shore of the Gulf of St. Lawrence. These gulls now nest at some 15 colonies, mostly along the northern coasts from Dalhousie to Murray Harbour, P.E.I. A few small nesting groups became established in the 1980s on islands in the St. John River and Grand Lake, N.B. Competition by larger gulls limits its access to the few island nest-sites in some suitable areas with agriculture nearby, e.g., north shore and Minas Basin, N.S. Some further spread and increases are plausible, but pest status is not anticipated here.

Ref. 84

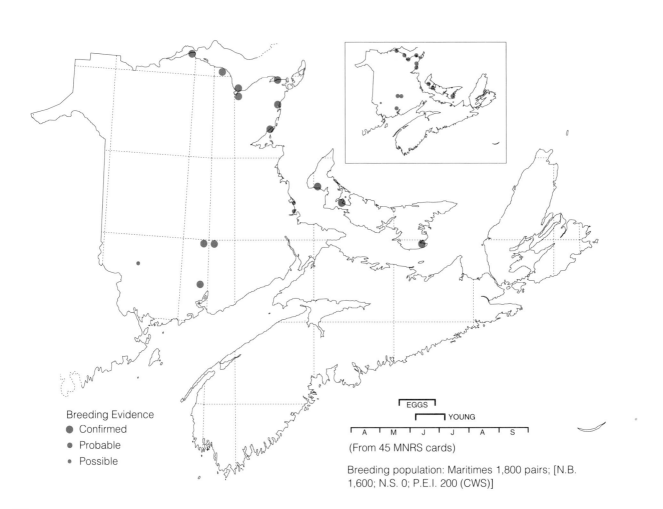

Breeding Evidence
● Confirmed
● Probable
· Possible

EGGS
YOUNG
A M J J A S

(From 45 MNRS cards)

Breeding population: Maritimes 1,800 pairs; [N.B. 1,600; N.S. 0; P.E.I. 200 (CWS)]

Herring Gull
Goéland argenté

Larus argentatus

This so-called "seagull" may originally have been an inland bird. It breeds inland as well as coastally in North America, but its breeding in Europe is thought to be recent. Herring Gulls breed from Yukon Territory and northern British Columbia to Baffin Island, Newfoundland, south to New York and locally farther. In the Maritimes, coastal breeding, mainly in colonies of 20–200 pairs, is usual, and inland breeding was confirmed only locally in western Nova Scotia. Herring Gulls do not breed until four years old and wander widely, so most gulls away from colonies were counted as non-breeders, unless other evidence was obtained. Probably many records of "possible breeding" were only wandering birds. There is little breeding along the steep, unbroken coasts of the middle Bay of Fundy and western Cape Breton Island. Gaps along Northumberland Strait may reflect the scarcity of islands safe from predators and the intensive recreational use of most mainland shores in summer. Sixty-two per cent (177 of 282 squares with Herring Gulls) of our records showed confirmed breeding, mostly nests

(149 squares) or free-running young birds.

Before European settlement, the Herring Gull, bred locally here, wherever "beach-combing" and fishing allowed them to maintain a population, subject to intermittent exploitation of their eggs and young for food by native people.

Europeans also used gull eggs and young for food into the 20th century, and this predation helped to reduce the breeding birds to a remnant by 1900. With protection since 1916, their numbers have increased, to pest status in some areas, and they exert pressure on other breeding birds such as terns. The huge numbers formerly attributed to the Kent Island, N.B., colony, (still one of the largest in the Maritimes with under 2,000 pairs) are now believed erroneous. Numbers overall have declined since 1960, partly through competition with the Great Black-backed Gull, which increased in this period. The large gulls have increased in number and become more concentrated in large colonies, since garbage in centralized dumps and landfill sites, and fish offal around large packing plants, provided ample and predictable food, available even in winter. Gulls will remain abundant as long as these food supplies are made available to them.

Refs. 22, 66, 79, 80, 82

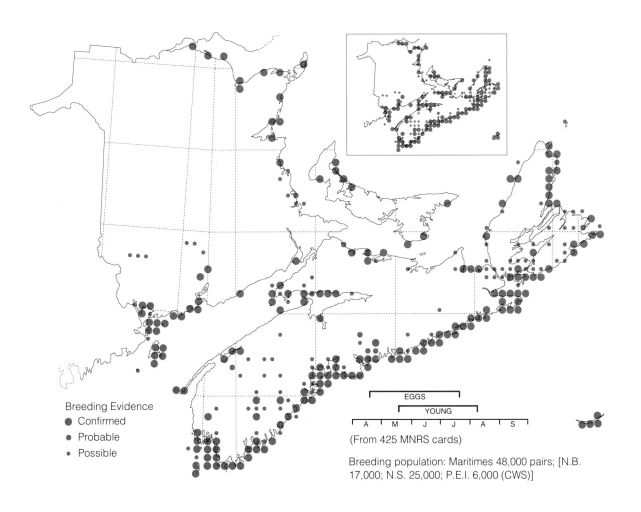

Breeding Evidence
- Confirmed
- Probable
- Possible

EGGS
YOUNG
A M J J A S

(From 425 MNRS cards)

Breeding population: Maritimes 48,000 pairs; [N.B. 17,000; N.S. 25,000; P.E.I. 6,000 (CWS)]

Great Black-backed Gull
Goéland à manteau noir

Larus marinus

The Black-back is the largest gull any-where. Its range spans the North Atlantic Ocean, from Greenland to New York and from Svalbard to France, with a few inland in eastern Canada. Here, it was mainly a coastal bird nesting on islands, with inland breeding by scattered pairs except at Lake George, Yarmouth Co., N.S. where about 150 pairs were breeding (in 1981). Many colonies were relatively small, but some, including Boot Island, N.S. (1,400 pairs in 1984), had several hundred pairs. Gaps in the coastal distribution mostly occurred along steep shores without islands, as in the upper Bay of Fundy and northwest Cape Breton Island. Most records of this species were of nesting birds and colonies (NY,NE,ON, in 176 of 323 squares with these gulls), other evidence (H,P,T) having been retained mainly to indicate possible inland sites.

Great Black-backed Gulls presumably nested along the coasts of the Maritimes in the distant past. The abo-riginal human population visited the coasts mainly in late summer, so exerted little pressure on breeding gulls, though they took birds and eggs for food opportunistically. Once perma-nent European settlements became estab-lished on the coasts and on many islands, gulls and other breeding seabirds rapidly disap-peared. By 1900, the Lake George colony was thought to include the only Black-backs nesting in the Maritimes. With protection under the Migratory Birds Convention of 1916, they increased gradually for the last 70 years. In recent decades, they have increased at the expense of Herring Gulls, as they overcame their suspicion of humans suffi-ciently to occupy areas where colonies would have been exploited until quite recently. Unless and until we reduce access by gulls to waste food at dumps and fish-packing plants, these birds will continue to prosper, and to exert pressure on all other seabirds breeding nearby. Local control of gull numbers, at least temporarily, may be unavoidable if we wish to retain breeding eiders, terns, or storm-petrels in some areas. We have helped and continue to help the gulls; can we refuse to help the species they victimize?

Ref. 82

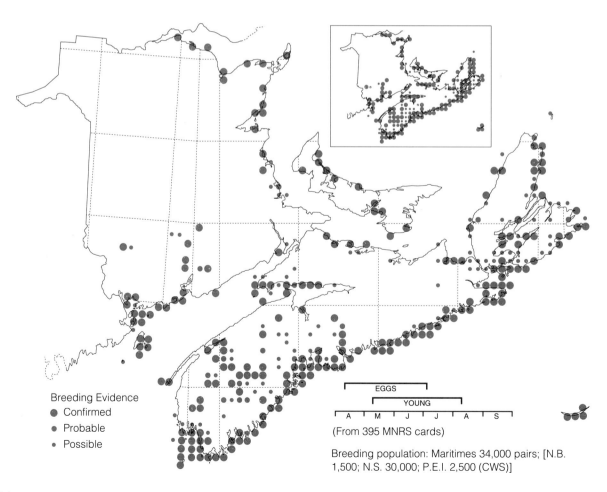

Breeding Evidence
● Confirmed
● Probable
● Possible

EGGS

YOUNG

A M J J A S

(From 395 MNRS cards)

Breeding population: Maritimes 34,000 pairs; [N.B. 1,500; N.S. 30,000; P.E.I. 2,500 (CWS)]

Black-legged Kittiwake
Mouette tridactyle

Rissa tridactyla

Kittiwakes are pelagic gulls that feed out at sea and come to shore only to breed. They nest colonially, sometimes in vast numbers, on cliffs around all the northern oceans as far north as open water prevails in summer, and south to the Gulf of St. Lawrence, Brittany (France), and the Kurile and Aleutian Islands (Pacific Ocean). The only colonies in the Maritimes, the southernmost anywhere, were on islets off Cape Breton Island. The Atlas records included only four (two in one square) of the five colonies known previously, plus two new ones nearby. The other known site, on a small rocky islet near Baleine, is not easily visited and cannot be viewed from the mainland.

Kittiwakes were first found breeding in the Maritimes in 1971, presumably by an extension southward from the rapidly expanding population on the island of Newfoundland. Additional colonies were found subsequently. The increase in Kittiwakes in Nova Scotia from 1971 to 1983 was near the maximum growth rate possible in a population, without assuming further immigration, which suggests that food was superabundant then. If a similar rate of increase (20 + per cent/year) has continued, the Nova Scotia population may have exceeded 2,000 pairs by 1990. Proposed human exploitation of the small fishes eaten by these birds may limit future growth in this population, if it has not already done so, and could jeopardize its existence here at the southern limit of the species' range.

Ref. 83

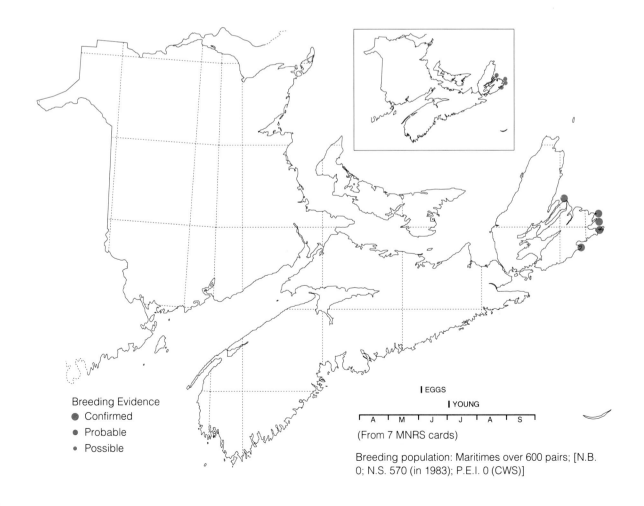

Breeding Evidence
● Confirmed
● Probable
● Possible

| EGGS
| YOUNG

A   M   J   J   A   S

(From 7 MNRS cards)

Breeding population: Maritimes over 600 pairs; [N.B. 0; N.S. 570 (in 1983); P.E.I. 0 (CWS)]

Roseate Tern
Sterne de Dougall

Sterna dougallii

This tern breeds locally in a few widely spaced areas: eastern America, western Europe, east Africa, Australasia, mainly in tropical to warm temperate regions. It is more pelagic than the Common Tern, and more tropical in distribution than the even more pelagic Arctic Tern. The Roseate Tern has declined to "threatened" or "endangered" status in many regions, through pressure by gulls at nesting areas and by humans in some wintering areas. In the Maritimes, it has declined along with the other coastal terns, of which it was always the scarcest species. Surveys in eastern Nova Scotia by government agencies in the early 1980s showed few more records than were obtained in the Atlas, and one in southwest Nova Scotia was the colony visited regularly by bird-listers. A few more colonies may still be active, as Roseate Terns are easily missed in mixed colonies.

Roseate Terns in the past were found sporadically in several locations where they have not been reported recently. They may have shared in the greater abundance of all terns, as at Sable Island around 1900, when all gulls had been greatly reduced, and they shared in the gradual decline of terns since World War II. A recent status report (1985) noted about eight sites recently occupied by Roseate Terns in Nova Scotia, with one pair having nested or tried to nest at Machias Seal Island, N.B., in several years. The species' foothold in the Maritimes, and in Canada, is precarious and depends for continuation on management to ensure more favourable conditions for terns in general. *Ref. 78*

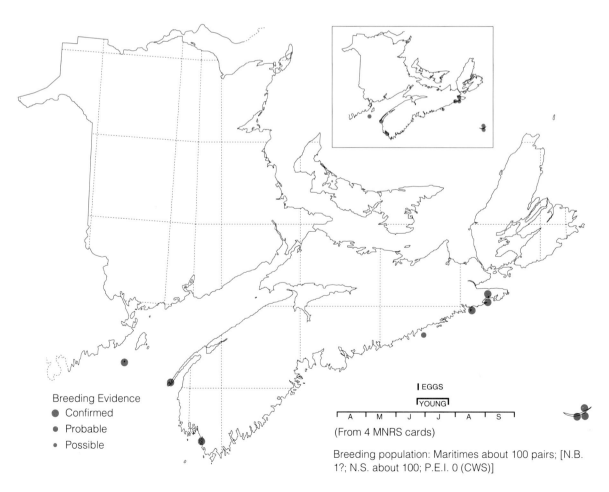

Breeding Evidence
- Confirmed
- Probable
- Possible

EGGS

YOUNG

| A | M | J | J | A | S |

(From 4 MNRS cards)

Breeding population: Maritimes about 100 pairs; [N.B. 1?; N.S. about 100; P.E.I. 0 (CWS)]

Common Tern
Sterne pierregarin

Sterna hirundo

The Common Tern is the most familiar small tern in temperate areas of the world. It breeds across Eurasia, and in North America, east of the Rockies south of the tundra, and southward along the Atlantic coast. Its range includes inland as well as coastal areas, where shallow waters for fishing occur near sandy or gravelly shores for nesting, but it does not forage far out to sea, unlike the Arctic Tern. In the Maritimes, Common Terns were mostly coastal birds, although inland breeding occurred locally along the St. John River and at lakes in southwest Nova Scotia. In the Bay of Fundy, muddy water prevents terns from finding fish in the upper reaches, and the steep middle coasts lack shallow water and beaches suitable for tern nesting. Terns were also scarce around northern Cape Breton Island. Over half of the squares (117 of 221) with Common Terns had breeding confirmed, mostly by finding nests, usually in colonies. Records of terns flying along a shoreline do not prove

breeding in a square, even on the coast, as terns may fly a long way to forage.

Terns were more common in earlier times. Although their eggs were collected from large colonies for human food, they were not an important food source except where very abundant. Killing of adult terns for decorating women's hats had adverse effects for a rather brief period late in the 19th century.

None of these pressures reduced the species below a "very common" level before protection was extended to gulls after 1916. Since then, gulls have increased enormously, encroaching on tern colonies and preying upon tern eggs and chicks, and now tern numbers are at levels arousing concern in management agencies. Present estimates of breeding Common Terns here total only a small fraction of the numbers that bred on Sable Island in 1900. The prospects for terns in the future, without relief from gull pressure, are not promising, but protection from gulls in a few key colonies would suffice to maintain tern populations here.

Ref. 85

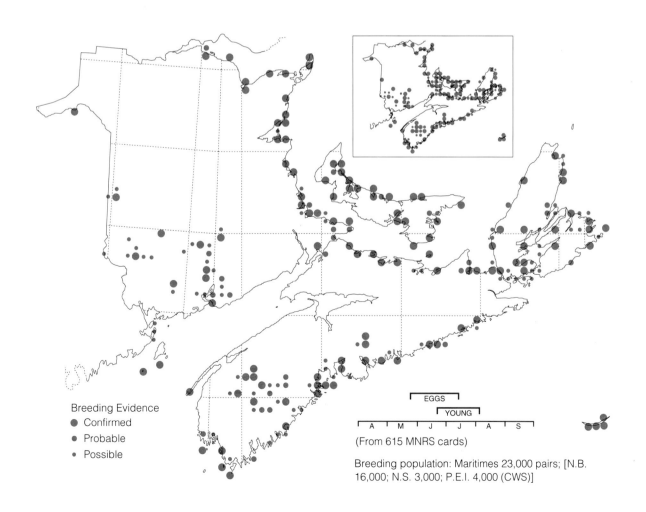

Breeding Evidence
- Confirmed
- Probable
- Possible

EGGS
YOUNG

A M J J A S

(From 615 MNRS cards)

Breeding population: Maritimes 23,000 pairs; [N.B. 16,000; N.S. 3,000; P.E.I. 4,000 (CWS)]

Arctic Tern
Sterne arctique

Sterna paradisaea

The Arctic Tern breeds across the low arctic and subarctic regions of the Earth, and south along coasts cooled by ocean currents flowing from the arctic. Its range overlaps in Europe and in the Maritimes with the temperate- and boreal-breeding Common Tern, with which colonies are often shared. Breeding here was on the cool outer coasts, from the lower Bay of Fundy to Cape Breton Island, with only a few small groups in warmer areas, on the north coast of Prince Edward Island and in northeast New Brunswick. Nearly all Maritimes nesting areas were on islands, facing the open sea, with foraging at all seasons typically offshore, unlike the Common Tern. The largest colony in our area (1,800 pairs in 1988) at Machias Seal Island, N.B., was on a 5-ha islet more than 20 km from the nearest coast, and the only other large colonies were on Sable Island. Because Arctic Terns are easily confused with the more widely distributed Common Tern, they were detected mainly at colonies, and most records (40 of 64) were of confirmed breeding (NE,NY,ON). No doubt some colonies were missed.

Terns in the past were exploited by humans wherever access to their colonies was feasible. Tern eggs were less sought-after than those of larger birds, but they were used where readily available, even into the 1980s, although such use has been illegal since 1916. The use of tern plumage for decorating hats in the late 1800s rapidly decimated tern numbers in eastern North America, but terns recovered rapidly once protection was legislated. More recently, increased numbers of gulls have affected terns adversely. Thus, numbers of all three coastal-breeding species of terns in the Maritimes have gone up and down several times over the past 150 years. At present, Arctic Terns breed in at least 4 colonies in New Brunswick, 30 in Nova Scotia, and 3 in Prince Edward Island. Prospects for the future remain obscure, even with proposals for intensive management, including gull control, in key colonies. Human exploitation of small fishes used by marine birds, such as the sand lance often taken by terns, is increasing as stocks of larger fishes are depleted. The future of seabirds, especially terns, depends on planned conservation of the marine ecosystem as a whole, not merely a few small aspects of it.

Refs. 65, 85, 110

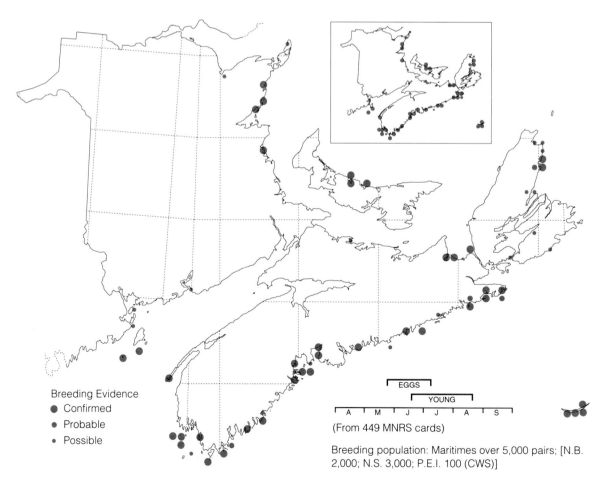

Breeding Evidence
● Confirmed
● Probable
● Possible

EGGS
YOUNG

A M J J A S

(From 449 MNRS cards)

Breeding population: Maritimes over 5,000 pairs; [N.B. 2,000; N.S. 3,000; P.E.I. 100 (CWS)]

Black Tern
Guifette noire

Chlidonias niger

The Black Tern is the most widespread of a group of small freshwater-marsh terns, breeding widely but discontinuously in Europe and western Asia and in North America. It is most general in the prairie marshes, but breeds from British Columbia and south Mackenzie to southern Ontario and Quebec, and south to California, Nebraska, and New York. Farther east it is very localized, and Maritimes Atlas records were restricted to the lower St. John River and New Brunswick–Nova Scotia border marshes. Its complete absence from wetlands in Prince Edward Island, the third major area of fertile wetlands here, is unexplained. Breeding at many sites is irregular and unpredictable, although the birds have been found annually for the last 50 years, and this erratic behaviour is characteristic of the species. The few records (in only 14 squares) did not allow generalization on types of evidence, but breeding was confirmed in 7 squares, with nests seen in 5 squares.

The Black Tern seems to have been a recent arrival as a breeding bird in the Maritimes. Before 1937, it was not known to breed east of Ontario and New York, and breeding in Maine was not proved until 1946. The species was first found breeding in our area in 1937, at Big Timber Lake, where it has been found on many later visits. Black Terns were first detected in the New Brunswick–Nova Scotia border area in 1966, and suitable habitat in new DUC impoundments and frequency of sightings here both increased in succeeding years. The areas first occupied, especially at Amherst Point, N.S., were largely abandoned after a few years, when the Black Terns apparently moved into more recently impounded areas nearby, a pattern also seen with Redhead, Ruddy Duck, and Common Moorhen. The species is marginal here, though persistent, and it would not take a major disaster to eliminate it from our breeding avifauna. Its use of marsh areas impounded for waterfowl ensures that habitat availability will not limit it here in future, but its small numbers and few breeding sites make its status precarious.

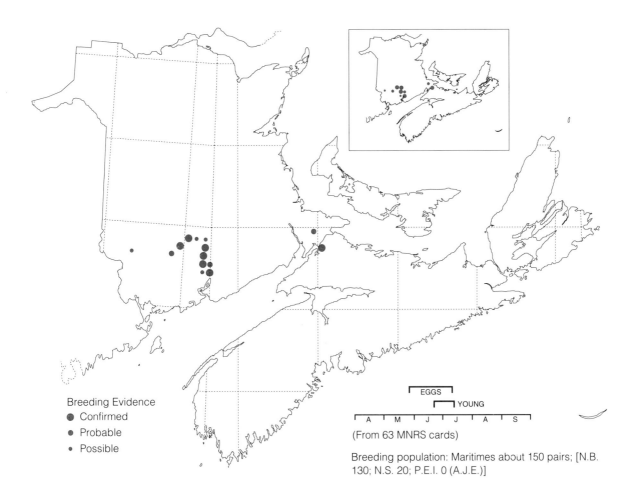

Breeding Evidence
- Confirmed
- Probable
- Possible

EGGS
YOUNG

A M J J A S

(From 63 MNRS cards)

Breeding population: Maritimes about 150 pairs; [N.B. 130; N.S. 20; P.E.I. 0 (A.J.E.)]

Razorbill
Petit Pingouin

Alca torda

The Razorbill is restricted to coasts of the North Atlantic Ocean, from Maine to Baffin Island and west Greenland, and from Iceland to the White Sea, the Baltic Sea, and Brittany, France. Most of the world population breeds in Iceland and the British Isles; only 2–3 per cent of the total breed in America, mostly in Labrador. Only five small colonies were found in the Maritimes, at the southern limit of the range. All are adjacent to cool coastal waters, in New Brunswick, at the mouth of the Bay of Fundy, on the outer coast of Nova Scotia, and around northern Cape Breton Island. As these birds do not breed until they are three to five years old, and they may forage at considerable distances from the colony, only records of Razorbills clearly associated with nesting sites (ON,NY,V) were counted.

The history of Razorbill populations in North America leaves no doubt that numbers have declined since European settlement began, through human exploitation of the birds and their eggs for food and human occupation of many former nesting islands. The decline continued into the 20th century. It seems likely that Razorbills formerly bred more widely on islands along the outer coast of Nova Scotia. The colonies at Hertford/Ciboux Islands, N.S., and Yellow Murr Ledge, N.B., were known before 1930, and that at Machias Seal Island, N.B., is also of long standing.

Evidence of breeding at Pearl Island and at Margaree Island, both in Nova Scotia, was not obtained until 1971 and 1981, respectively. These latter may be the first suggestions of re-occupation of former sites, but numbers at the larger colonies have not increased noticeably in 60 years. Four of the five known sites have some legal protection by federal, provincial, or private agencies, but difficulty of access to most locations is their main protection from incautious visitation, now a more serious threat than deliberate disturbance. Small breeding numbers, and susceptibility to oil spills at sea while birds are away from the colonies, make this one of the most vulnerable members of our avifauna.

Ref. 109

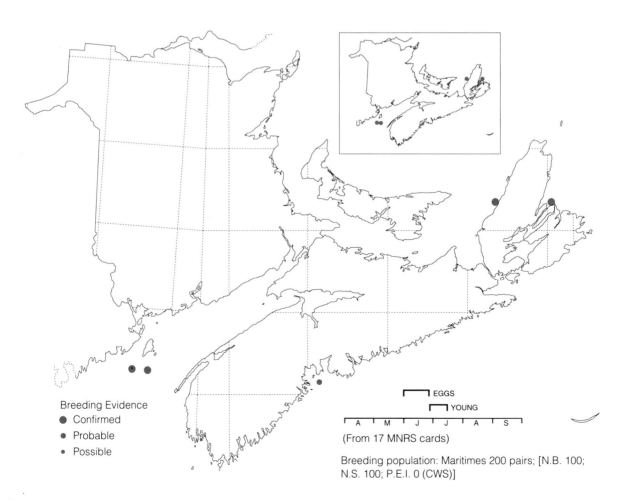

Breeding Evidence
● Confirmed
● Probable
• Possible

EGGS
YOUNG

A   M   J   J   A   S

(From 17 MNRS cards)

Breeding population: Maritimes 200 pairs; [N.B. 100; N.S. 100; P.E.I. 0 (CWS)]

Black Guillemot
Guillemot à miroir

Cepphus grylle

This, the most widely distributed member of the auk family, breeds around the high arctic and south to temperate waters in Europe and eastern North America, with closely related forms in the Pacific Ocean. The range in the Maritimes is discontinuous, with concentrations around Grand Manan Island N.B., and off Yarmouth County and around northern Cape Breton Island, N.S. Probably many more than were detected breed on islands along the Atlantic coast of Nova Scotia, where suitable habitat—broken, bouldery shore—is abundant. Scarcity of such habitat limits breeding by Black Guillemots in Prince Edward Island and along the Gulf shores of Nova Scotia and New Brunswick. This bird, known to fishermen as the "seapigeon," is present year-round, but moves farther offshore in winter when coasts are icebound. They re-appear in April around breeding areas, where their nests are in rock crannies and among boulder piles. The black-and-white plumage and red feet make this a striking bird, easily detected where present. Most records (53 of 85 squares with the species) were of confirmed breeding, usually occupied nests (ON, 25 squares) or adults carrying food to suitable cliffs (AY, 12 squares).

Whereas the other auks nest in large colonies vulnerable to predation by mammals, the Black Guillemot's dispersed breeding pattern and well-hidden nest sites saved it from much human exploitation in the past. Although some were taken for food during the period of subsistence economy in the Maritimes, which extended into the present century locally, its small size made this species less attractive than most other marine birds, so its numbers never declined far below those when Europeans first came here. The decimation of other seabirds may even have reduced competition for food, allowing Guillemots to increase above former levels. Nearly 1,000 pairs were estimated to breed in the Grand Manan archipelago in 1936, but fewer bred there by 1950. We have no evidence that suggests a decline is in progress now or expectable in the immediate future.

Refs. 28, 109, 116, 119, 158

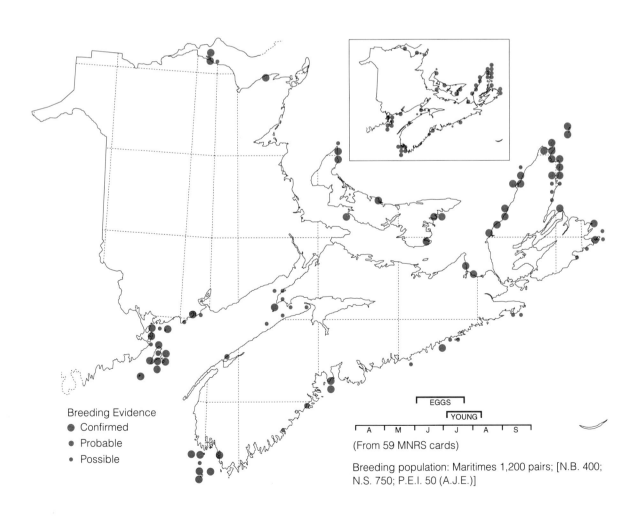

Breeding Evidence
- Confirmed
- Probable
- Possible

EGGS
YOUNG

A M J J A S

(From 59 MNRS cards)

Breeding population: Maritimes 1,200 pairs; [N.B. 400; N.S. 750; P.E.I. 50 (A.J.E.)]

Atlantic Puffin
Macareux moine

Fratercula arctica

Everyone knows the Puffin from its picture, but few people ever see one alive. This small auk with its grotesque beak breeds around the North Atlantic Ocean, from Svalbard and northwest Greenland south to Maine and Brittany. The world population centre is in Iceland, where several millions breed, and most of the North American birds breed in southeast Newfoundland (270,000 pairs), so the Maritime population is only a peripheral remnant. Atlas work confirmed breeding at the three previously known sites: Machias Seal Island, N.B., Pearl Island, N.S., and Hertford/Ciboux Islands, N.S. Evidence was received by the Atlas of unconfirmed breeding in Nova Scotia in the Tusket Island and Mud Island groups, both near Seal Island where breeding occurred up to about 1912; with the few birds at Pearl Island, these probably represent partial re-occupation of former

breeding sites, whereas the larger, less-disturbed colonies had persisted throughout.

Presumably Puffins were formerly much more numerous in the Maritimes. Predation by dogs, cats, and rats, and also by humans, eliminated them as breeding birds from many islands around our coasts in the centuries following European settlement here. The Puffin is at the southern limit of its range in our area and is everywhere vulnerable to disturbance at its nest-sites and to disruption of its food supply. Incautious human exploitation of food fish, here usually sand lance or capelin, could have a disastrous effect on these and other seabirds in the future. Predation and stealing of food by large gulls may also hinder recolonization by Puffins of some former breeding areas. We should not expect major numerical increases or expansion of breeding range by Puffins in our area, under existing circumstances.

Ref. 109

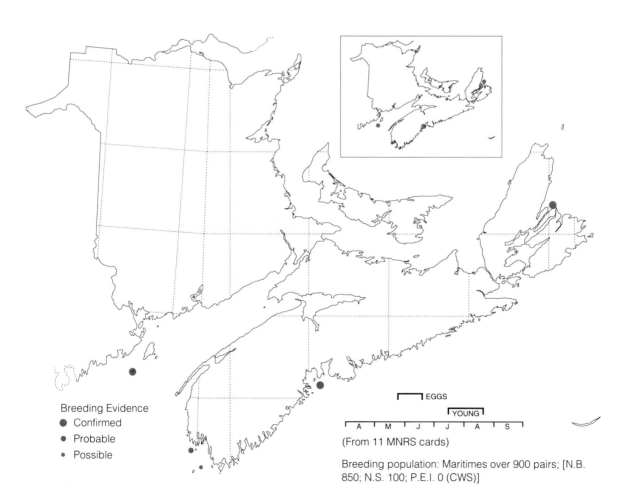

Breeding Evidence
● Confirmed
● Probable
● Possible

EGGS
YOUNG

| A | M | J | J | A | S |

(From 11 MNRS cards)

Breeding population: Maritimes over 900 pairs; [N.B. 850; N.S. 100; P.E.I. 0 (CWS)]

Rock Dove
Pigeon biset

Columba livia

Pigeons were brought to North America from Europe as free-flying domestic birds. Their original range probably extended from western Europe and north Africa to China. Pigeons have been established by human actions in all continents, for food, sport, and show, and their breeding almost everywhere is now associated with human buildings and agriculture, whether in the original or the extended range. Their distribution in the Maritimes includes most towns and cities, many villages, and much farmland, but the map shows striking gaps in forested regions such as northern New Brunswick and Cape Breton Island. Rock Doves breed almost throughout the year, breeding evidence (both extreme dates involving copulations) during our Atlas study extending from 11 January 1986 to 19 December 1990! Their nest-sites on buildings and, increasingly, on highway overpasses are easily detected. Breeding was confirmed in 328 squares (58 per cent of 562 with

pigeons), with nests (ON,NE,NY) found in 223 squares (40 per cent). As these birds, especially "racing" pigeons, often stray (e.g., to Sable Island and northern Cape Breton Island), some isolated single sightings (H) likely were vagrants rather than "possible breeding."

The earliest introduction of Rock Doves to Canada was in 1606–07, in Nova Scotia (Saunders 1935), but abandonment or repeated destruction of most settlements here in wars doomed most introductions before 1750. Present stocks presumably stem from later arrivals. Feral birds bred away from dovecotes as soon as suitable nest-sites on buildings and feeding areas on cleared lands became available nearby. Numbers peaked with the maximum extent of agricultural land early in the 20th century and probably declined somewhat since then. Efforts to reduce fouling by urban pigeons have lowered numbers in some towns and cities, but the BBS shows no overall change in the Maritimes since 1966. As with the other "feathered rats" that successfully follow human habitations, the future of Rock Doves here seems assured, whether we like them or not.

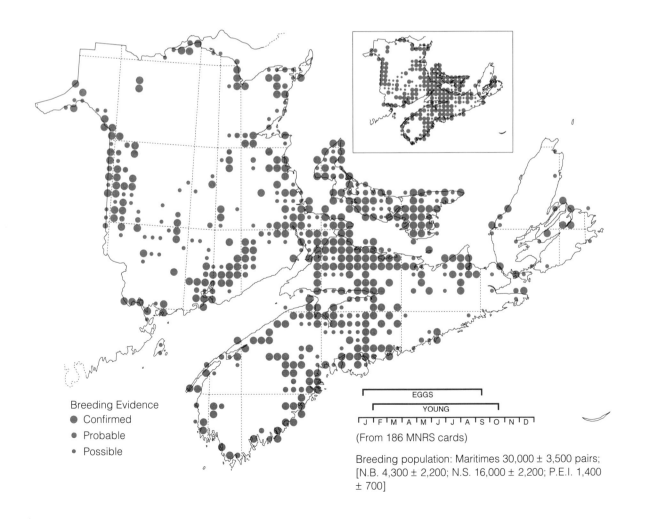

Breeding Evidence
● Confirmed
● Probable
· Possible

EGGS

YOUNG

J F M A M J J A S O N D

(From 186 MNRS cards)

Breeding population: Maritimes 30,000 ± 3,500 pairs; [N.B. 4,300 ± 2,200; N.S. 16,000 ± 2,200; P.E.I. 1,400 ± 700]

Mourning Dove
Tourterelle triste

Zenaida macroura

The Mourning Dove breeds across southern Canada and throughout the United States and Central America. It frequents open areas for foraging, but nests in trees, most commonly open-grown conifers as in windbreaks or pine plantations; thus it is absent as a breeding bird from both the treeless plains and the northern forests. The Mourning Dove reaches its northeast limit in the Maritimes, where its range in the St. John River Valley and Charlotte County, N.B., is continuous with that in Maine. It also breeds widely in other farming areas through southern and eastern New Brunswick and the Annapolis Valley, N.S., and in settled coastal areas near Yarmouth, Bridgewater, and Halifax. Although not common birds here, their habit of perching on roadside wires and their distinctive call "Oh, woe-woe-woe!" make them easy to detect. Single sightings (H) or continued presence (T) were the most frequent breeding evidence (40 per cent and 16 per cent, respectively, of 498 squares with doves reported), and fledged young (FL, in 106 squares) were the most com-

mon confirmation. Nests were reported in only 34 squares. Mourning Doves often winter around grain storage and feed-lot areas in the Maritimes, so not all spring records were of migrants returning to former breeding areas, although nesting may start as early as late March.

In the largely forested Maritimes before European settlement, Mourning Doves cannot have been common, and they were easily confused with the much more numerous Passenger Pigeons (now extinct). Doves seem to have occurred and bred in New Brunswick from early times, but they became more widespread and numerous in the last 40 years. Wintering began to be frequent in Nova Scotia in the 1950s and in New Brunswick from 1970. Breeding was proved in Nova Scotia in 1964, though no nest was found until 1979, and in Prince Edward Island in 1975. Present estimates of Mourning Dove numbers represent the largest population ever in our area. Projections of land use, with clear-cutting of forests followed by plantations, often of pines, should favour the species in future, and climatic warming from the "greenhouse effect" would also improve the marginal status of these birds. They are here to stay.

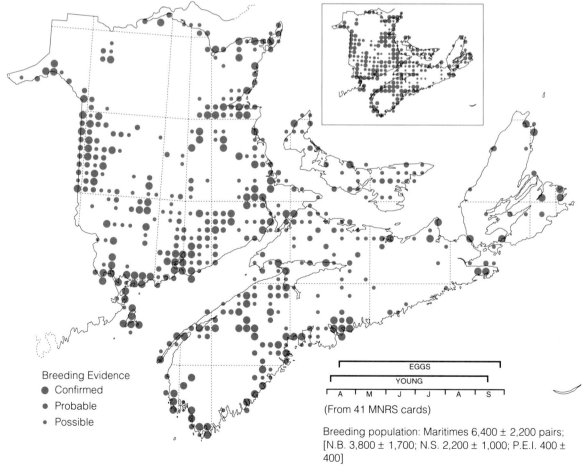

Breeding Evidence
● Confirmed
● Probable
● Possible

EGGS
YOUNG
A M J J A S

(From 41 MNRS cards)

Breeding population: Maritimes 6,400 ± 2,200 pairs; [N.B. 3,800 ± 1,700; N.S. 2,200 ± 1,000; P.E.I. 400 ± 400]

Black-billed Cuckoo
Coulicou à bec noir

Coccyzus erythropthalmus

This cuckoo is the more northern of the two North American species, breeding across southern Canada and most of the United States. In the northern parts of its range, its numbers vary greatly from year to year in response to outbreaks of both the forest and orchard species of tent caterpillars, on which it feeds. It is associated with open woodland and forest edge and nests in small trees and tall shrubs. The far-carrying calls, (grouped in threes thus: "cooc-cooc-cooc," in long series) aid in its detection, as cuckoos are skulkers in the underbrush except in flight. No widespread tent caterpillar outbreak occurred during the Atlas period, and the mapped distribution of the species under-represents its intermittently used range in the Maritimes. Only a few well-scattered records (15, of 167 squares with the species) were of confirmed breeding, mostly of adults carrying food to young (AY); the majority of records (68 per cent and 13 per cent, respectively) were of calling birds detected once (H) or more often (T) in an area. Opening-up of the formerly more continuous forests of the Maritimes probably allowed the species to increase, but replacement of mixed forest by conifers had a contrary effect. The suggestion of a decline in Nova Scotia, attributed to arsenical sprays (Tufts 1962), could only have been local, as orchards requiring spraying were never widespread in most areas of the Maritimes and have become much scarcer since World War II. An overall decline might reflect the gradual disappearance of marginal and abandoned farms and their reversion to forest unsuitable for this species, but the wide variation in numbers between tent caterpillar outbreaks would mask anything except a major decrease. Our estimates of present numbers, owing to the cuckoo's under-representation during the Atlas period, are minimal, and very approximate. If forestry and agriculture continue to become more intensive where practised, the future for this species is not likely to improve, but global warming could make the area less marginal for cuckoos.

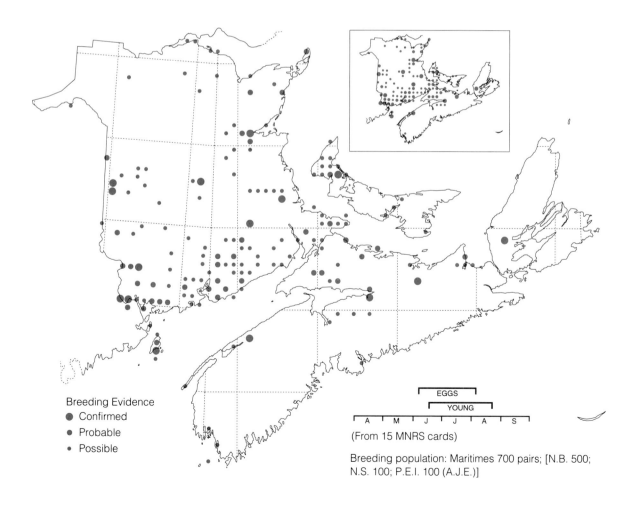

Breeding Evidence
● Confirmed
● Probable
• Possible

EGGS
YOUNG
A | M | J | J | A | S

(From 15 MNRS cards)

Breeding population: Maritimes 700 pairs; [N.B. 500; N.S. 100; P.E.I. 100 (A.J.E.)]

95

Great Horned Owl
Grand-duc d'Amérique

Bubo virginianus

The Great Horned Owl occurs throughout the Americas, from the arctic tree-line south to the Strait of Magellan. In the Maritimes, it is our most widely distributed and possibly our most numerous owl, although Barred Owls exceed its numbers in Nova Scotia. Like all night birds, it was often missed by atlassers, and it probably occurs more generally than our records show. The period of most frequent calling in spring coincides with the "mud season" here, when many rural roads are all but impassable for conventional vehicles; this difficulty, coupled with relatively low densities of all owls, greatly reduced the attractiveness of night surveys, and few observers spent much time listening for owls. These large birds are detected almost as often by sight as by sound, and most records of confirmed breeding were of nests (NY,NE,ON, 44 squares) or newly fledged young (FL; 55 squares); these comprised 24 per cent of 414 squares with Great Horned Owls. These owls do not build a nest, but take over nests of large hawks or crows; they readily accept nest-platforms erected for their use, as discovered inde-

pendently here (though known earlier in other areas). Large owls were shot on sight in the Maritimes for many years, as they were considered harmful to human interests. This species, which feeds on almost any available animals, is more likely than most raptors to have taken domestic poultry or desirable game species, but its effects, even at undisturbed population levels, would not have matched those of human predators. Tufts (1986, also noted in 1962 & 1973 editions) considered that this species had decreased from former abundance in Nova Scotia, but this was not a widespread consensus; as a boy in Nova Scotia in the 1940s I never saw one, but nearby in New Brunswick I now see a few every year while spending less time in the field. Shooting of hawks and owls is less general than formerly, though it still persists, so their numbers might have been recovering from earlier persecution 40 years ago. True numbers may be several times greater than our minimal estimates. Although numbers of Great Horned Owls elsewhere fluctuate with those of the Snowshoe Hare, one of their main prey species, this relationship has not been evident in the Maritimes. There seem to be no serious threats to its continued abundance here.

Ref. 147

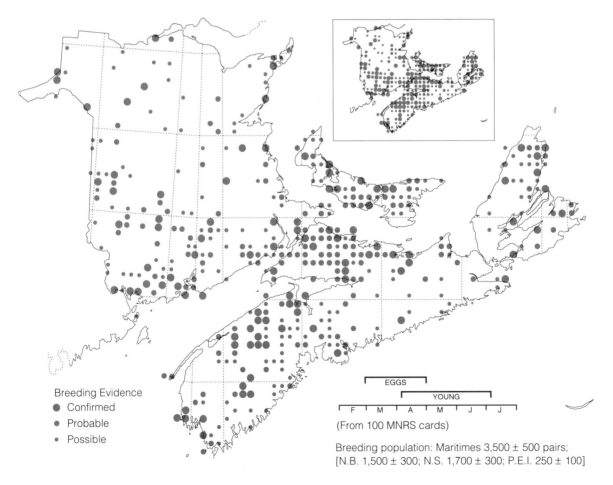

Breeding Evidence
● Confirmed
● Probable
· Possible

EGGS
YOUNG
F M A M J J

(From 100 MNRS cards)

Breeding population: Maritimes 3,500 ± 500 pairs;
[N.B. 1,500 ± 300; N.S. 1,700 ± 300; P.E.I. 250 ± 100]

Barred Owl
Chouette rayée

Strix varia

One of the two common large owls of temperate North America, the Barred Owl is a permanent resident, breeding throughout the eastern hardwood and mixed forests of Canada and the U.S.A., south at higher elevations into Central America, and recently westward into British Columbia. The Maritimes is near its northeast limit, as it scarcely penetrates beyond the Gulf of St. Lawrence. Like other owls, this nocturnal species was certainly more generally distributed than the Atlas data showed. Its scarcity in New Brunswick outside the southwest quarter and in eastern Nova Scotia may reflect some avoidance of the more coniferous forests there. Most records (80 per cent of 377 squares with Barred Owls) were of single (H) or repeated (T) sightings or calls, and only 19 squares had nests reported, mostly in nest-boxes. These birds originally nested in large, hollow trees, and their ready adoption of nest-boxes suggests that they are sometimes limited by availability of suitable cavities.

The greater extent in former times of mature forests, with more large old trees and more frequent tree cavities, indicates that Barred Owls were more abundant when Europeans first came to the Maritimes than at present. Subsequent selective logging, taking the largest trees first, reduced availability of nest-sites at the same time that indiscriminate shooting of large birds, and especially raptorial species, reduced the numbers of Barred Owls. Though always considered one of the most common owls in the Maritimes, this bird was probably at its lowest numbers early in the 20th century, and it may have increased somewhat since then. Atlas records provided a weak sampling base for estimating populations. Shorter cycles in the cutting of managed forests in future will limit the old-growth trees that provide natural cavities, but many suitable trees remain in neglected areas. This is hardly a common bird, but its numbers seem unlikely to change greatly in the next few decades.

Ref. 29

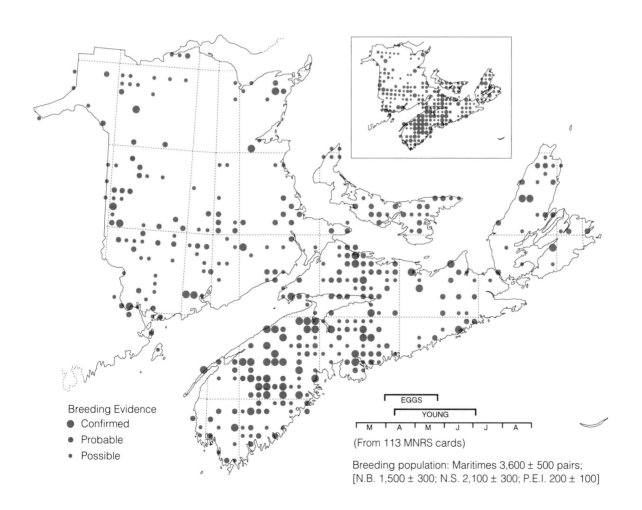

Breeding Evidence
- Confirmed
- Probable
- Possible

EGGS
YOUNG
M A M J J A

(From 113 MNRS cards)

Breeding population: Maritimes 3,600 ± 500 pairs;
[N.B. 1,500 ± 300; N.S. 2,100 ± 300; P.E.I. 200 ± 100]

Long-eared Owl
Hibou moyen-duc

Asio otus

This owl occurs across the boreal and temperate regions of Eurasia and North America; related but larger species occur in tropical America and Africa. It is a permanent resident in milder areas, but withdraws from more northern parts of its range in winter. It frequents woodlands large or small, dense or open, conifer or broad-leafed, at all seasons, but it also forages over open areas. Long-eared Owls are seldom seen except at nests or roosts, as they usually stay in dense cover except during their nightly foraging for small mammals, which make up their main prey. Their calls are varied, including squawks and screams not easily recognized as belonging to an owl. Long-eared Owls are easily missed, and this may well be the most poorly represented species in the Maritimes Atlas. During the five years of field-work, it was recorded in only 35 squares, in all three provinces, including the first ever found on Cape Breton Island. It is probably much more widely distributed than our records suggest, but we can only guess at its real frequency. In the

Ontario atlas, with more intensive field work, Long-eared Owls were found in about one-third as many squares as Barred Owls, vs. one-tenth here; in Maine, with less intensive effort, they were found in only 6 squares vs. 160 with Barred Owls. Long-eared Owls probably occur, at low density, all over the Maritimes, as our records were scattered in 14 of the 23 atlassing regions.

Historic changes in the forest cover of the Maritimes seem unlikely to have much altered its suitability for this species. The former widespread persecution of raptorial birds is unlikely to have affected this strictly nocturnal species. The Atlas data provide no direct basis for estimating their total numbers here, but comparison with the Barred Owl allows us a guess. Although Long-eared Owls often move in autumn to roosts near favoured foraging areas (e.g., Kings Co. dykelands in Nova Scotia), our birds are not known to leave the Maritimes in winter; thus any changes, in the future as in the past, would result from persecution or environmental changes within the Maritimes. The fragmentation of our forests in recent years seems likely to affect the numbers of these owls more by influencing the availability of their prey than by reducing their habitat.

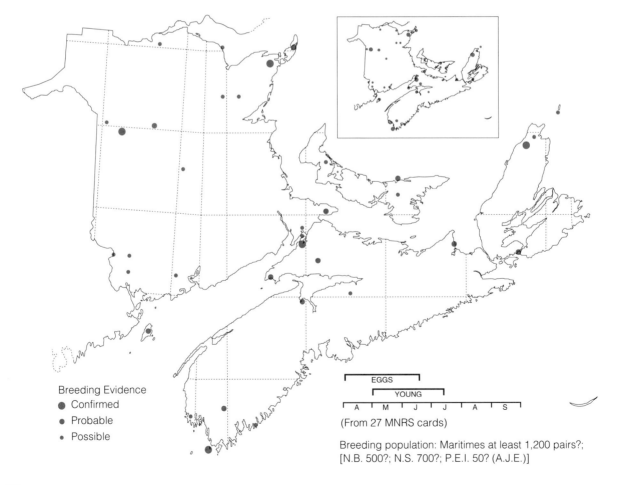

Breeding Evidence
● Confirmed
● Probable
· Possible

EGGS
YOUNG
A M J J A S

(From 27 MNRS cards)

Breeding population: Maritimes at least 1,200 pairs?;
[N.B. 500?; N.S. 700?; P.E.I. 50? (A.J.E.)]

Short-eared Owl
Hibou des marais

Asio flammeus

The Short-eared Owl occurs erratically in open grassy areas on most continents and some oceanic islands. In North America, it ranges in low arctic tundras from Alaska to northern Labrador, and in boreal and temperate grass- and shrub-lands south to California and New Jersey, but its presence and numbers in any area vary from year to year. These owls have bred in the Maritimes in dyked wet meadows and marshes and in coastal bogs and grasslands in the past, and they were found locally in such habitats during the Atlas, with no confirmed breeding away from the coasts. Their numbers, and those of the meadow voles on which they fed, were unusually high in 1980, when R. Simmons found many nests near Jolicure, N.B., and B. Forsythe found several nests near Grand Pré, N.S.; no such abundance occurred during the Atlas period, but nests were found on the Grand Pré dykelands also in 1987 and 1988. As Short-eared Owls hunt at dawn and dusk in open country, they are more easily detected than other owls, although they seldom vocalize. The scarcity of mapped records indicates that little time was spent searching suitable habitats in the twilight hours, except in squares where observers lived. The data probably under-represent the usual status for this species, but perhaps less so than for most owls.

Although European settlement of the Maritimes resulted in extensive destruction of marshes, the dyking of salt marsh, by excluding tidal effects, may have helped breeding by Short-eared Owls. Their numbers vary so erratically, independently of human actions, that there are no firm grounds for believing that Short-eared Owls were generally much more or less common in the past than at present. We surmise that in years with peaks in vole abundance, such as 1980, their numbers may increase to more than double the usual population, but we do not know how much the numbers in favoured sites spill over into other potential areas. Presumably these birds will continue to fluctuate in distribution and numbers here in the future. Their nomadic habits outside of nesting season mean that they can easily re-occupy suitable habitats when food is available, even though present numbers here are low.

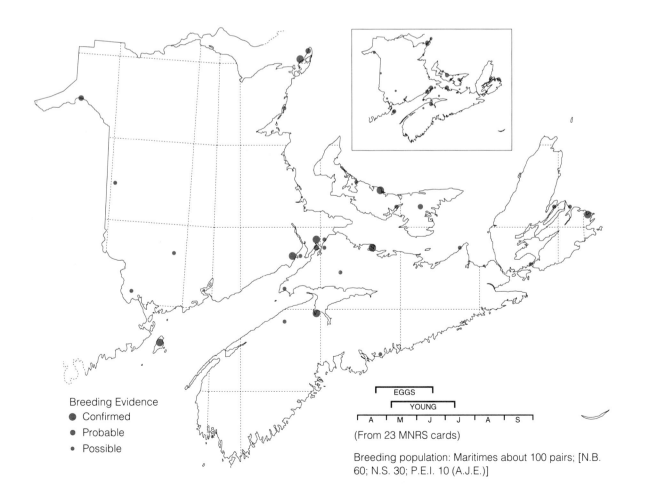

Breeding Evidence
- Confirmed
- Probable
- Possible

EGGS
YOUNG

A M J J A S

(From 23 MNRS cards)

Breeding population: Maritimes about 100 pairs; [N.B. 60; N.S. 30; P.E.I. 10 (A.J.E.)]

Boreal Owl
Nyctale boréale

Aegolius funereus

This small owl is poorly known in North America, although it has been well-studied in Europe. It breeds across the boreal forests of Eurasia and North America, and locally farther south in the mountains, nesting in woodpecker holes and other tree cavities. Boreal Owls are active only at night, and they call early in the spring when few observers are afield, so they are easily missed. Until recently, inaccurate descriptions of the territorial call in most North American bird books hampered recognition of Boreal Owls throughout their range on this continent. They were meagerly represented in the Atlas records in 11 squares, which were restricted to northeastern New Brunswick and Cape Breton Island, except for one report in eastern Prince Edward Island. Our data certainly under-represent the species, but the scarcity of records at other seasons suggests that we do not have a large unseen population of Boreal Owls.

With so few records anywhere on this continent, we have little perspective from which to assess its past or present status in our area. No Boreal Owls were found in the Atlas study in Maine, and only 22 squares in the Ontario Atlas had this species. It seems plausible that the old records (1924–32) from the Grand Manan archipelago were a temporary phenomenon only, and that the Atlas records from the northern periphery of our region represent the southern limit of its regular range in the east; the sparse Quebec records support this interpretation. Boreal Owls may have bred in our northern bogs all along, though they were not recognized to do so until the Atlas period. Those habitats, and presumably also the birds in them, have changed rather little over the period of European settlement here. Our estimates are certainly minimal, but undetected birds are unlikely to push the total above a few hundred pairs. There are no obvious human threats to the species, but a tendency towards climatic warming could lead to a northward retreat of its range limit.

Ref. 13

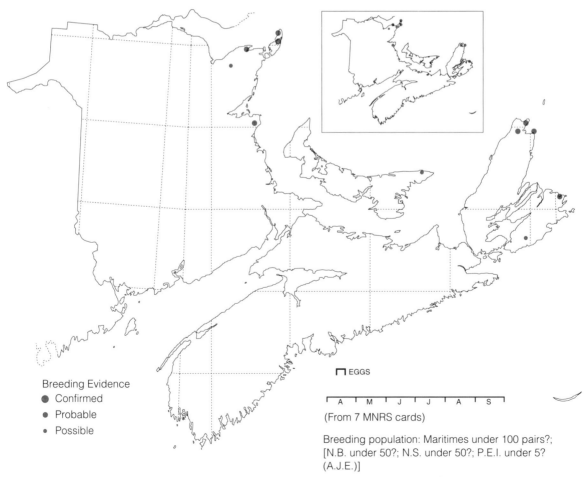

Breeding Evidence
- Confirmed
- Probable
- Possible

⊓ EGGS

| A | M | J | J | A | S |

(From 7 MNRS cards)

Breeding population: Maritimes under 100 pairs?; [N.B. under 50?; N.S. under 50?; P.E.I. under 5? (A.J.E.)]

Northern Saw-whet Owl
Petite Nyctale

Aegolius acadicus

This small owl breeds across southern Canada, south to northern Mexico and to Missouri and Maryland, U.S.A. It is partly migratory, as numbers have been banded in autumn at concentration areas along shorelines, including in Nova Scotia, but some others winter here. Like other owls, it is seriously under-represented by the Atlas data in the Maritimes, largely owing to a scarcity of observations when the birds are calling, at night in spring when back-country roads are best avoided. The data here were too sparse to provide a convincing pattern of absence vs. presence, but they suggested that Saw-whet Owls occur throughout our area, as indicated by earlier summaries based on even fewer records. Most reports were of single (H) or repeated (T) sightings or calls (in 86 per cent of 168 squares with the species). Nests in woodpecker holes or nest-boxes were found in

only 11 squares, and fledged young in 10 squares.

Historical data on trends in numbers of Saw-whet Owls are lacking completely. We infer that suitable nesting cavities were more frequent in the more generally mature forests before European settlement than at present, but we do not know if nest-sites have limited their numbers, recently or in the more distant past.

Cold winters with deep snow, which may have been more frequent during the Little Ice Age, still cause losses by starvation of wintering Saw-whet Owls in some years. Reports are neither more nor less widespread or frequent now than 30–40 years ago; this need not point to a decline, as the greatly increased numbers of observers in recent years do not spend much, if any, more time afield at night. Our estimates of breeding numbers are little better than guesses. There are few obvious threats to this inconspicuous species, which is likely to persist, unobserved, in our forests for years to come.

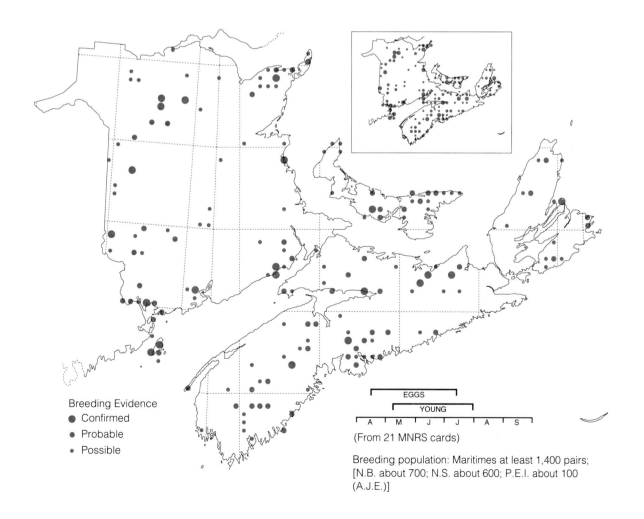

Breeding Evidence
- ● Confirmed
- ● Probable
- · Possible

(From 21 MNRS cards)

Breeding population: Maritimes at least 1,400 pairs; [N.B. about 700; N.S. about 600; P.E.I. about 100 (A.J.E.)]

Common Nighthawk
Engoulevent d'Amérique

Chordeiles minor

Nighthawks breed across Canada from the Yukon to the Maritimes, and south through the U.S.A. into Mexico and the West Indies. We found the bird to be widespread but nowhere really common. There were records in all regions, but notably fewer in northwest New Brunswick (cool highlands), Prince Edward Island (farmland), and eastern Nova Scotia, especially Cape Breton Highlands (cooler); no one factor explains scarcity in all these areas. Nighthawks nest on sparsely vegetated or bare ground in open "wastelands" such as pine barrens, forest cut-overs, or burns, and secondarily on flat roofs of buildings. Most records (443 out of 539 squares with Nighthawks, 82 per cent) were of birds seen and heard while foraging and calling/courting, high over the breeding area. Nests were found in only 25 squares (5 per cent) and newly flying young in 20 squares. Nighthawks were under-represented, as only observers who camped in the breeding areas (usually wilderness) were there at the dusk and dawn calling periods. Low representation, with breeding seldom confirmed, was typical for this species in Maine and Ontario also.

The more continuous forest cover, with fewer fires and minimal clearing, in the Maritimes before European settlement provided fewer opportunities for Common Nighthawks to nest, even though the flying insects that make up their food were more plentiful then. The species probably increased up into this century. Most long-term observers share the impression that its numbers have declined over the past 50 years, but the BBS suggests an upward trend in New Brunswick since 1966, based on small samples. The apparent decline may be mainly in urban-nesting Nighthawks, as great efforts have gone into reduction of flying insects in cities and towns since World War II. For example, Nighthawks had been regular, though becoming scarcer, in summer over Sackville, N.B., but none at all were seen there since about 1984. Clear-cutting in forest harvest operations continues to provide plenty of nesting habitat, but spraying against insects harmful to forests may reduce some insects eaten by Nighthawks. Whether or not the decline reported in this species is real, its decline elsewhere suggests that we need to monitor its numbers more carefully in future.

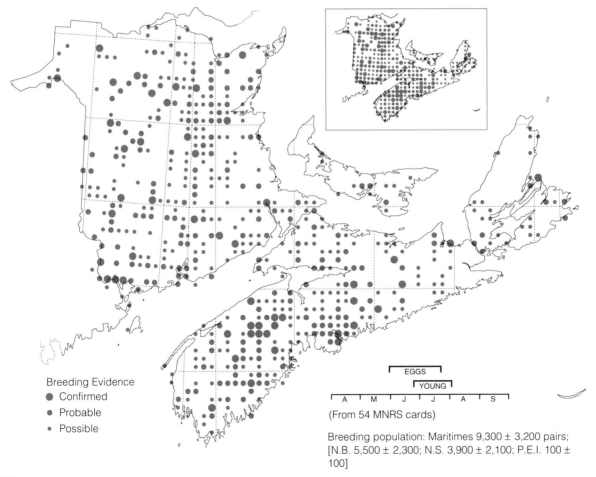

Breeding Evidence
- Confirmed
- Probable
- Possible

EGGS
YOUNG
A   M   J   J   A   S

(From 54 MNRS cards)

Breeding population: Maritimes 9,300 ± 3,200 pairs; [N.B. 5,500 ± 2,300; N.S. 3,900 ± 2,100; P.E.I. 100 ± 100]

Whip-poor-will
Engoulevent bois-pourri

Caprimulgus vociferus

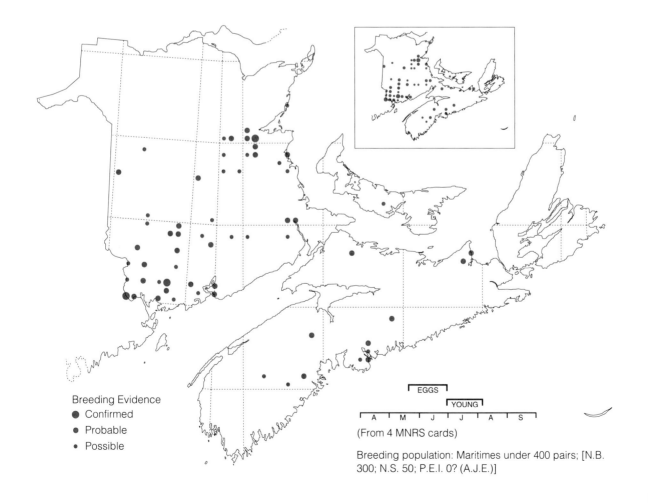

The Whip-poor-will is known by name to many who have never heard it, and by voice to many who have never seen one. This species belongs to the eastern broad-leafed forest, a generally warmer region, breeding from the eastern edge of the prairies south to Texas and east to Virginia and, very sparsely, to the Maritimes. In our area, there is a suggestion of general distribution in southwest New Brunswick, with a cluster in the lower Miramichi Valley and scattered records along other valleys of New Brunswick. Most Nova Scotia records were in the south-central part, where nesting was first proved in 1930. A scarcity of night surveys in most areas means the Atlas seriously under-represents the occurrence of Whip-poor-wills, but this surely is a scarce species everywhere in the Maritimes. Of 62 squares with these birds reported, only three had breeding confirmed, one with newly flying young (FL); only 3 nests were known in the region, none since 1942, before the one in the Atlas files. Other records were mainly of calling birds (H,T).

Whip-poor-wills were probably at the limits of their breeding range here even when mature broad-leafed forests were more general. We have no evidence that a change in either numbers or range has occurred in historic times. The Atlas data are not sufficient for extrapolated estimates of the populations, and our guesses are as likely to be generous as low. The Whip-poor-will is sparsely distributed in adjacent parts of southern Maine, with which the southwest New Brunswick range is contiguous. It is unlikely to become more common here, unless under the influence of global warming, and it could disappear without most people even noticing its absence; it is not regular enough that we can define areas in which it needs special protection.

Breeding Evidence
- ● Confirmed
- ● Probable
- • Possible

EGGS
YOUNG

| A | M | J | J | A | S |

(From 4 MNRS cards)

Breeding population: Maritimes under 400 pairs; [N.B. 300; N.S. 50; P.E.I. 0? (A.J.E.)]

Chimney Swift
Martinet ramoneur

Chaetura pelagica

This is the only swift in eastern North America, where it breeds from eastern Saskatchewan and the Maritimes south to Florida and Texas. These birds are seen only on the wing and while entering their nesting places; these are often in chimneys or old cabins in the forest, but most swifts originally nested, and still nest, in hollow trees. Swifts were seen in most regions of New Brunswick and Nova Scotia. Very few were seen in Prince Edward Island, and they also were notably scarce in the regions of New Brunswick and Nova Scotia adjoining Northumberland Strait. Breeding was confirmed in only 92 squares (20 per cent of 471 squares with the species) with single (H) or repeated (T) sightings comprising over two-thirds of the records. Surprisingly, most confirmed breeding was based on nests, found in 74 squares (16 per cent), mostly widely scattered in rural areas, with some urban centres such as Saint John and Shediac, N.B., completely unrepresented.

Although large hollow trees may have been more common in the pre-settlement forests of the Maritimes than at present, they seem unlikely ever to have supported very large populations of Chimney Swifts. The former reports of "many thousands" entering individual chimneys, e.g., in Fredericton, in the early 1900s, suggest that human settlements improved nesting opportunities for these birds for a while. They have decreased in some parts of the Maritimes, notably in the New Brunswick-Nova Scotia border area, in recent decades. BBS data for these aerial wanderers are too variable to represent trends in their numbers. Chimney Swifts were termed "common" in New Brunswick and "uncommon" in Prince Edward Island in earlier summaries, but they are certainly less numerous in those provinces now. Aerial spraying against spruce budworm, ongoing in New Brunswick since 1952, may have reduced flying insects below the levels needed to support earlier swift numbers, especially in areas downwind of the spray zone. Projected trends in forest management are unlikely to benefit swifts in future, with old trees becoming scarcer and pesticide spraying continuing. Large urban chimneys decreased with the shift from wood or coal to oil or electric heating. The Chimney Swift has declined markedly in the last 30 years, and its numbers should be monitored carefully.

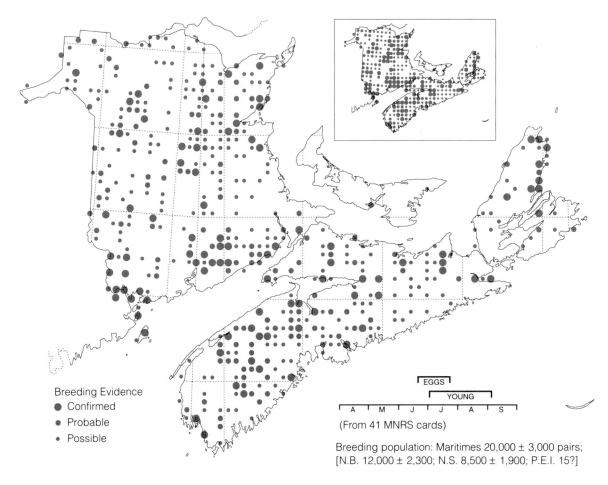

Breeding Evidence
● Confirmed
● Probable
● Possible

EGGS
YOUNG
A M J J A S

(From 41 MNRS cards)

Breeding population: Maritimes 20,000 ± 3,000 pairs;
[N.B. 12,000 ± 2,300; N.S. 8,500 ± 1,900; P.E.I. 15?]

Ruby-throated Hummingbird
Colibri à gorge rubis

Archilochus colubris

The hummingbirds are mostly native to Central and South America, and this is the only species found regularly in eastern North America. It breeds from the Gulf of Mexico north to southern Alberta and the Gaspé Peninsula. Hummingbirds range throughout the Maritimes, from the forests of New Brunswick's northwestern highlands to lowland farms and towns, nowhere abundant, everywhere unique. These exotic creatures, scarcely bigger than a moth but with gleaming metallic plumage, flying like a bumblebee rather than a bird, are unmistakable, if your eyes focus fast enough to recognize them. The often-sparse Atlas data reflect their moderate density and elusiveness, as single (H) or repeated (T) sightings accounted for records in two-thirds of the squares (525 of 784) in which hummingbirds were found. Breeding was confirmed in 145 squares, mainly through fledged young (FL) seen by resident observers, who could make prolonged observations,

i.e., in settled areas. Nests were found in 26 squares, often by seeing the female bringing nest material (NB). Hummingbirds feed mainly on nectar from flowering plants and on the tiny insects found around flowers, and they are typical of open woods and settled lands where flowering is more prolonged than in the forest. Many forest flowers bloom only for brief periods in the spring before leaves appear on the trees. It seems plausible that hummingbirds were somewhat less common in the primaeval forests of the Maritimes, but it is unlikely that they increased greatly following European settlement. This is a species that makes use of our gardens and orchards without becoming dependent on them. Our estimate of their present numbers is certainly low, because they were missed in many squares that received only minimal coverage. There seems no reason to doubt that hummingbirds will be around the Maritimes in the future, and global warming, if it occurs, might improve their prospects.

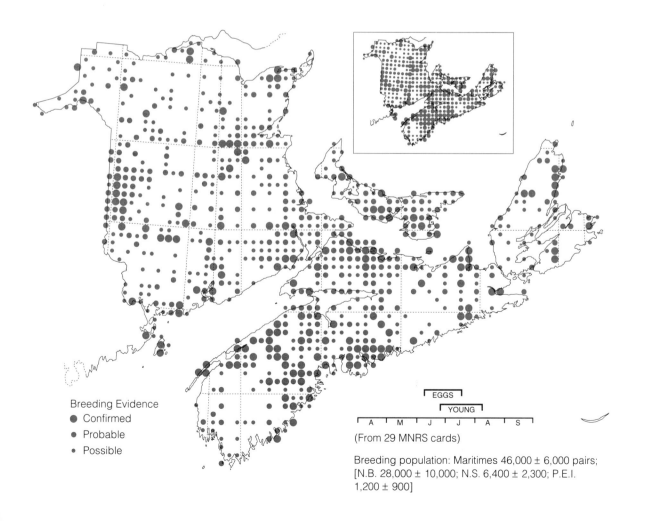

Breeding Evidence
- Confirmed
- Probable
- Possible

EGGS
YOUNG

A M J J A S

(From 29 MNRS cards)

Breeding population: Maritimes 46,000 ± 6,000 pairs; [N.B. 28,000 ± 10,000; N.S. 6,400 ± 2,300; P.E.I. 1,200 ± 900]

Belted Kingfisher
Martin-Pêcheur d'Amérique

Ceryle alcyon

Kingfishers of various sizes and colours occur on every continent except Antarctica, but this is the only one regular in America north of Mexico. It breeds from the southern U.S.A. to Alaska and Newfoundland and is common throughout the Maritimes. Kingfishers occur around both salt and fresh waters, wherever suitable earth banks for nests, and trees and other elevated perches from which to watch for fish, are available. Atlassers found them almost everywhere that an adequate search was made of suitable places by rivers, lakes or sea-shores. Kingfishers are easily seen and heard in the open water-side habitats they frequent, so the Atlas records may represent their occurrence quite well, better than for many other species. Nests were easily found, but the contents are difficult to inspect; nests (mainly ON) were reported in 24 per cent of 841 squares with Kingfishers, and newly fledged young in family groups were also frequent (82 squares).

These fish-eating birds were excluded from protection under the Migratory Birds Convention (1916) as supposedly harmful to human interests. For years, they were shot regularly "to protect fishery interests," but this persecution was sporadic and evidently ineffective in reducing their numbers. Although cut banks suitable for their nesting are more widely available now than before European settlement, such sites were always common in natural situations; on balance, it seems unlikely that their numbers have changed much over the last four centuries. The BBS shows no significant population trend since 1966. A possible threat to their future occurrence and breeding here may be the effect of acid rain in reducing fish stocks in our already acidic inland waters. Forecast changes in land use and climate seem unlikely to have important effects on future Kingfisher numbers here.

Ref. 155

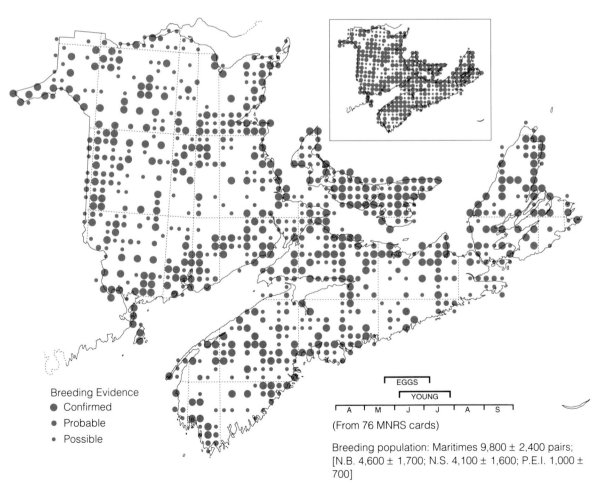

Breeding Evidence
● Confirmed
● Probable
• Possible

EGGS
YOUNG
A M J J A S

(From 76 MNRS cards)

Breeding population: Maritimes 9,800 ± 2,400 pairs; [N.B. 4,600 ± 1,700; N.S. 4,100 ± 1,600; P.E.I. 1,000 ± 700]

Yellow-bellied Sapsucker
Pic maculé

Sphyrapicus varius

These woodpeckers breed all across boreal Canada, from southern Yukon to Newfoundland, and south into the transitional forests of New England, and in mountains to Georgia. Closely related forms occur in the west from British Columbia to California. Everywhere these are birds of broad-leafed woodlands, nesting in live poplars and birches in the boreal range but in various tree species (usually dead or decaying) farther south, including in the Maritimes. The "sap-sucking" habit is general, but seems unlikely to provide a major part of their energy budget, as their diet is mostly of insect larvae, including spruce budworms when these are abundant. In the Maritimes, Sapsuckers were widespread, but scarcer along the Atlantic slope of Nova Scotia, on Cape Breton Island, and along the Bay of Fundy coastline, these areas having smaller proportions of hardwoods suitable for these birds. They seemed not to use the floodplain forests of the St. John and other

rivers as much as upland hardwoods. This is our only woodpecker that may be identified reliably by its "drumming," an irregular rhythm in this species, so these birds are easily identified. Also, young in the nests are noisy, so breeding is easily confirmed. Nests were noted in 166 squares, 23 per cent of 728 with Sapsuckers, and 48 per cent of those with breeding confirmed, higher proportions than even for the more widely known Flickers.

Presumably Sapsuckers were always common in the Maritimes forests, changes resulting from settlement and lumbering not having inconvenienced them greatly. Present numbers are substantial, but the BBS indicates a significant decline since 1966. This decrease may be related to increasingly widespread clear-cutting of forests and suppression of successional hardwoods, but other factors may be implicated. These forestry practices are likely to continue, but they seem unlikely to threaten the Sapsucker's status as a common bird in many parts of the Maritimes.

Refs. 42, 54

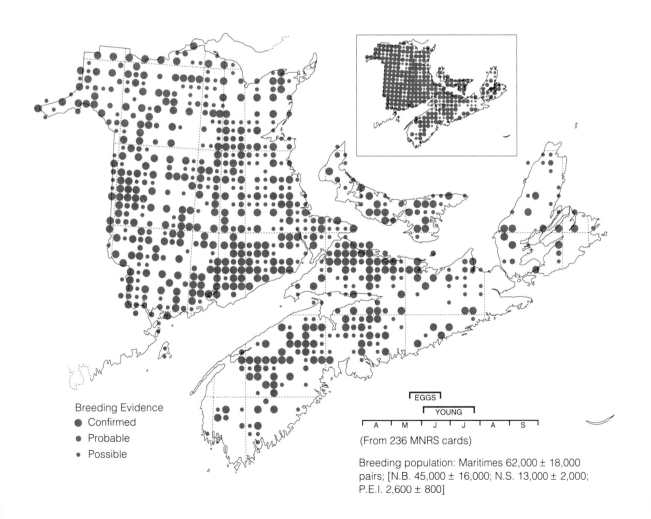

Breeding Evidence
- Confirmed
- Probable
- Possible

EGGS
YOUNG
A M J J A S

(From 236 MNRS cards)

Breeding population: Maritimes 62,000 ± 18,000 pairs; [N.B. 45,000 ± 16,000; N.S. 13,000 ± 2,000; P.E.I. 2,600 ± 800]

Downy Woodpecker
Pic mineur

Picoides pubescens

The Downy is the smallest widespread wood-pecker species in North America, where it is a permanent resident all across Canada and the U.S.A. The woodpeckers in Eurasia that are ecological equivalents seem less similar to ours than in other bird groups, and they are clearly distinct species. These small woodpeckers, in both New and Old Worlds, are largely restricted to woodlands with appreciable proportions of broad-leafed trees. Downy Woodpeckers seldom excavate their nests in evergreen trees, and they nearly always choose dead trees, or dead parts of live trees. They readily use habitats opened or modified by human actions, such as gardens, orchards, and city parks, in which they are more conspicuous than in undeveloped areas. Downy Woodpeckers were found across the Maritimes, with few obvious gaps, the most

conspicuous of which was in northeast New Brunswick. Breeding was confirmed in 296 squares (38 per cent of 789 squares with the species), and nests were found in 96 squares (12 per cent), proportions equivalent to those for other woodpeckers.

As Downy Woodpeckers have adapted well to settled areas and forest edge habitats, they are likely to have maintained or increased their numbers since the start of white settlement and the opening-up of the forests in the Maritimes. Their effect in seeking out and eating insect pests has been long-recognized as more important than any minor damage to wood caused by their excavations. We have no indications of trends in their numbers. There seem to be no good reasons to expect that they will increase or decrease here in the near future, as they are not narrowly restricted to any one habitat or type of food.

Ref. 89

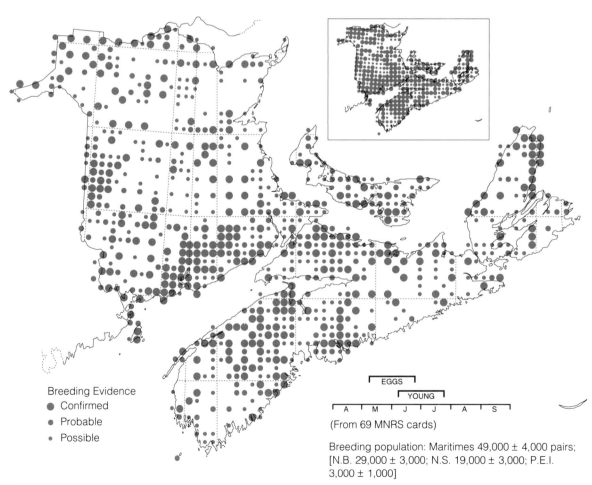

Breeding Evidence
- Confirmed
- Probable
- Possible

EGGS

YOUNG

A M J J A S

(From 69 MNRS cards)

Breeding population: Maritimes 49,000 ± 4,000 pairs; [N.B. 29,000 ± 3,000; N.S. 19,000 ± 3,000; P.E.I. 3,000 ± 1,000]

Hairy Woodpecker
Pic chevelu

Picoides villosus

The larger of our two familiar black-and-white woodpeckers, the Hairy occurs and breeds throughout North America's woodlands, from near the arctic tree-line to Panama and the West Indies. As they forage mainly on trunks of trees and stubs, they are quite catholic in habitat choice as long as trees are present, though they are more common in open woodlands than in either dense forest or settled lands with few trees. In the Maritimes, the Hairy Woodpecker was widespread, but the fact that it was found in a larger proportion of surveyed squares in Prince Edward Island than in much of New Brunswick tends to confirm that continuous forests are less used than more open regions. It was nowhere really abundant, as its numbers are limited to those the local habitat can sustain through the winters; species that migrate south for the winter achieve higher breeding densities. As a rather scarce bird, its detection depended to some extent on the duration of coverage; the areas with few Hairy

Woodpeckers recorded generally had received only brief surveys, even if by experts. As with other woodpeckers, breeding by Hairies was confirmed (in 280 squares, 35 per cent of those with the species) as often by finding the nest (in 110 squares, especially NY) as by other means (e.g., FL in 123 squares). Young Hairy Woodpeckers in the nest are noisy, and thus the nests can be detected from considerable distances.

The species has presumably increased since the pre-settlement era, but perhaps not greatly; Hairy Woodpeckers depend neither on closed forests nor cleared lands, and small openings in wooded areas must always have existed. Even the recent provision of winter feeding stations occurs on such a limited scale as to have only local influence on distribution or numbers. The BBS suggests a non-significant increase since 1966. Present breeding populations are very similar to those of Downy Woodpeckers. Prospects for the future are obscure but optimistic, with no obvious threats to the species' security and no clearly beneficial trends in land use or climate forecast.

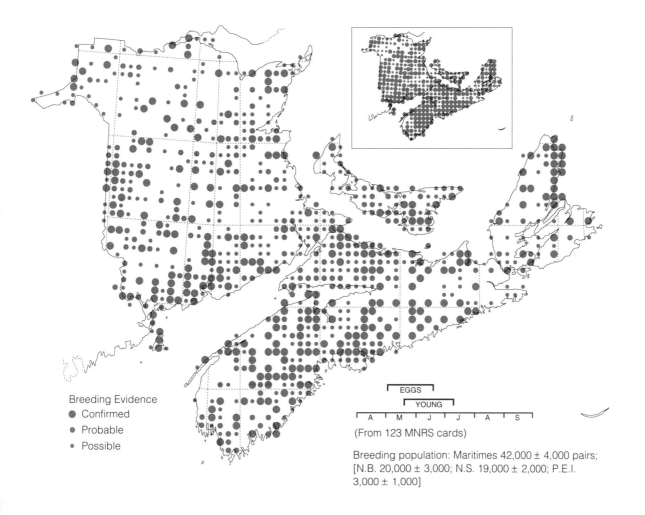

Breeding Evidence
- Confirmed
- Probable
- Possible

EGGS
YOUNG
A   M   J   J   A   S

(From 123 MNRS cards)

Breeding population: Maritimes 42,000 ± 4,000 pairs;
[N.B. 20,000 ± 3,000; N.S. 19,000 ± 2,000; P.E.I.
3,000 ± 1,000]

Three-toed Woodpecker
Pic tridactyle

Picoides tridactylus

This is the most northerly woodpecker species, breeding in the boreal conifer forests nearly to the arctic tree-line from Scandinavia across Siberia and from Alaska to Labrador. There are isolated populations in mountains farther south, in Eurasia and also in North America, south to Arizona in the west but only to New York and New England in the east. In the Maritimes its range is largely restricted to New Brunswick, where it is more general in higher areas of the northwest. Breeding in Nova Scotia was first proved during the Atlas period and was limited to Cape Breton Island; the one record in Prince Edward Island was the first known breeding there. Three-toed Woodpeckers were seen so seldom that special efforts were made whenev-

er one was found, and breeding was confirmed in 12 of 42 squares with this species, 7 involving nests with young heard or seen (NY,ON).

Although breeding by Three-toed Woodpeckers in New Brunswick was not confirmed until 1965, small numbers of these birds presumably bred here in the more distant past, even before the start of European settlement. As they forage especially on dead or dying conifer trees resulting from lumbering and from insect damage, the changes wrought on our forests over the past four centuries should have helped as much as they harmed these birds' opportunities. We cannot demonstrate any population change in the short or the long term. For the future, forest management seems unlikely to affect the species much. Any shift towards more temperate forest types as a result of global warming could, in time, eliminate this northern bird from the Maritimes.

Ref. 53

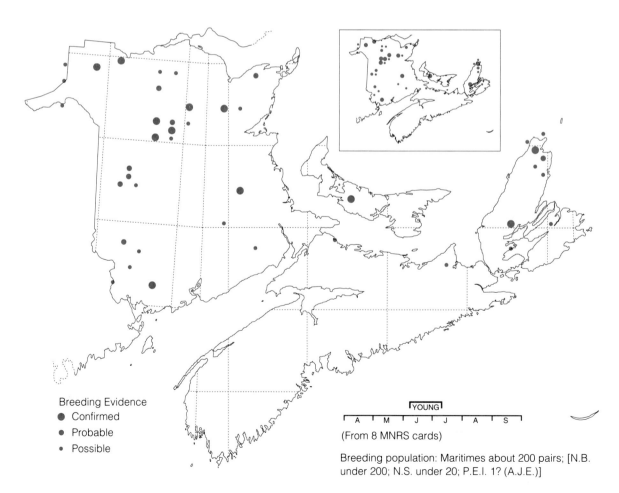

Breeding Evidence
● Confirmed
● Probable
• Possible

┌─YOUNG─┐
| A | M | J | J | A | S |

(From 8 MNRS cards)

Breeding population: Maritimes about 200 pairs; [N.B. under 200; N.S. under 20; P.E.I. 1? (A.J.E.)]

Black-backed Woodpecker
Pic à dos noir

Picoides arcticus

This is slightly the larger and more southern species of the three-toed woodpecker group and, unlike the Three-toed Woodpecker, it is restricted to North America. It breeds across the boreal regions from Alaska to Labrador, and south in the mountains to California and New England. In the Maritimes, it was widely but thinly distributed in conifer forests throughout, becoming more common farther north. The Black-backed Woodpecker was scarce in western Prince Edward Island and was very local in southwest Nova Scotia. These birds forage on trees damaged by forest insects, especially bark beetles, and their characteristic flaking-off of bark fragments in search of food can be an aid in detecting them. Nests here are often in quite open situations, such as cut-over areas, open jack pine stands, and the edges of woodland gardens. The young sometimes call continuously in the nest for minutes at a time, so nests (NY,ON) were quite easy to locate (in 16 per cent of 279 squares with the species) and were the most frequent evidence of confirmed breeding, though single birds (H) were seen more often (147 squares).

These woodpeckers were originally scarcer in parts of the Maritimes, before the primaeval hardwood and mixed forests were cleared or burned and replaced by fir and spruce. Elsewhere, where conifers prevailed from earlier times, their numbers have probably changed little since European settlement began. This is the scarcest of our breeding woodpeckers except for the preceding species. Management of forests for pulpwood (conifer) species is likely to ensure a continuing place for these birds to breed, but shorter cycles between timber cutting will reduce the time available for insect damage to weaken trees to the stages especially favoured by the three-toed woodpeckers. They are unlikely to increase here.

Ref. 52

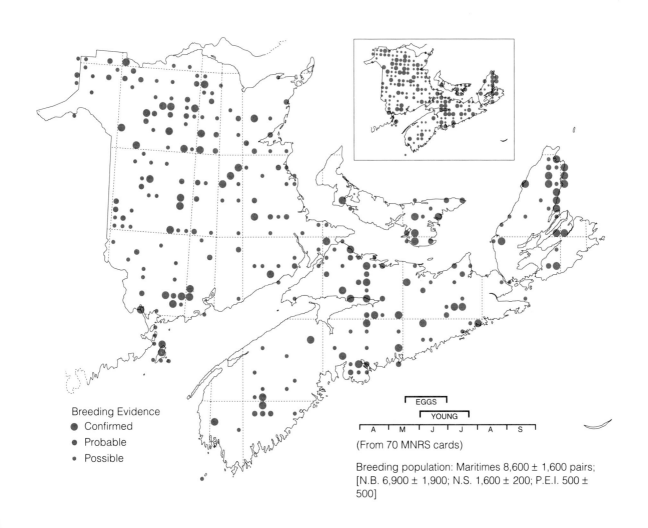

Breeding Evidence
● Confirmed
● Probable
• Possible

(From 70 MNRS cards)

Breeding population: Maritimes 8,600 ± 1,600 pairs; [N.B. 6,900 ± 1,900; N.S. 1,600 ± 200; P.E.I. 500 ± 500]

111

Northern Flicker
Pic flamboyant

Colaptes auratus

One or another form of the common Flicker breeds almost everywhere south of the tree-line in Canada and through the U.S.A. into the West Indies. This is a migratory species in northern areas so, its numbers not being limited by food in the harsh winters, it attains much higher breeding densities here than the resident woodpecker species. Flickers frequent relatively open areas, where they often feed on the ground on ants, but there must be some trees, or posts or utility poles, nearby for roosting and nesting. They were found throughout the Maritimes, in almost every square that received more than a casual visit, but confirmation of breeding was more dependent on intensive coverage. These birds are easily detected, both by sound and by sight, being large, vocal, and colourful and often frequenting relatively open areas. Breeding was most often confirmed by seeing family groups after fledging (FL, in 256 squares, 22 per cent of 1162 with Flickers),

and nests (NY,NE, and especially ON, but not including nest-building, N) were noted in 207 squares (18 per cent).

Presumably the opening-up of the formerly more continuous forests of the Maritimes after the start of European settlement increased the habitat suitable for Flickers, which thus increased in numbers up into this century. More recently, the introduction of the European Starling to North America and its spread to the Maritimes by the 1920s added a formidable competitor to the community of birds and mammals using Flicker cavities. A new equilibrium between cavity-makers and users has developed since then; Flickers are markedly scarcer than before around settled areas, where Starlings, which also use holes in buildings and thus are not limited by availability of Flicker cavities, are more abundant. In more wooded areas, Flicker numbers probably were little affected. There are no obvious trends in land use or climate that are thought likely to cause these birds to decline in future, and the BBS indicated that their numbers here have been stable since 1966.

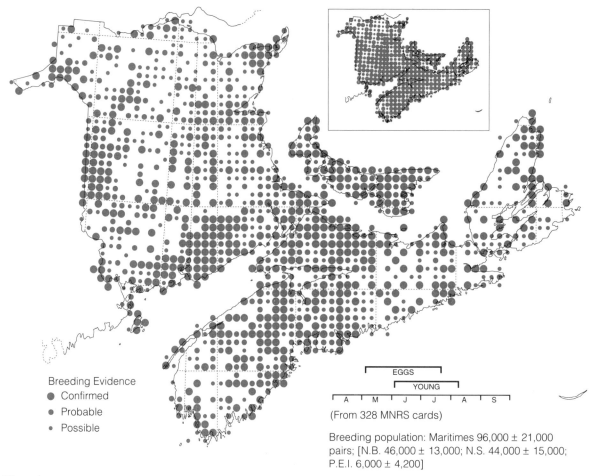

Breeding Evidence
- Confirmed
- Probable
- Possible

EGGS
YOUNG

A M J J A S

(From 328 MNRS cards)

Breeding population: Maritimes 96,000 ± 21,000 pairs; [N.B. 46,000 ± 13,000; N.S. 44,000 ± 15,000; P.E.I. 6,000 ± 4,200]

Pileated Woodpecker
Grand Pic

Dryocopus pileatus

This woodpecker is a permanent resident, breeding throughout forested regions of Canada and the United States. Owing to its large size, its nesting is restricted to extensive forest areas with old, large-diameter trees. In the Maritimes, its local distribution seems focused in stands dominated by broad-leafed trees, perhaps because more large trees remain in the hardwood stands. The entrances to its nests, and to many of its feeding excavations, are distinguishable from those of other woodpeckers by having squared corners. Despite its large size, striking crested appearance, and loud far-carrying calls, this bird is inconspicuous as, unlike the smaller woodpeckers, it seldom comes out into open woodland near habitations. Only one-fifth (99 out of 507 squares with Pileated Woodpeckers) of Atlas records reported confirmed breeding, and little over half of these (60 squares) were of nests (NY,NE,ON). Inclusion of recent excavations, which were not accepted as breeding evidence during the first two years, would have increased substantially the number of squares in which the species was recorded.

We can infer that the Pileated Woodpecker declined, in parallel with cutting of the older-growth forests of the Maritimes, since European settlement began. This loss of mature trees still continues, with a trend to shorter cutting cycles in the forest industry, and with re-growth of trees slower here than in warmer areas with more fertile soils. The Pileated Woodpecker was extirpated from Prince Edward Island, largely through loss of nesting habitat, before 1900; the records there during the Atlas period, starting in 1987, were the first on the Island for many years, but breeding was not confirmed there in 1986–90. With reversion to forest of much farmland in Prince Edward Island during the last 40 years, and subsequent maturation of the trees, the species seems likely to re-establish a small population there soon. Reforestation of cut-over areas using jack pines will never provide homes for Pileated Woodpeckers, as is evident from the large gaps in the New Brunswick range. In general, these birds were found more widely than expected. The long-range survival of the species in the Maritimes depends on maintaining enough large tracts of older forests for this, the only large hole-nesting species that is not readily attracted to nesting boxes.

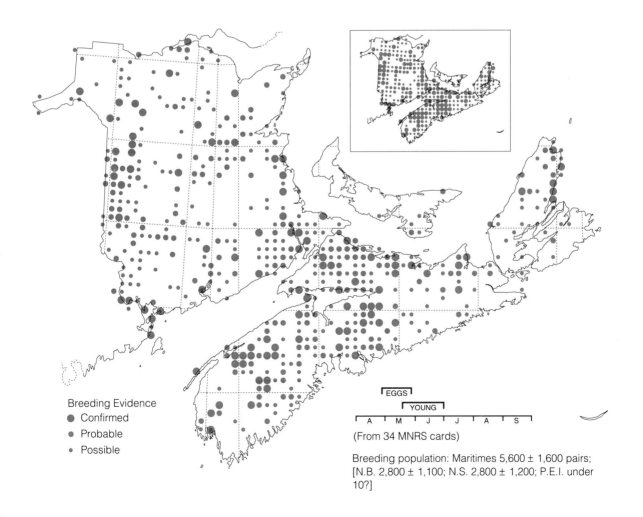

Breeding Evidence
● Confirmed
● Probable
● Possible

EGGS
YOUNG
A M J J A S

(From 34 MNRS cards)

Breeding population: Maritimes 5,600 ± 1,600 pairs; [N.B. 2,800 ± 1,100; N.S. 2,800 ± 1,200; P.E.I. under 10?]

Olive-sided Flycatcher
Moucherolle à côtés olive

Contopus borealis

This is one of our larger flycatchers, found in open woodlands and other places where scattered trees remain after cutting or fire in forested regions. It breeds from Alaska to Labrador, south across Canada, and beyond in mountains to Arizona and North Carolina. The Olive-sided Flycatcher was found throughout the Maritimes, never abundantly, and most sparsely in eastern New Brunswick and western Prince Edward Island where dense young second-growth forest, in succession after fires or farm abandonment, is prevalent. It was less general in the regions where hardwoods predominated, but no large areas lacked it altogether. Breeding was confirmed less often than for some other flycatchers, in 101 (15 per cent) of 677 squares with these birds, and 73 per cent of squares had records only of presence (H) or territory (T). Nests were noted in only 13 squares.

It seems probable that Olive-sided Flycatchers were rather less widespread in the former, more closed forests of the Maritimes, and they became more abundant during the settlement phase when cutting and clearing first became general and wildfires more frequent. With abandonment of much marginal farmland scattered with old trees and stubs, and its reversion to low, closed forest, much formerly suitable habitat was lost in some areas in the past century, and this species was thought to have declined in Nova Scotia between 1900 and 1940. The small samples obtained in the BBS suggested no change since 1966. Clear-cutting in recent forest exploitation probably leaves fewer isolated trees than in the past, but forestry practices seem likely to ensure continued availability of suitable forest-edge habitats for the Olive-sided Flycatcher in the Maritimes. Forest spraying is unlikely to benefit insect-eating birds in general, but there seems to be no reason to consider this species as threatened here.

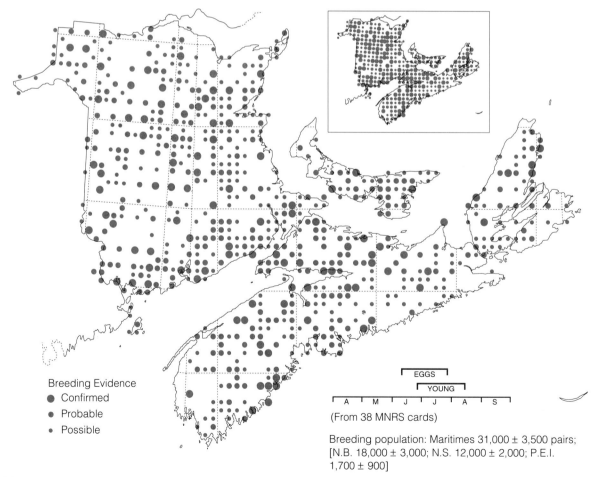

Breeding Evidence
- Confirmed
- Probable
- Possible

EGGS
YOUNG
A M J J A S

(From 38 MNRS cards)

Breeding population: Maritimes 31,000 ± 3,500 pairs; [N.B. 18,000 ± 3,000; N.S. 12,000 ± 2,000; P.E.I. 1,700 ± 900]

Eastern Wood-Pewee
Pioui de l'Est

Contopus virens

This small flycatcher breeds across the eastern U.S.A. from Texas and Florida north, extending into southern Canada from Manitoba to the Maritimes. It reaches its northeastern limits hereabouts, as pewees are not known to breed in the Gaspé Peninsula or Newfoundland. It was found in all regions of our area, but with obvious thinning to the north and east where broad-leafed trees are scarcer. Pewees were conspicuously absent in parts of the highest uplands of northwest and north-central New Brunswick and in northern Cape Breton Island, and on the Atlantic slope (Guysborough Co. east) in eastern Nova Scotia. Breeding was confirmed in similar proportions to other small flycatchers, in 148 (19 per cent) of 789 squares with pewees, and with nests (ON,NE,NY) in only 22 squares (3 per cent). Three-quarters of all squares had only single (H) or repeated (T) "sightings," usually of singing birds, as pewees are easily recognized by voice. They are unlikely to be missed if present in an area.

The forests of the Maritimes before Europeans came here were probably suitable for this species, although the cooler climate during the Little Ice Age (ca.1350–1850) may have restricted its numbers near the present limit of its breeding range. The subsequent opening-up of the forests made them more attractive, as pewees are birds of openings and edges more than of closed forest in our region, and they readily use well-spaced shade trees in rural and urban settlements. There is no indication, from BBS data or general impressions, of recent population changes. Forestry practices in the future seem unlikely to limit pewee numbers generally, although widespread emphasis on conifer production might restrict opportunity locally. Any tendency towards global warming should encourage a more temperate mixed forest and is more likely to help than harm this species.

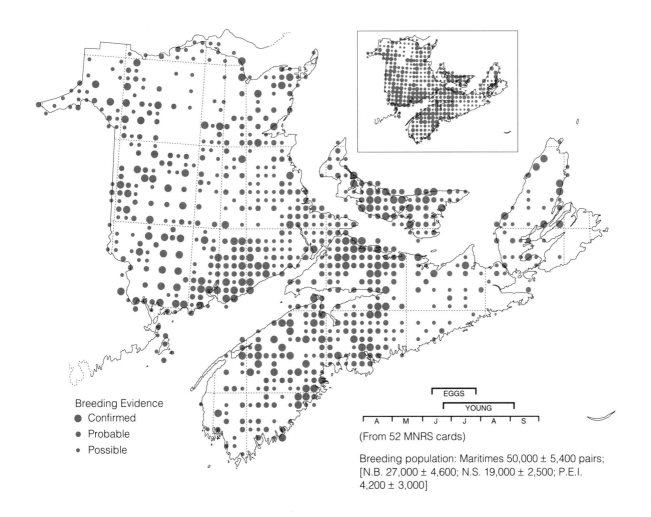

Breeding Evidence
- Confirmed
- Probable
- Possible

EGGS
YOUNG

A M J J A S

(From 52 MNRS cards)

Breeding population: Maritimes 50,000 ± 5,400 pairs; [N.B. 27,000 ± 4,600; N.S. 19,000 ± 2,500; P.E.I. 4,200 ± 3,000]

Yellow-bellied Flycatcher
Moucherolle à ventre jaune

Empidonax flaviventris

Like all our flycatchers, this is an American species, associated mainly with the eastern boreal forest in breeding season, scarcer west of Ontario. It usually frequents stands where conifers, especially spruces, predominate, with up to one-third broad-leafed trees not deterring the species. It frequents both mature and successional stands where trees exceed 4–5 m in height, but where shrubs and herbs are not dense. Atlas work showed it occurring throughout the Maritimes, but with gaps in the hardwood-dominated regions in southwest New Brunswick, southwest and northern Nova Scotia, and in Prince Edward Island. Some observers may have missed it through misidentification: its usual single "shleck" or "killick" calls, under-emphasized in Peterson field guides, are often mistaken for the rapidly repeated "che-bec" calls of the Least Flycatcher, which is confined to broad-leafed stands. Yellow-bellies probably breed in most squares with extensive areas of conifers. Its nests are on the ground, usually in a thick moss mat, but only 12 nests were reported during the Atlas. Records of confirmed breeding were usually adults carrying food (AY, in 79 of 673 squares with the species), and most reports (in 506 squares, 75 per cent) were of singing birds (H or T).

We have no strong evidence of historical changes in status, in the Maritimes or elsewhere in its range. Presumably clearing of forests across Canada somewhat reduced its range along the southern edge, but former farms reverting to forest restored much of the habitat lost in the Maritimes. The BBS suggested a non-significant increase here since 1966, which might arise from improved recognition of its calls. As the species is under-represented by the Atlas data, our population estimates are minimal. This may be the most numerous flycatcher in the Maritimes, as its habitats exceed in area those used by most other species in the family; the use of edges by Alder Flycatchers makes the latter more obvious to people travelling by road, but that apparent greater abundance may not be representative. The maintenance of extensive conifer forests for the wood-fibre industry should ensure breeding habitat for Yellow-bellied Flycatchers here in the foreseeable future.

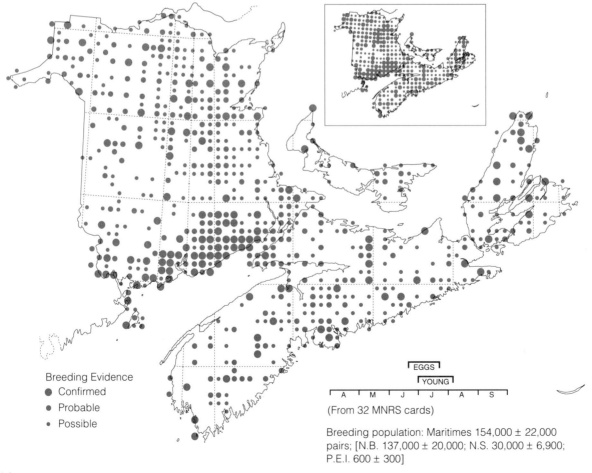

Breeding Evidence
● Confirmed
● Probable
● Possible

EGGS
YOUNG

A M J J A S

(From 32 MNRS cards)

Breeding population: Maritimes 154,000 ± 22,000 pairs; [N.B. 137,000 ± 20,000; N.S. 30,000 ± 6,900; P.E.I. 600 ± 300]

Alder Flycatcher
Moucherolle des aulnes

Empidonax alnorum

This is the northern member of the "Traill's Flycatcher" complex (see also Willow Flycatcher), breeding from Alaska across Canada south of the tree-line, but reaching only the northernmost United States (Michigan to Maine). It was found almost throughout the Maritimes, with an unexplained gap in the northwest, mostly in highlands. Its relative abundance is probably over-estimated, because its habitat preference for shrubby edges such as roadsides makes it very easy to detect. Away from roads, it is more frequent in damp shrubby swales of alders and willows than in upland areas reverting to forest. Its nests are low in shrubbery and are difficult to find; only 23 of 230 squares with breeding confirmed had nests seen, one of the lowest proportions for any flycatcher. Most records (72 per cent of 1008 squares with these birds) were of singing birds (H or T), as with most other flycatchers.

Habitat for Alder Flycatchers in the Maritimes was scarcer, and mostly confined to edges of streams and beaver meadows, before Europeans began clearing the forests for settlement 400 years ago. We infer that numbers of this species increased greatly up to around 1900 and may have declined somewhat since then as sub-marginal and abandoned farms reverted to woodland. The BBS indicated a significant increase since 1966; those counts are confined to roadsides, so would not measure a decline away from roads, where most potential habitat losses occurred. Although abandoned farmland is still reverting to forest, new habitat is being created along the edges of areas clear-cut in the forest industry. These trends seem likely to cancel each other, and habitat for Alder Flycatchers should not limit their numbers here in the foreseeable future. If these birds also use early successional habitats in their wintering areas, the sweeping forest clearances in Latin America may not pose undue problems for them.

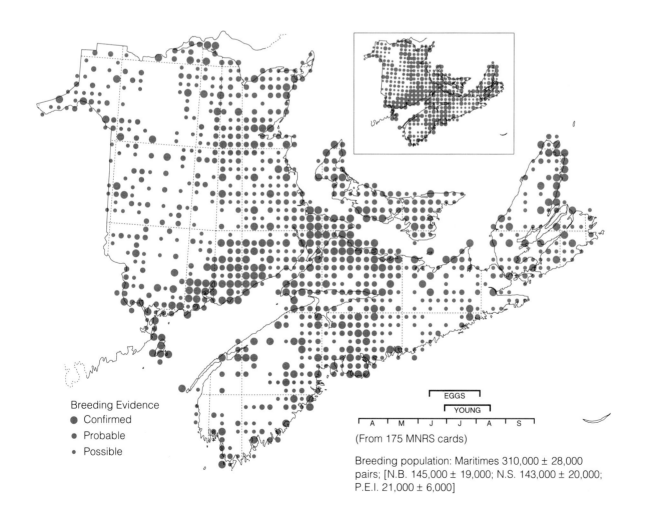

Breeding Evidence
- ● Confirmed
- ● Probable
- • Possible

EGGS
YOUNG

A M J J A S

(From 175 MNRS cards)

Breeding population: Maritimes 310,000 ± 28,000 pairs; [N.B. 145,000 ± 19,000; N.S. 143,000 ± 20,000; P.E.I. 21,000 ± 6,000]

Willow Flycatcher
Moucherolle des saules

Empidonax traillii

Willow and Alder Flycatchers were combined as "Traill's Flycatchers" until 1973. Before then, the "fitz-bew" song-type of Willow Flycatcher had not been reported in the Maritimes, and Traill's here were assumed to be all Alder Flycatchers. The Willow Flycatcher breeds across the U.S.A., and north into southern Canada from British Columbia to Quebec. Since it was accepted as a distinct species, only a few reports have been received in the Maritimes, mostly from southern New Brunswick. The Atlas received records from 23 squares, with breeding confirmed in both Prince Edward Island and in New Brunswick in 1989. Breeding had been apparently confirmed in Nova Scotia in 1980, but the Atlas received only two records from Nova Scotia during 1986–90, neither in a location suggesting range continuous with the New Brunswick records. Willow Flycatchers were undoubtedly often confused with other small flycatchers, so this species is under-represented by our data. However, the species was found only sparsely across southern Maine in 1977–83, so it is unlikely to be of general occurrence even in southern New Brunswick.

As a southern species and one frequenting shrubby habitats, we can infer that the Willow Flycatcher was scarcer, if it occurred here at all, in 1600, when more of the region was forested and the climate was cooler. Its arrival in the Maritimes was probably very recent, as the species is thought to have increased and spread all along the northern limits of its range farther west. In view of the mapped distribution, the numbers breeding are probably many times greater than the few reported, though it is still a scarce species. Global warming in future may tend to aid the spread of this southern bird in our area, but increased familiarity with its calls should fill in some of the apparent gaps in its range.

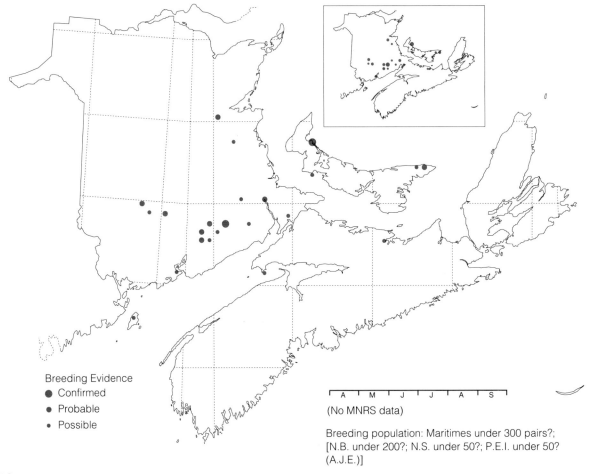

Breeding Evidence
● Confirmed
● Probable
• Possible

(No MNRS data)

Breeding population: Maritimes under 300 pairs?; [N.B. under 200?; N.S. under 50?; P.E.I. under 50? (A.J.E.)]

Least Flycatcher
Moucherolle tchébec

Empidonax minimus

This is the flycatcher of broad-leafed wood-lands across most of Canada and the northern-most U.S.A. It breeds from the Gulf of St. Lawrence to southern Yukon and interior British Columbia, and south to Wyoming, Missouri, and North Carolina, the widest range of any in this group of flycatchers. In the Maritimes, earlier accounts emphasized its occurrence in shade trees in suburban and farming areas, but Atlas workers found it more generally in forests, as is typical across boreal Canada. It was found somewhat less frequently in areas with more conifers, particularly on the Atlantic slope of Nova Scotia, and was almost absent in western Prince Edward Island. Possibly this species was sometimes reported in error instead of the Yellow-bellied Flycatcher, whose "killic" or "shlick" calls (under-emphasized in field guides) are often confused with the clearly two-syllabled and oft-repeated "che-bec che-bec che-bec" of the Least,

although Yellow-bellies seldom call in rapid series. Atlas statistics for the Least Flycatcher were typical of the flycatcher family. Breeding was confirmed in 185 squares (22 per cent of 826 with these birds), although nests were reported in only 32 squares. Nearly 70 per cent of squares (559) had only single (H) or repeated (T) records, mostly of singing birds.

It seems likely that declines in northern hard-woods (maples, beech) since the start of European settlement in the Maritimes were largely balanced by increases in successional white birch and aspen stands, which are equally acceptable to Least Flycatchers as long as conifers are scarce therein. There seems no good reason to believe that these birds have changed greatly in overall numbers since 1600, and no change was found by the BBS in the last 25 years. Although forest management nowadays attempts to suppress broad-leafed growth in commercially exploited forests, much forest is still left to regenerate on its own in small private woodlots, so there is no shortage of mixed and open hardwood stands. Least Flycatchers will probably continue to do well here.

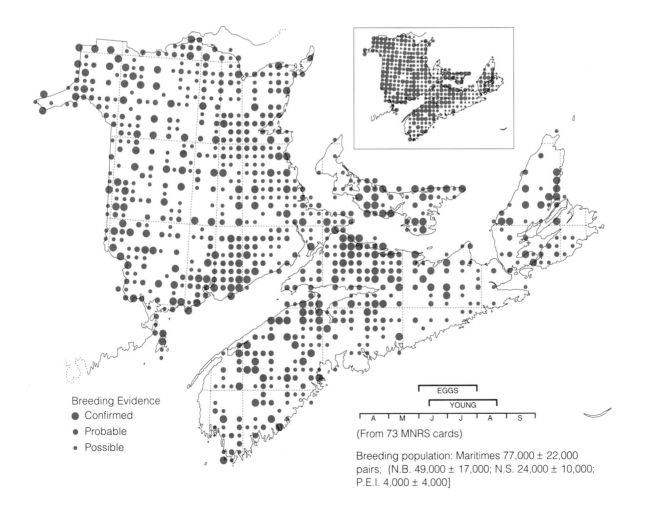

Breeding Evidence
● Confirmed
● Probable
• Possible

EGGS
YOUNG
A　M　J　J　A　S

(From 73 MNRS cards)

Breeding population: Maritimes 77,000 ± 22,000 pairs; [N.B. 49,000 ± 17,000; N.S. 24,000 ± 10,000; P.E.I. 4,000 ± 4,000]

Eastern Phoebe
Moucherolle phébi

Sayornis phoebe

This Phoebe breeds across Canada and the United States east of the Rockies. In the west it extends into the northern forests beyond latitude 58° N, but in the east it belongs to the temperate hardwoods region, reaching its northeast limits in the Maritimes. Phoebes are commonly found near water, and the distribution in our area was largely confined to inland valleys except in the southwest half of New Brunswick. The significance of isolated records beyond the relatively continuous range is uncertain. Singing males (H,T) often appear far beyond the known range, as in Prince Edward Island, and may call there for weeks in apparently suitable habitats; evidently some of these birds secure mates and breed, as confirmed on Cape Breton Island for the first time during the Atlas period. The Phoebe is not shy and often nests on buildings and bridges, as well as under bank and cliff overhangs, but it is not conspicuous. We suspect that some inexperienced observers confused the call of the more common Alder Flycatcher with this species, so the Phoebe is as likely to be over- as under-represented in the Atlas data. The finding of nests (NY,NE,ON, in 78 squares) was the most frequent breeding evidence, and breeding was confirmed in just over half of the 243 squares in which Phoebes were reported, by far the largest proportion among the flycatchers.

Probably Phoebes already bred in New Brunswick before the start of European settlement. Though they were long established in western New Brunswick, the known range seems to have expanded over the past 30 years, with nesting first reported in Nova Scotia in 1963 but still not confirmed in Prince Edward Island. How much this apparent extension is an artifact of more and better informed observers remains uncertain. The species is still too scarce to provide trends through the BBS. Population estimates here exclude the outlying records of "possible breeding" in Prince Edward Island. Projected habitat and climatic trends suggest that the Eastern Phoebe is more likely to expand than decrease in the Maritimes in future.

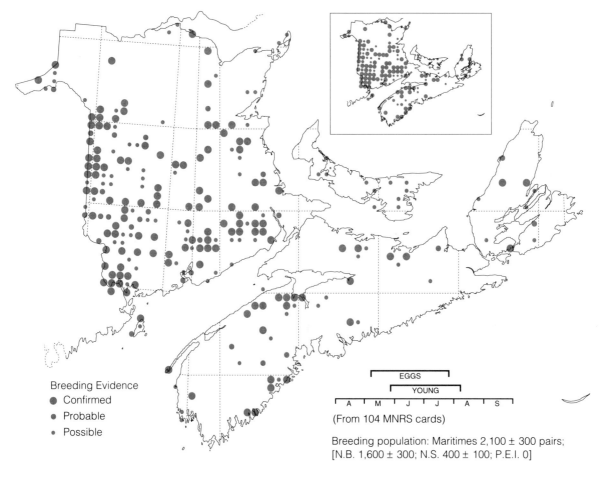

Breeding Evidence
- Confirmed
- Probable
- Possible

EGGS
YOUNG
A   M   J   J   A   S
(From 104 MNRS cards)

Breeding population: Maritimes 2,100 ± 300 pairs;
[N.B. 1,600 ± 300; N.S. 400 ± 100; P.E.I. 0]

Great Crested Flycatcher
Tyran huppé

Myiarchus crinitus

This is a bird of the eastern broad-leafed region from Texas and Florida north to southern Canada, where it breeds sparsely, west to Alberta and east into the Maritimes. In our area, the Great Crested Flycatcher had a range, in open woodlands and settled areas, centred in valleys in the southwestern half of New Brunswick, with only scattered records (some breeding confirmed) farther east and in southwest Nova Scotia. As this robin-sized bird nests in tree cavities and nest-boxes, competition for suitable nest sites by Starlings and failure to find a mate mean that many birds seen beyond the main range did not breed. The Crested Flycatcher was considered accidental in Prince Edward Island, where breeding was confirmed during the Atlas period, and in eastern Nova Scotia. Nearly half the records of confirmed breeding (17 of 35) involved nests (NY,NE,ON), which are easy to find in nest-boxes. The relative frequency of sightings (H,T, in two-thirds of 148 squares with the species) indicates that it was easy to detect, but also that some detections did not involve nesting.

The historical record sheds no light on whether this southern species was already in the Maritimes before European settlement, when suitable forest-edge habitat was present but scarcer than recently. Great Crested Flycatchers were known by the late 1800s to nest in extreme western New Brunswick, but records elsewhere in the Maritimes were much later (first noted in Nova Scotia in 1931, first breeding there in 1956), perhaps reflecting the increase in observers in recent decades. The spread of the Starling here since the 1920s made nest-sites less available and slowed any tendency towards breeding range extension by the flycatcher. A trend towards global warming would favour southern species such as this one. Other land-use trends seem unlikely to pose major threats to its continuing to breed uncommonly in the Maritimes.

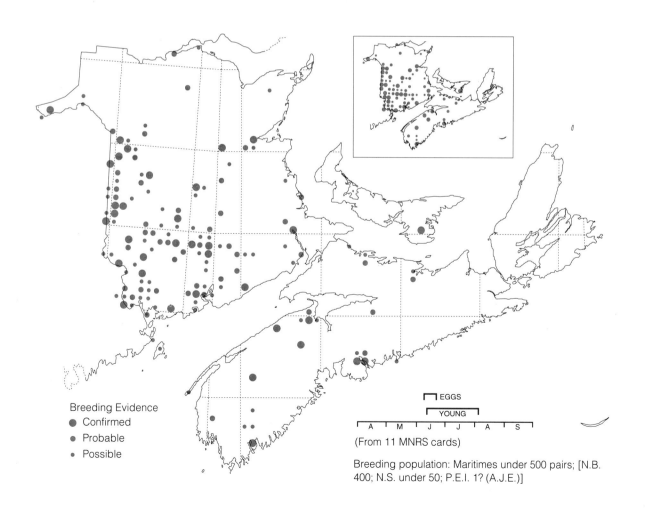

Breeding Evidence
- Confirmed
- Probable
- Possible

EGGS
YOUNG

A   M   J   J   A   S

(From 11 MNRS cards)

Breeding population: Maritimes under 500 pairs; [N.B. 400; N.S. under 50; P.E.I. 1? (A.J.E.)]

Eastern Kingbird
Tyran tritri

Tyrannus tyrannus

The Kingbirds are large flycatchers, members of a family centred in tropical America. This species breeds through most of the U.S.A. and Canada, from California to southern Mackenzie and from Florida to the St. Lawrence estuary. The Maritimes is near the northeast limit of the range, and their distribution here is less continuous than suggested by previous publications. These birds of open areas were found mostly in farming districts, especially the St. John River Valley in New Brunswick, Kings and north Cumberland Counties in Nova Scotia, and widely in Prince Edward Island, though they also occurred, less conspicuously, by rivers, swamps, and cut-over areas in forested regions. Kingbirds were notably scarce and local on the Atlantic slope and on Cape Breton Island. They were detected more often by sight than sound, and breeding was confirmed in 41 per cent (294) of 714 squares with

Kingbirds. Nests, easy to find on (often low) branches in open situations, accounted for over one-third of confirmed breeding records, in 105 squares (15 per cent).

Kingbirds were probably rather scarce and localized breeders in the Maritimes 400 years ago, restricted to riverside and meadow-edge situations in a largely forested region. European settlement, by creating openings and fields, allowed these birds to spread and increase. Old orchards and abandoned fields were much used during the decline of marginal farming here, but Kingbird numbers may have declined in some areas in recent decades as abandoned farmlands finally reverted to forest. The BBS detected no change in the Maritimes since 1966. Environmental trends predicted for the coming years, including global warming, acid rain, and land-use patterns in the Maritimes and in wintering areas, reveal no threats bearing particularly on this species. Ongoing efforts to reduce flying insects around settled areas may reduce feeding opportunities locally. Kingbirds should persist at levels far above those of the distant past.

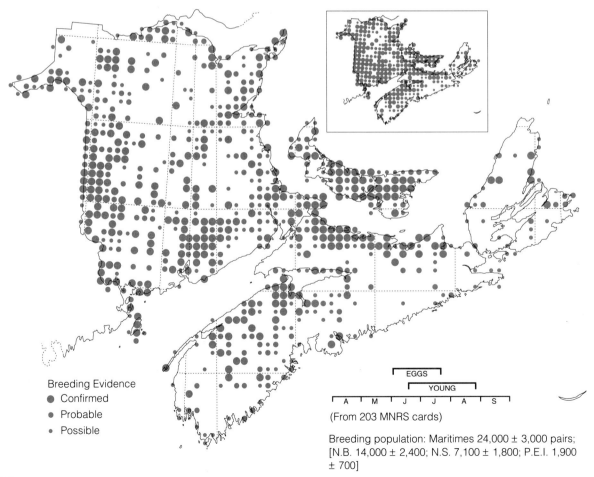

Breeding Evidence
- Confirmed
- Probable
- Possible

EGGS
YOUNG
A M J J A S

(From 203 MNRS cards)

Breeding population: Maritimes 24,000 ± 3,000 pairs; [N.B. 14,000 ± 2,400; N.S. 7,100 ± 1,800; P.E.I. 1,900 ± 700]

Horned Lark
Alouette cornue

Eremophila alpestris

The Horned Lark is a widespread species of open areas in the northern hemisphere, in Eurasia as well as North America. It frequents arctic tundras, alpine barrens, prairies, and semi-deserts, as well as various farmlands, but it is absent from woodland regions. Its Maritimes breeding range, given the generally forested aspect of our area, is quite restricted, with treeless farmlands in the upper St. John River Valley and in southeast New Brunswick having the main concentrations. Coastal grasslands in eastern New Brunswick had Horned Larks too, but similar habitats in western Prince Edward Island, where they were regular in the 1970s, had few reports. Airfields were the main habitats used in Nova Scotia, with records at or near Greenwood, Waterville, Halifax and Sydney airports; I heard singing larks at Stanley and Trenton airports (N.S.) in 1961, but none were reported there in 1986–90. These birds are easily detected and identified, so it is unlikely that they were under-represented if their habitats were visited.

They were confirmed as breeding in 25 of 69 squares with records of larks, and nests were found in 7 squares.

Horned Larks were absent from southeast Canada 100 years ago, except as migrants from the arctic in autumn and winter. They spread as breeding birds eastward from the prairies with the widespread clearing of land for settlement during the 1800s and reached the Maritimes early in the 20th century. Probably their numbers peaked with the greatest extent of cleared land here, between 1920 and 1950. Since then, their numbers have decreased, and they have disappeared from some areas formerly occupied. The few records in the Atlas allow only rough extrapolation to present populations. The species' future here is uncertain, as heavy use of agricultural chemicals has reduced opportunities for breeding in potato-growing areas, which were used regularly in the past. Many formerly open areas are reverting to woodland, and others are used too intensively to be available for Horned Larks. Larks have dwindled elsewhere in eastern Canada for similar reasons, and we cannot assume that they will persist indefinitely here under present uses of open lands.

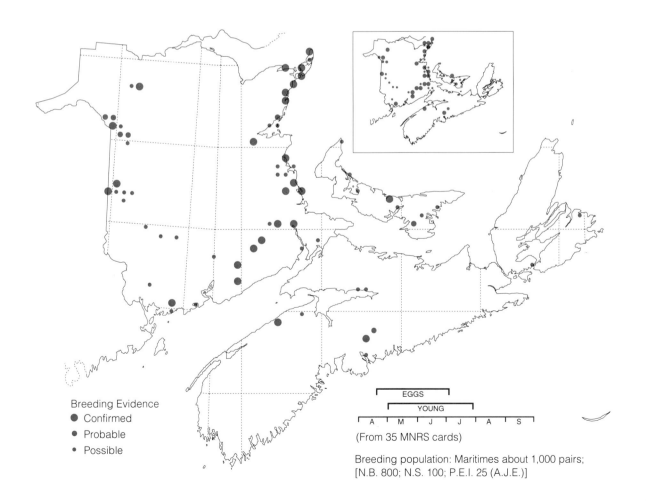

Breeding Evidence
● Confirmed
● Probable
· Possible

EGGS
YOUNG
A M J J A S

(From 35 MNRS cards)

Breeding population: Maritimes about 1,000 pairs;
[N.B. 800; N.S. 100; P.E.I. 25 (A.J.E.)]

Purple Martin
Hirondelle noire

Progne subis

Our largest swallow, and the one longest associated with people in America, the Purple Martin is the most northerly of a tropical group. It breeds across the eastern U.S.A. from Texas and Florida north into southern Canada, and west to Alberta, with a separate population along the west coast from south British Columbia to Mexico. In the Maritimes, its range comprised interior valleys of southern New Brunswick, with a few colonies in northwest Nova Scotia. No breeding was detected in the Miramichi Valley, where a colony at Boiestown was active until 1963, but other sites at the north and northeast limits are of long standing. Purple Martins are seldom seen far away from their colonies, and breeding was confirmed in 69 of 82 squares (84 per cent) with the species. As all known colonies here are in nest-boxes, the data for martins, as with other swallows, are quite representative of their status, which is over-represented relative to other common birds.

Although the agricultural Indians of the eastern U.S.A. placed gourds as nest-sites for martins before Europeans came there 400 years ago, these birds presumably nested only in tree-holes in the Maritimes, if they were present here in that era. It is unlikely that nest-boxes influenced their local numbers much before the mid-1800s, but their range formerly extended farther into Nova Scotia than at present, including sites in Truro and Windsor until the 1940s. Competition from House Sparrows and especially Starlings eliminated several former colonies, and these alien species continue to disturb nesting martins. Here at the limits of the species' range, Purple Martins sometimes suffer heavy losses during cold weather in May, after the pairs have occupied their nest-boxes; many colonies lost nearly all their breeding birds in 1990, but some of the non-territorial yearlings survived to continue the local traditions. Given the small population and the ever-present threats of Starlings and House Sparrows, Purple Martins depend heavily on human assistance for their survival here. Global warming seems unlikely to make their status much less precarious, although the region's coolness is often limiting.

Refs. 39, 71

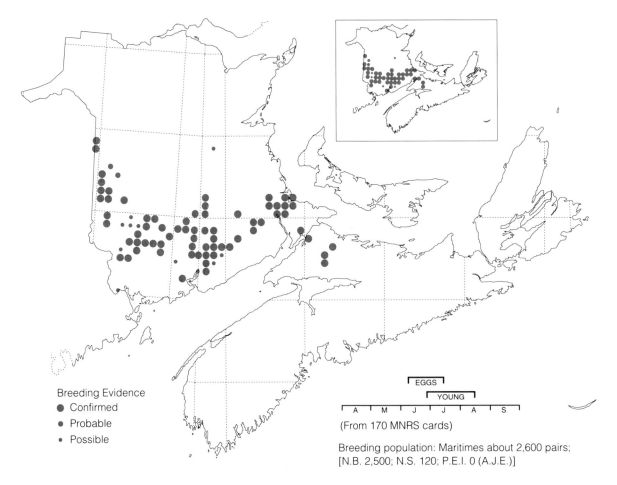

Breeding Evidence
● Confirmed
● Probable
· Possible

EGGS
YOUNG
A M J J A S
(From 170 MNRS cards)

Breeding population: Maritimes about 2,600 pairs;
[N.B. 2,500; N.S. 120; P.E.I. 0 (A.J.E.)]

Tree Swallow
Hirondelle bicolore

Tachycineta bicolor

The "one swallow to make a summer" was a different species, but this is the bird that convinces us that spring has come, at last, to the Maritimes. Tree Swallows breed across Canada and Alaska, north to near tree-line and south through most of the United States, except the southeast. This is the earliest to return as well as the most widespread of our swallows in the Maritimes, occurring in forested as well as open areas. They are somewhat less obvious and general in forests, where they are mainly found near lakes and ponds over whose surfaces they forage for insects. Their use of nest-boxes, erected for them around human dwellings, makes them familiar to nearly everyone and makes it easy both to detect them and to confirm their breeding. Nests (NY,NE,ON) comprised half (586 of 1175 squares) of the breeding evidence for the species, with recently fledged young (FL, 162 squares) also frequent. This species should not have been missed in any square that was studied seriously, except in 1990 (see below).

The erection of nest-boxes may have increased this species' numbers locally in the past, e.g., in the well-studied "colony" at Kent Island, N.B. However, most pairs of the species still nest in woodpecker holes and other cavities in trees, in natural situations. The loss of suitable nest trees through forestry and the clearing away of dead stubs and snags everywhere is likely to have counter-balanced any population gains attributable to nest-boxes. Tree Swallows in the Maritimes were markedly reduced in numbers following losses in eastern wintering areas in 1958 and in breeding areas in 1959; another large die-off occurred here in cold wet weather during May 1990, resulting in relatively fewer Atlas records in squares covered only in that year. Such losses probably recur at intervals and are recouped within a few years. There seems no reason to doubt that they will continue as one of our more abundant birds in future.

Refs. 39, 64, 114, 141

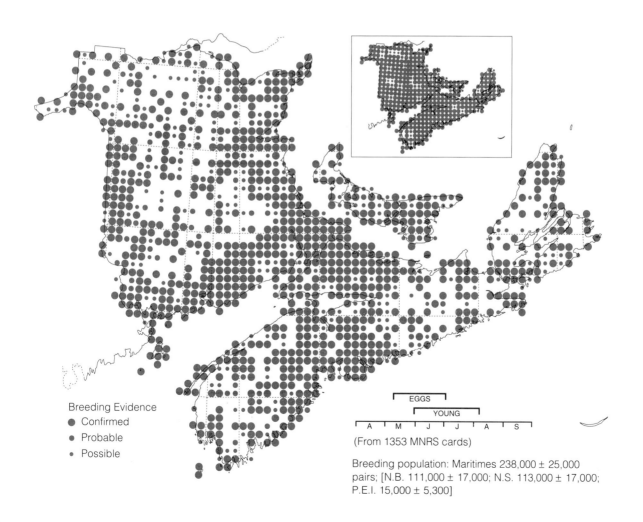

Breeding Evidence
- Confirmed
- Probable
- Possible

EGGS

YOUNG

A M J J A S

(From 1353 MNRS cards)

Breeding population: Maritimes 238,000 ± 25,000 pairs; [N.B. 111,000 ± 17,000; N.S. 113,000 ± 17,000; P.E.I. 15,000 ± 5,300]

Northern Rough-winged Swallow
Hirondelle à ailes hérissées

Stelgidopteryx serripennis

This swallow is widespread through the U.S.A. and Mexico, with a closely related species in Central and South America. It breeds only a short distance into Canada, mainly in south British Columbia and south Ontario. The Rough-winged Swallow is sparsely distributed across south Maine, with several records of confirmed breeding in the southeast and one in the extreme north. Breeding along the Meduxnekeag River near Woodstock (N.B.) was first proved in 1969, and breeding also occurred at Browns Flats in 1972 and at Mill Settlement in 1973. During the Atlas period, confirmed breeding records included flying young (FL) fed by an adult at St. George in 1988, and birds entering a nest-burrow (ON) near Lorneville in 1986, with probable breeding (P,T) northeast of Nackawic in 1989 and at Grand Falls Dam in 1988, all in southwest New Brunswick. With the earlier records, these suggest that Rough-winged Swallows breed sparsely all across the southwest quarter of New Brunswick, this species often being overlooked through confusing it with the similarly brown-backed but much more common Bank Swallow. Although both nest in cut banks, Rough-wings seldom if ever excavate burrows; their nests are more often in crevices in rocky cliffs than in the softer earth or mudstone banks where Bank Swallows nest.

Although Rough-winged Swallows were not detected except as vagrants before they were found breeding in the Maritimes in 1969, they probably were present in small numbers earlier, but undetected, as elsewhere near the periphery of their range. The Maritimes breeding population, all in New Brunswick, is small, but probably a lot larger than is recognized. The pattern from other areas does not suggest that this is likely to change much in future, unless global warming makes a major difference in the regional climate. The Rough-winged is likely to remain our scarcest swallow, as it is elsewhere in Canada.

Ref. 39

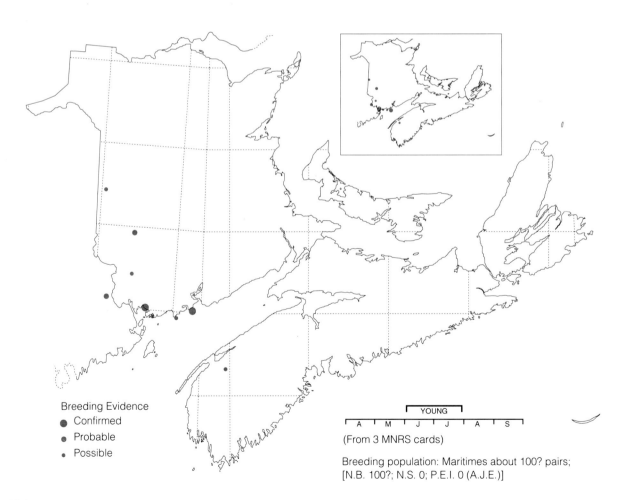

Breeding Evidence
- Confirmed
- Probable
- Possible

YOUNG

| A | M | J | J | A | S |

(From 3 MNRS cards)

Breeding population: Maritimes about 100? pairs; [N.B. 100?; N.S. 0; P.E.I. 0 (A.J.E.)]

Bank Swallow
Hirondelle de rivage

Riparia riparia

This bird breeds in North America, Europe, and Asia, scattered wherever suitable nesting sites in banks and cliffs are available. Its breeding requirements restrict it in many flat areas, as in much of the prairies and the Florida Peninsula, but its continental range extends from Alaska and Labrador to the southern U.S.A. Bank Swallow nesting colonies usually are found near open areas, and often close to water. Where all conditions are satisfied, as in many coastal farming areas of the Maritimes, this may be locally the most abundant swallow species. Although nests in artificial banks such as road cuttings and gravel pits are frequent here, the abundance of suitable banks along the coasts has resulted in a larger proportion of Bank Swallow nests being in natural sites here than elsewhere in Canada. During the Atlas project, this species was found in all regions of the Maritimes, but it was scarce in many inland forested areas. As these birds nest in obvious colonies in open areas, breeding was very easily confirmed, in 79 per cent of the 799 squares with Bank Swallows, with nests in 519 squares (65 per cent).

Like other swallows, this species benefited from the advance of human settlement here. The creation of cut banks and of cleared lands enlarged its opportunities for breeding in the Maritimes continuously over the last 300 years. BBS data since 1966, though not well suited for monitoring colonial species, suggest that Bank Swallows are still increasing. People like to have swallows of any kind around, and Bank Swallows have adapted well to habitats altered through human action in the past. Unless and until environmental insults harmful to these birds emerge, here or in their wintering areas, we can expect Bank Swallows to continue to enliven our shores in future summers.

Ref. 39

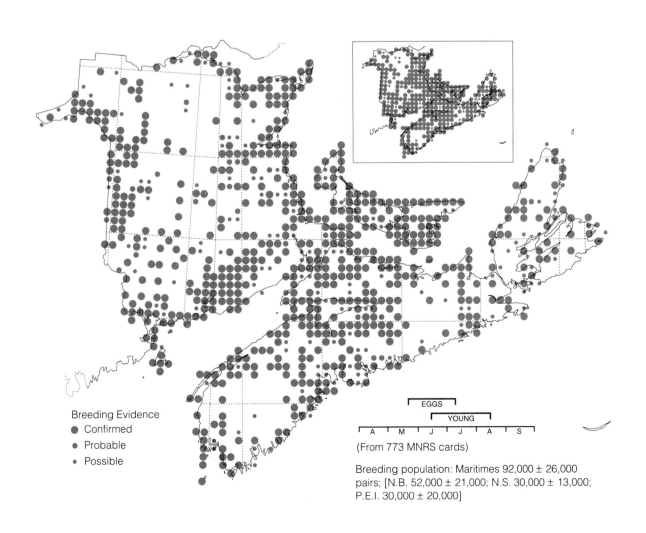

Breeding Evidence
● Confirmed
● Probable
● Possible

EGGS
YOUNG
A M J J A S
(From 773 MNRS cards)

Breeding population: Maritimes 92,000 ± 26,000 pairs; [N.B. 52,000 ± 21,000; N.S. 30,000 ± 13,000; P.E.I. 30,000 ± 20,000]

Cliff Swallow
Hirondelle à front blanc

Hirundo pyrrhonota

The Cliff Swallow was and remains a bird mainly of western North America. Although it breeds from Alaska to Mexico in the west, its eastern range extends only from southern Quebec and the Maritimes to North Carolina. This bird proved more widespread than had been expected before the Maritimes Atlas project began, except in Prince Edward Island; the one record there, south of Mt. Stewart in 1988, was the first reported in over 20 years, although the species was regular earlier. Cliff Swallows were found in most settled areas of New Brunswick, with obvious gaps in forested regions, but they were sparser in Nova Scotia, especially farther east and towards the Atlantic coast, where the humid climate may make the mud nests less stable than in drier areas. Confirmation of breeding, by finding the distinctive and persistent mud nests, was very easy for this species. No less than 88 per cent of squares with Cliff Swallows (526 of 598) had breeding confirmed (the highest proportion for any common species except Starling), with nests seen in 386 squares (65 per cent).

The Cliff Swallow was probably local or absent in the Maritimes when explorers and settlers first came. The species was formally described, from specimens taken in South America, in 1817, but anecdotal records later extended knowledge of its nesting in Maine to before 1800. Its spread through the eastern U.S.A., around the same time, followed land-clearing and erection of buildings, which latter freed the species from dependence on overhanging cliffs for nest-sites. Evidently it flourished in the Maritimes only for a few decades, as the increase of House Sparrows after the 1880s provided serious competition for these swallows. Decreases were noted in most previous regional summaries, but Cliff Swallow numbers may have stabilized later as the BBS showed no change since 1966, although this colonial species is not well sampled by those counts. Cliff Swallows in the Maritimes nest almost entirely on man-made sites, with concrete bridges and other structures being used more often as houses increasingly are covered with vinyl or metal siding to which the mud nests will not adhere. The species is surviving better than was feared until recently, and the continuing decline in House Sparrows may help it to prosper.

Refs. 39, 43

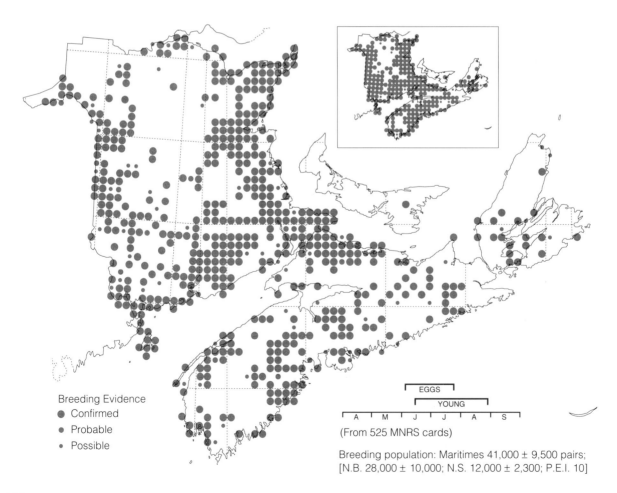

Breeding Evidence
- Confirmed
- Probable
- Possible

EGGS
YOUNG
A M J J A S
(From 525 MNRS cards)

Breeding population: Maritimes 41,000 ± 9,500 pairs;
[N.B. 28,000 ± 10,000; N.S. 12,000 ± 2,300; P.E.I. 10]

Barn Swallow
Hirondelle des granges

Hirundo rustica

The Barn Swallow occurs all across the northern hemisphere, with related species in Africa, southeast Asia and Australia. In North America, it breeds from Alaska to Mexico in the west, but less widely in the east, where it reached the eastern Gulf of Mexico only in the 1970s, and it is scarce north of the Gulf of St. Lawrence. In the Maritimes it breeds everywhere there are buildings and other structures that provide sheltered, dry nest-sites, even nesting on isolated cabins in deep woodland and on fishing shacks on offshore islands. A recent innovation, in remote logging areas with no alternatives, has been their basing nests on bolt-heads low in the sides of large corrugated metal culverts. However, nests in natural situations, in caves or under overhanging cliffs, usually close to water, are very rare, with only 5 among 1093 nests in the MNRS to 1977. Like several other swallows, it associates with human habitations, so is easily detected, and its nests are obvious, noted in more squares (693) than any other species. Barn Swallows were confirmed as breeding in 85 per cent of squares (988 out of 1159) with the species, the highest percentage for any common species except Starling and Cliff Swallow.

In the absence of permanent buildings before European settlement, Barn Swallows must have been rare and local birds in the Maritimes and throughout eastern North America 400 years ago. Their subsequent spread and increase continued into this century, but BBS data suggest that numbers in the Maritimes and generally across Canada have been stable since 1966. Although occasional pairs of Barn Swallows build nests in places where the fallout is unwelcome, this species is generally popular and well-adapted to a world managed for human convenience. Some recent building innovations (e.g., vinyl and metal siding) reduce nesting opportunities for these birds, and efforts to control flying insects affect their food supply locally. These adverse influences seem unlikely to interfere seriously with their remaining abundant birds here.
Ref. 39

Breeding Evidence
- Confirmed
- Probable
- Possible

EGGS
YOUNG

A M J J A S

(From 2234 MNRS cards)

Breeding population: Maritimes 242,000 ± 26,000 pairs; [N.B. 94,000 ± 16,000; N.S. 128,000 ± 19,000; P.E.I. 20,000 ± 6,000]

Gray Jay
Geai du Canada

Perisoreus canadensis

The Gray Jay is the North American representative of the northernmost group of jays, which breed all around the boreal regions of Eurasia and North America. It occurs year-round in the conifer forests from near the arctic tree-line of Alaska and Canada south to California and New Mexico, the treeless prairies, and the eastern hardwood forests of the Great Lakes and New England states. These birds were found all over the Maritimes except where extensive conifer forests are lacking, as in the western half of Prince Edward Island. The gap in northern New Brunswick was not obviously for that reason. Gray Jays seldom leave the spruce and fir forests where they nest, and this, with sparse visitation of such habitats, leads to under-representation of the species in our data. Family groups including sooty-plumaged fledged young (FL) are quite noisy and easily seen, where present, and these were by far the most frequent evidence of breeding obtained in the Atlas (60 per cent of 694 squares with these jays), outnumbering all other types of evidence. Nests were reported in only 4 squares.

When Europeans first settled the Maritimes, conifer forest habitat for Gray Jays may have been less extensive than in recent decades. Their populations now may be as large as in any period in historic times, but the ongoing logging of mature conifer stands in New Brunswick is likely to reduce the amount of suitable habitat in future, if it has not already done so. Forest management favouring conifers should ensure enough habitat that their numbers do not fall to levels requiring concern in the future, even if present high numbers are not maintained.

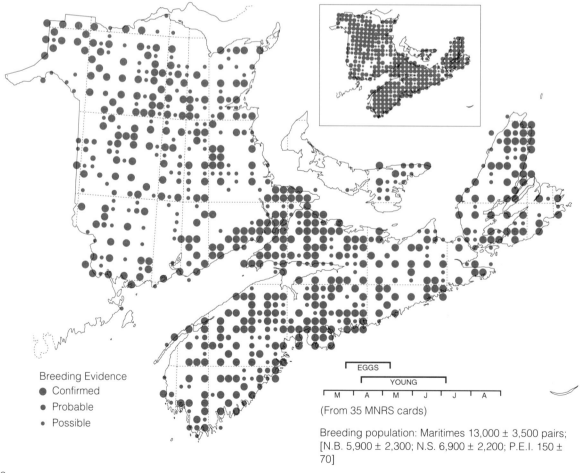

Breeding Evidence
- ● Confirmed
- ● Probable
- ● Possible

EGGS

YOUNG

M A M J J A

(From 35 MNRS cards)

Breeding population: Maritimes 13,000 ± 3,500 pairs; [N.B. 5,900 ± 2,300; N.S. 6,900 ± 2,200; P.E.I. 150 ± 70]

Blue Jay   Geai bleu

Cyanocitta cristata

The Blue Jay breeds throughout eastern North America from the Gulf Coast to southern Canada, and westward, except in the treeless prairies, to the Rockies. This is a bird of deciduous and mixed woodlands, though in northern areas it usually nests in conifers. All jays are opportunistic feeders, often preying on eggs and nestlings of other birds as well as on all kinds of invertebrates, and even taking sunflower seeds at bird feeders. This species' bold and conspicuous habits, except during nesting season, suggest that people never treated it as a harmful bird, in contrast to the shy and wary European Jay and the less-closely related crows and ravens. In the Maritimes, Blue Jays were found widely; they were sometimes missed in well-forested regions where field work was less extensive, as they were less easily seen during their secretive nesting period when much atlassing was done in such areas. Probably the Atlas data under-represent their occurrence to some degree in less well-worked

areas. Once the young leave the nest, they are more obvious, and family groups of flying young were the most frequent evidence confirming breeding (in 416 squares, 39 per cent of 1065 with Blue Jays). Nests, in only 25 squares, were rarely noted.

The primaeval mixed forests of the Maritimes, with larger proportions of mature hardwoods than at present, were good habitat for Blue Jays, but these birds adapted readily to environments altered by human actions in succeeding centuries. Their winter numbers here vary widely, as greater or smaller proportions of the population migrate south, presumably in anticipation of less or more food availability in the coming months. We have no indications of long-term decrease or increase in their numbers in summer, from the BBS or anecdotal data. This species seems likely to be able to adapt to any foreseeable environmental changes in the future, whether these are caused directly by people, as in land use, or only indirectly, as with global warming.

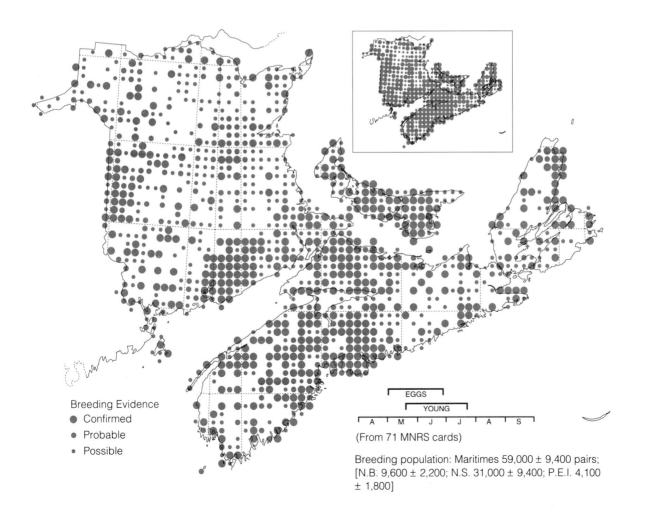

Breeding Evidence
- ● Confirmed
- ● Probable
- • Possible

EGGS
YOUNG

| A | M | J | J | A | S |

(From 71 MNRS cards)

Breeding population: Maritimes 59,000 ± 9,400 pairs; [N.B. 9,600 ± 2,200; N.S. 31,000 ± 9,400; P.E.I. 4,100 ± 1,800]

American Crow
Corneille d'Amérique

Corvus brachyrhynchos

The Crow is familiar across Canada, and south through the United States to the Gulf of Mexico, but it does not occur quite everywhere even though it eats almost anything. Crows nest in trees, but they forage mainly on the ground in the open; this combination restricts their use of forested regions to the edges of bogs, marshes, and other openings, and it excludes them from tundra and alpine areas and from deserts. In the Maritimes, farmlands are Crows' main foraging areas, though they also use shores at all seasons, especially in winter; the largest winter roost here, with over 50,000 birds at times, is on Boot Island, in the main agricultural area of Nova Scotia. The main foods overall are probably ground insects, except in winter, but opportunistic predation on small mammals, birds and their eggs, and frogs, and use of carrion and garbage at all seasons, all contribute to their diet. Crows are so conspicuous and noisy that they were detected, if present, almost everywhere that atlassing was undertaken; breeding was confirmed in 58 per cent of 1115 squares with Crow records, usually as newly flying young (FL, 481 squares), although nests (ON,NE,NY) were found in 114 squares.

Crows benefited from the clearing of forests here since European settlement began, as well as from human wastes. Wintering in the Maritimes must have become more practicable for them in the last century, although some are thought still to migrate southward. Their numbers are greater now, and their range more extensive, than before settlement, and the BBS indicates a continuing significant increase in numbers in the Maritimes since 1966. Their present range corresponds well with the distribution of agriculture and urban settlements in the Maritimes. Despite persecution by people and continuing lack of protection, the Crow has always flourished, and it will no doubt thrive in the future.

Ref. 104

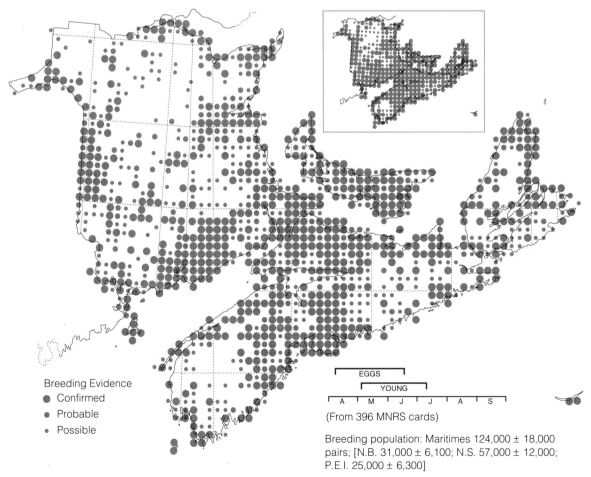

Breeding Evidence
- Confirmed
- Probable
- Possible

EGGS
YOUNG
A M J J A S

(From 396 MNRS cards)

Breeding population: Maritimes 124,000 ± 18,000 pairs; [N.B. 31,000 ± 6,100; N.S. 57,000 ± 12,000; P.E.I. 25,000 ± 6,300]

Common Raven
Grand Corbeau

Corvus corax

The Raven is one of the most adaptable and wide-ranging birds. It breeds from the high arctic of Canada and Alaska to the mountains of Central America, and all across Eurasia. Its nests are placed in trees, on cliffs, and in old buildings, and its food includes almost anything it can secure and swallow. However, Ravens long suffered from humans' prejudice against black birds (e.g., cormorants, crows, starlings), and they retreated into the wilderness before the guns of "civilized" humanity. Once they learned that settled areas were again safe, they rapidly returned to exploit the edible wastes in garbage dumps and along roadsides. In the Maritimes they were found in all regions, with no apparent restrictions. Many people with only a casual interest in birds confuse Ravens with Crows, but the latter are much more closely associated with farmlands and settled areas. Although Ravens range widely while foraging, over both forests and open areas, rather few squares include so little woodland, for nesting area, that sightings there need be assumed to involve visitors from another square. Most confirmed breeding was based on newly flying young (FL, in 464 squares, 41 per cent of 1137 with Ravens), and nests (NY,NE,ON) were found in 88 squares (8 per cent).

Before European settlement of the Maritimes, availability of food probably held Raven numbers below present levels. Subsequently, persecution reduced them still farther, as these birds were considered uncommon in New Brunswick and virtually absent in Prince Edward Island at the turn of the century. They evidently regained former numbers, and more, since then and may be more numerous now than ever before. Like the large gulls, Ravens combine predation, scavenging, and omnivorousness. As long as we assemble edible wastes in public dumping grounds and continue to kill animals on the highways with our vehicles, Ravens will thrive here. The proposed consolidation in New Brunswick of public dumps into fewer but larger regional landfill sites may reduce the number of such feeding areas, but as to whether it will reduce greatly the overall population, "Quoth the Raven, 'Nevermore!'" (E.A. Poe).

Ref. 148

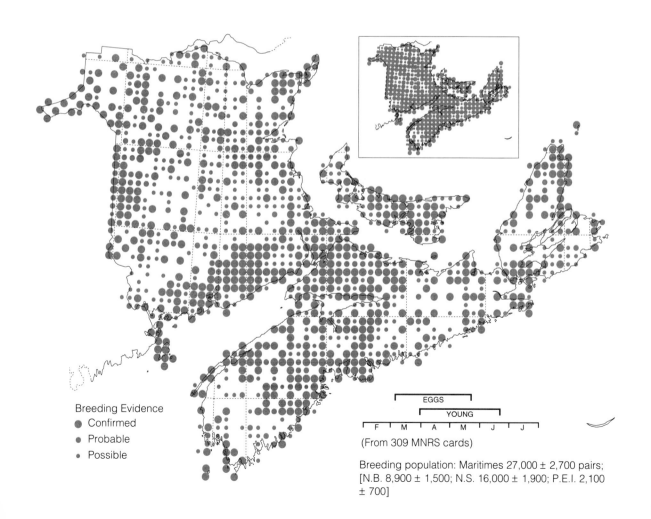

Breeding Evidence
- Confirmed
- Probable
- Possible

EGGS
YOUNG
F | M | A | M | J | J

(From 309 MNRS cards)

Breeding population: Maritimes 27,000 ± 2,700 pairs; [N.B. 8,900 ± 1,500; N.S. 16,000 ± 1,900; P.E.I. 2,100 ± 700]

Black-capped Chickadee
Mésange à tête noire

Parus atricapillus

This chickadee is recognized by almost every-one with even a casual interest in birds, as the members of this family (chickadees, titmice, tits) have few equals in their readiness to approach and accept food from people. The Black-capped Chickadee breeds across Alaska and Canada in the southern parts of the boreal forest, and south in temperate forests and wood-lands to northern California and New Jersey, as it uses conifer and mixed stands as well as hardwoods. In the Maritimes, it was found almost everywhere with ade-quate coverage, being missed mainly in treeless coastal squares. The Atlas data should represent its status ade-quately. Chickadees come to winter feeding stations even in cities and towns well removed from continuous tree cover, and they often stay to breed in urban areas, using nest-boxes; in the wild, however, they excavate their nest-cavities in rotted tree stubs. Nests were quite often found and reported to the Atlas (in 75 of 1124 squares with the species, 7 per cent), but the most frequent confirmations of breeding were by sightings of family groups (FL, 393 squares).

The wide habitat usage by this species, including its acceptance of urban areas, suggests that the changes wrought by humans on the pre-settle-ment forests of the Maritimes have not altered appre-ciably the breeding opportunities for chickadees here. With winter feeding, which should improve their sur-vival during the period when food is most limiting, their present numbers are probably higher than before European settlement, and both the BBS and Christmas bird counts indicate increases over the last 25 years. The unusually abundant spruce cone crop in 1988–89 evidently provided food for chickadees as well as cross-bills; late in 1989, with the cone crop exhausted, the chickadees, and also red squirrels, appeared at nearby feeders in droves. Their general adaptability makes it likely that future land-use and climate changes will not limit their numbers in the Maritimes.

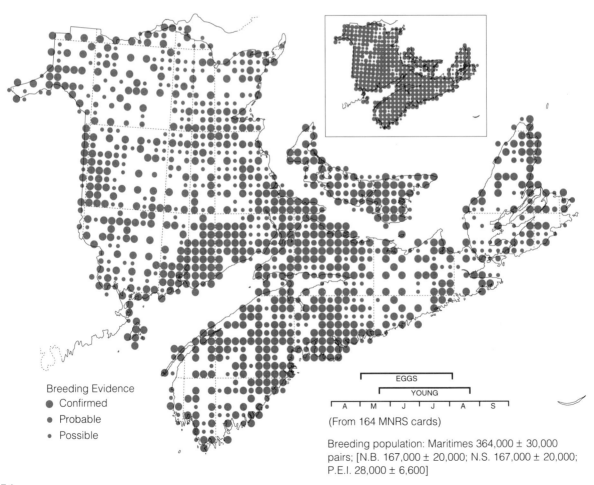

Breeding Evidence
● Confirmed
● Probable
● Possible

EGGS

YOUNG

A M J J A S

(From 164 MNRS cards)

Breeding population: Maritimes 364,000 ± 30,000 pairs; [N.B. 167,000 ± 20,000; N.S. 167,000 ± 20,000; P.E.I. 28,000 ± 6,600]

Boreal Chickadee
Mésange à tête brune

Parus hudsonicus

The Boreal is the chickadee of conifer, and especially of spruce, forests all across the northern regions of Canada. It breeds from Alaska to Labrador, and south only to the northernmost states of the U.S.A. This species seldom leaves the closed forest to visit winter feeding stations in suburban areas, so it is much less familiar to casual bird-watchers than is the Black-capped Chickadee and may have been missed by some atlassers. Boreal Chickadees were found in all parts of the Maritimes, though a general scarcity in the hardwood-dominated areas of western New Brunswick may reflect lack of suitable habitat (see below). Atlas statistics were closely parallel to those for Black-capped Chickadees, but Boreals were found in only two-thirds as many squares (748 vs. 1124). Breeding by this species was confirmed in 346 squares (46 per cent), but nests were found in only 20 squares (3 per cent).

Until recently, changes in the Maritimes forests since the start of European settlement seemed unlikely to have affected Boreal Chickadee numbers greatly. Rotting trees and stubs, of the small sizes in which these birds excavate their nests and roost-holes, were always numerous. Numbers of Boreal Chickadees often vary greatly between years, and it is likely that winter survival sets the main limit on their populations, past and present. Extensive clear-cutting of pure softwood stands, including "salvage" after major budworm kills since 1970, has recently reduced available habitat across much of southern New Brunswick. The BBS, though working with small samples, suggests a decline is general in the Maritimes, and Christmas Bird Counts in New Brunswick since 1970 support this impression. Continued emphasis on forest management for fibre production from conifers seems likely to ensure that Boreal Chickadees can persist here in future, but any trend toward global warming will make conditions for conifers here less favourable.

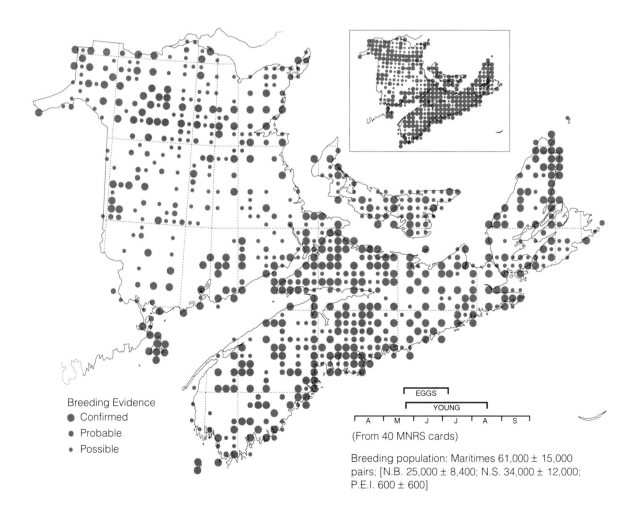

Breeding Evidence
- Confirmed
- Probable
- Possible

EGGS
YOUNG

A M J J A S

(From 40 MNRS cards)

Breeding population: Maritimes 61,000 ± 15,000 pairs; [N.B. 25,000 ± 8,400; N.S. 34,000 ± 12,000; P.E.I. 600 ± 600]

Red-breasted Nuthatch
Sittelle à poitrine rousse

Sitta canadensis

This species breeds across the North American boreal forest region from Alaska to Newfoundland, and south in mountains to California and North Carolina. Other nuthatches of similar size and pattern occur as relict populations scattered from Algeria and Corsica, France, to Yunnan, China, and some workers consider them all the same species. These are small birds of conifer forests, where they forage in the outer foliage of spruce and fir trees rather than on tree-trunks as in the larger nuthatches (e.g., White-breasted Nuthatch.). Although some Red-breasted Nuthatches are here year-round, others migrate, particularly when populations are high after seasons with abundant food. In the Maritimes, these birds were found throughout, but the mapped distribution was probably affected by an unprecedented spruce cone crop, from late summer 1988 well into 1989, when they were very abundant; see also White-winged Crossbill, affected similarly. Nests, self-excavated in dead trees or stubs, were seldom found, in only 41 (4 per cent) of 995 squares with this species. Family groups (FL), in 291 squares, were the most frequent confirmation of breeding.

Changes in the forests of the Maritimes since the start of European settlement have benefited this species more than they hindered it. The increase of conifer stands, as secondary succession after logging of the original, more hardwood-dominated, forest, provided much suitable habitat for both foraging and nesting, and the ravages of spruce budworm and clear-cut logging in more recent decades have not reversed that trend. The present numbers of this species are probably no fewer, and likely more, than those present 400 years ago. Future needs of the forest industry will tend to perpetuate conifer forests suitable for this species, but global warming, if it continues over a long period, would tend to encourage a temperate forest with more broad-leafed trees. Red-breasted Nuthatches are likely to remain regular birds of our forests in our lifetimes.

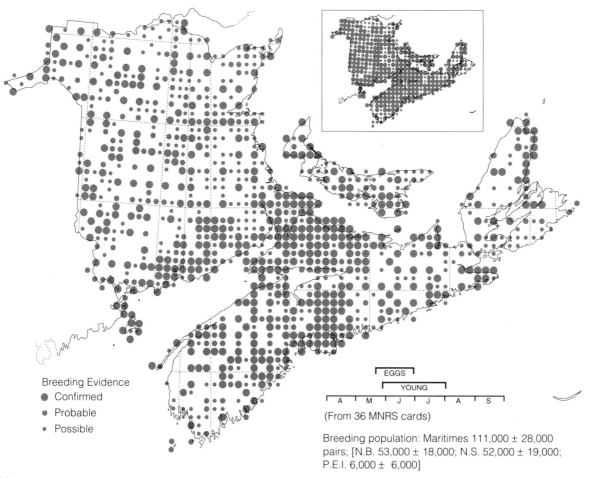

Breeding Evidence
- Confirmed
- Probable
- Possible

EGGS
YOUNG

A M J J A S

(From 36 MNRS cards)

Breeding population: Maritimes 111,000 ± 28,000 pairs; [N.B. 53,000 ± 18,000; N.S. 52,000 ± 19,000; P.E.I. 6,000 ± 6,000]

White-breasted Nuthatch
Sittelle à poitrine blanche

Sitta carolinensis

This is a more southern species than the preceding one, and it breeds from southern Canada through the United States to Mexico. As a tree-cavity nesting bird, it frequents open woodlands of all kinds and also urban areas with suitable large trees. The White-breasted Nuthatch is a year-round resident, but it may move some distance to winter feeding stations, which it commonly frequents. In our area, it is associated with broad-leafed trees rather than conifer forest, but its distribution does not fit any obvious pattern. Its largely southwesterly distribution reflects the greater prevalence of hardwoods in those areas. However, breeding in northeastern Cape Breton Island, here confirmed, was already inferred from summer sightings in the early 1960s. Numbers, and thus breeding range, are limited by survival through the winters, and feeding stations may have some influence on

their summer distribution. Many squares with possible breeding were scattered among better documented records and were likely valid, but some may have been birds lingering near winter feeders. Confirmations, mostly FL or AY, made up a rather small proportion (59 of 202 squares with the species) of all the Atlas records, and only eight nests were reported.

There is no evidence regarding historic abundance or scarcity of the White-breasted Nuthatch in the Maritimes. The assumption of more continuous and more generally broad-leafed forest here before European settlement suggests that its status is unlikely to have changed much. The data from the Atlas project were too scattered to allow extrapolation to total populations with much conviction. The species is unlikely to become abundant here, but there are no obvious threats to its continuing survival. If global warming becomes a reality, a trend towards more temperate forests would favour these birds.

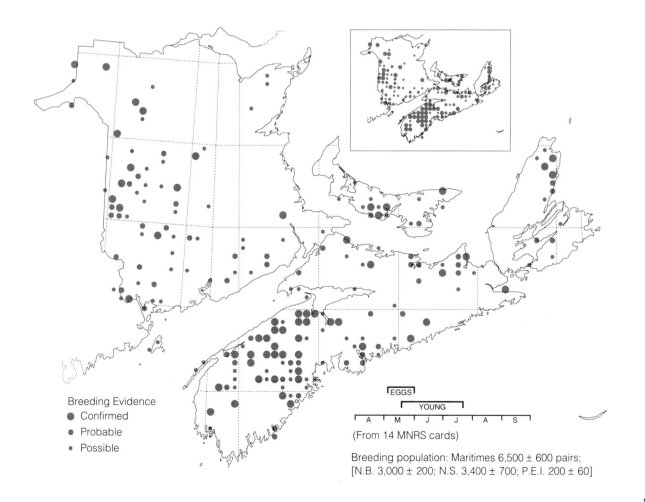

Breeding Evidence
- Confirmed
- Probable
- Possible

EGGS
YOUNG

A M J J A S

(From 14 MNRS cards)

Breeding population: Maritimes 6,500 ± 600 pairs;
[N.B. 3,000 ± 200; N.S. 3,400 ± 700; P.E.I. 200 ± 60]

Brown Creeper
Grimpereau brun

Certhia americana

This is the only tree-creeper in North America, the others being native to Europe and Asia. It breeds in conifer and mixed forests in the southern boreal and temperate zones of the northern U.S.A. and Canada, from central British Columbia to Newfoundland, and south in mountain regions to North Carolina and through Mexico to Nicaragua. Like other forest creepers, it is inconspicuous, with protective coloration and high-pitched faint call-notes and song. Although often termed a year-round resident, many creepers migrate southward, appearing every spring and fall south along the Atlantic coast in areas where they do not breed. Most migration in the Maritimes passes unnoticed, except on offshore islands, and the species must also be seriously under-represented in the Atlas breeding records. The creeper was noted in all regions, but in only about one-third of the squares with adequate coverage. Over half of the records (195 of 365 squares) were of single sightings (H). Creepers probably occur in most squares with mature or mid-

aged forests, as their nests (noted in 12 squares) are typically under patches of loose bark on old or dead tree-trunks.

It seems plausible that Brown Creepers were more widespread in the more often old-growth, continuous forests that clothed the Maritimes when European settlement began. Clearing of forests for settlement, and replacement of mature mixed forests by conifer stands, with increased frequency of forest fires, had negative influences on creeper numbers. Extrapolation of abundance indices to provincial populations, given the low detectability of the species, gives only a suggestion of numbers of these distinctive birds; census plot data suggest that creeper densities are similar to those of Winter Wrens and Canada Warblers, which often frequent the same habitats, but for which our population estimates are much higher. The BBS also detected too few Creepers to allow trend analysis, but there is no impression that they are declining. Forestry practices, with shorter intervals between cutting in future, are likely to reduce creeper habitat further, but their ability to use mature trees in suburban areas may partially offset such losses.

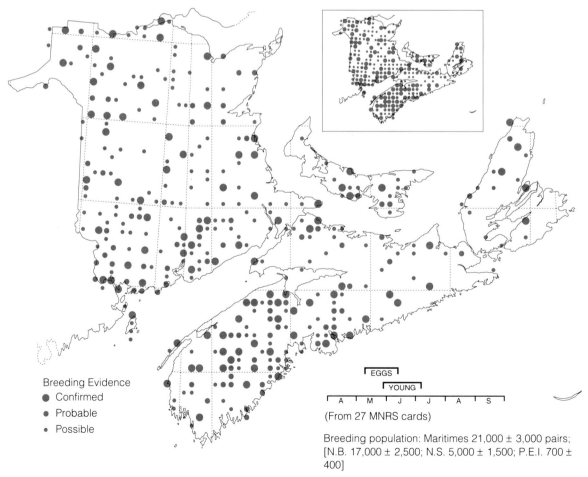

Breeding Evidence
● Confirmed
● Probable
● Possible

EGGS
YOUNG
A M J J A S
(From 27 MNRS cards)

Breeding population: Maritimes 21,000 ± 3,000 pairs; [N.B. 17,000 ± 2,500; N.S. 5,000 ± 1,500; P.E.I. 700 ± 400]

Winter Wren
Troglodyte des forêts

Troglodytes troglodytes

This is the northernmost of the wrens, an American family of small but voluble birds, of which only this species breeds also in Eurasia. Unlike its status as a familiar garden and farmland-hedgerow bird (the "Jenny Wren") in western Europe, in Canada it is found mainly in dense, damp, usually conifer forests, from coast to coast. However, the slash piles and brushy edges left by most logging operations also provide acceptable habitat, with many potential nest sites among the roots of fallen trees. The Winter Wren was formerly considered uncommon in New Brunswick and Nova Scotia and in many other parts of Canada, but it was found in two-thirds of the Atlas squares here that were adequately sampled (833 in all). Breeding was seldom confirmed, with records of fledged young (FL) or adults carrying food (AY) in only 14 per cent of squares with the species. The long and elaborate "tinkling" song was by far the most frequent evidence of its presence

in summer (H, 411 squares; T, 229 squares). Despite its name, only a very small number of our wrens attempt, seldom if ever successfully, to remain year-round in this area; Winter Wrens in Canada east of the British Columbia Coast Mountains are fully migratory. They return here in late April and early May, when insects again become available.

The changes in Maritimes forests in the past 400 years, from mature, closed, mixed forests to (in many areas) mainly conifers and then to logging slash and clear-cut areas, have benefited Winter Wrens as often as they disadvantaged them. There seem to be no intuitive reasons to believe that the species was formerly much more or less common than at present, although lack of familiarity with the bird and its song may have created an impression of former scarcity. However, the BBS suggests a decline, mainly in New Brunswick, since 1966. Tidier forestry practices may have some adverse effects in future, but maintenance of conifer stands for fibre production is likely to ensure a continuing place for the Winter Wren in the Maritimes.

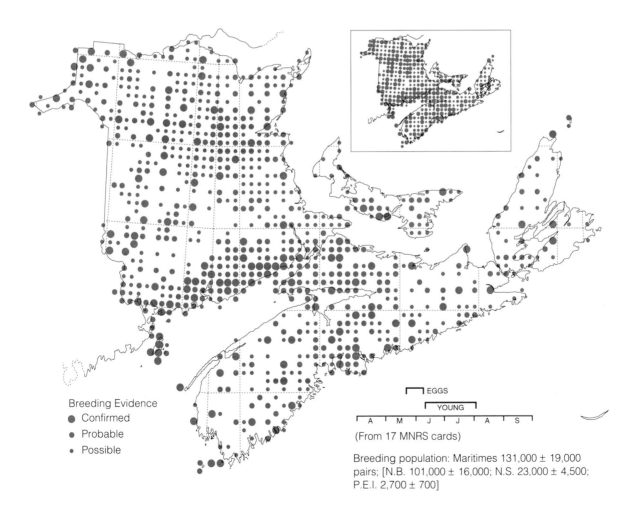

Breeding Evidence
- Confirmed
- Probable
- Possible

EGGS
YOUNG

A M J J A S

(From 17 MNRS cards)

Breeding population: Maritimes 131,000 ± 19,000 pairs; [N.B. 101,000 ± 16,000; N.S. 23,000 ± 4,500; P.E.I. 2,700 ± 700]

Marsh Wren
Troglodyte des marais

Cistothorus palustris

The (Long-billed) Marsh Wren frequents marshes across Canada from British Columbia to southern Quebec, and south to California and Florida. Although it also uses salt marsh farther south, it is strictly a bird of freshwater marshes in Canada, with cattail and bulrush stands in permanent waters the preferred habitats. It was quite widespread in south-central Maine, but it has been and remains very local in the Maritimes, where only four nest records have been received since 1968, although breeding was suspected as early as 1938. The Atlas received records in 11 squares, mostly from artificial impoundments created for waterfowl. All confirmed breeding was in areas where the species was already known: Red Head Marsh, N.B., and the New Brunswick–Nova Scotia border area, with dummy nests reported in similar habitat at the Germantown Marsh in New Brunswick. The sightings in Prince Edward Island, where its status was "accidental," and eastern Nova Scotia were probably of vagrants. In their main range, Marsh Wrens are vocal and easily detected; entry to nesting habitat, except by boat or wading, is less convenient. The Atlas data may be representative, as the species is absent or irregular in many oft-visited marshes in the New Brunswick–Nova Scotia border region and in Prince Edward Island.

The scarcity of suitable fertile breeding habitat in the Maritimes must always have made this area marginal for Marsh Wrens, and their status has fluctuated with the progress of marsh drainage and restoration here. After scattered records from 1938 to 1955, very few were noted until the recent DUC impoundments appeared after 1970. Its colonial habit means that even fewer locations were involved; it is unlikely that substantial unknown concentrations exist elsewhere. Although Marsh Wrens have persisted in the border area since 1938, the future for such a small population is uncertain, despite sustained efforts to maintain its preferred habitat for waterfowl.

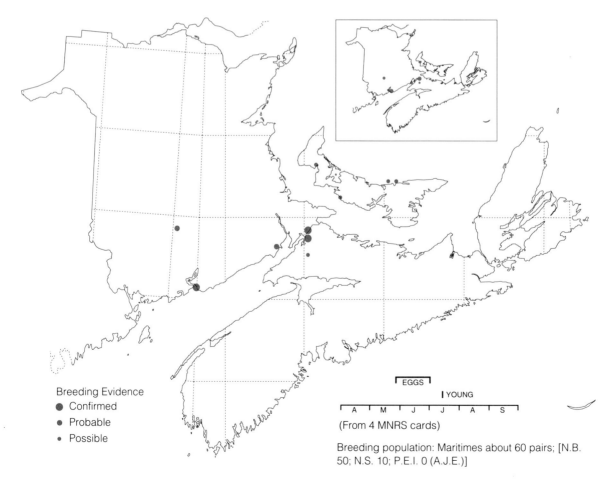

Breeding Evidence
● Confirmed
● Probable
● Possible

EGGS

| YOUNG

A   M   J   J   A   S

(From 4 MNRS cards)

Breeding population: Maritimes about 60 pairs; [N.B. 50; N.S. 10; P.E.I. 0 (A.J.E.)]

Golden-crowned Kinglet
Roitelet à couronne dorée

Regulus satrapa

The kinglets are among the smallest birds of the northern conifer forests, with different species occupying the same niches in Eurasia and America. Golden-crowned Kinglets breed in cooler coniferous forests all across Canada from Alaska to Newfoundland, and south in mountains to South Carolina and to Guatemala. Golden-crowned Kinglets are migratory in northern parts of their range, but farther south, including the Maritimes, many are permanent residents. In our area, these kinglets were found in all regions, and little less generally where the preferred conifer forests were less prevalent, as in the St. John River Valley and southwest Nova Scotia. An important influence on the mapped distribution was the low detectability of this bird, whose song is high-pitched, weak, and infrequent compared to many species. Golden-crowned Kinglets were missed in some briefly sampled squares for this reason, but they were found in almost every square that was visited for 30 hours or more, and appreciably more generally than

had been foreseen at the start of the Atlas project. They were confirmed as breeding in 482 squares (52 per cent of those with these birds), most often by detecting groups of newly fledged young (FL, in 338 squares, 36 per cent). Nests were rarely found (in 4 squares).

Reduction of mature conifers, especially white pines and eastern hemlocks, since the start of European settlement probably had a negative influence on kinglet numbers, but may have been balanced by the proliferation of balsam fir forests, until these were ravaged by spruce budworm or clear-cut for pulp logs. Overall, the long-term change, whether downward or upward, seems unlikely to have been great. The BBS suggested an increase since 1966, but that was based on small samples because the faint songs of these birds are seldom heard from roadsides. The future seems rather less promising, with the forest industry aiming for shorter cutting rotations, thus reducing the proportions of mature trees, and global warming perhaps reducing conifers relative to hardwoods. Golden-crowned Kinglets are unlikely to change much in the short term, but their long-range future here is unclear.

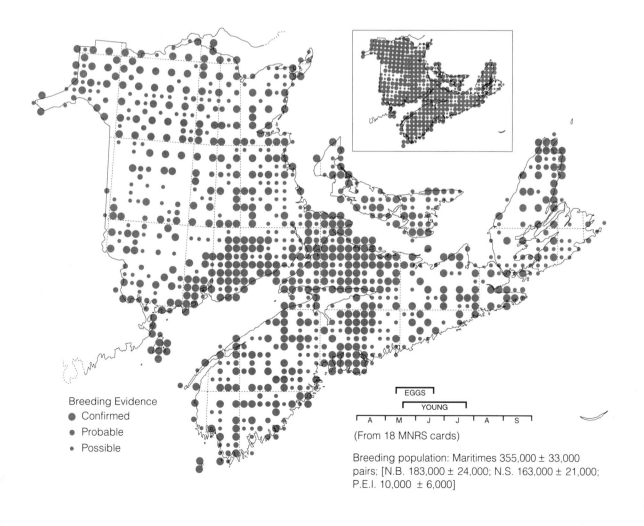

Breeding Evidence
● Confirmed
● Probable
● Possible

EGGS
YOUNG

A M J J A S

(From 18 MNRS cards)

Breeding population: Maritimes 355,000 ± 33,000 pairs; [N.B. 183,000 ± 24,000; N.S. 163,000 ± 21,000; P.E.I. 10,000 ± 6,000]

Ruby-crowned Kinglet
Roitelet à couronne rubis

Regulus calendula

The Ruby-crowned Kinglet breeds across this continent from Alaska to Labrador, and south in the mountains to California, but only north of the prairies and the temperate forests farther east. In the Maritimes it was found commonly throughout, except sparsely in the more hardwood-dominated woodlands of southwestern New Brunswick. Ruby-crowns are more tolerant of mixed woods than are Golden-crowned Kinglets, though both nest only in conifers. This is a small and drab-coloured bird, its ruby "crown" being usually concealed, but its loud emphatic song makes it conspicuous out of proportion to its size and numbers, and it is seldom missed if present. Records (in 1067 squares) were fairly equally divided between confirmed (36 per cent), probable, and possible breeding, as adults carrying food (AY) are easily attracted by squeaking or spishing sounds. The higher frequency of confirmed breeding in southern New Brunswick and central Nova Scotia reflected variation in the observers rather than in the birds. Nests were found in only five squares, although nests previously recorded here were situated low (1–3 m) in spruces as often as higher up (8–15 m).

Ruby-crowned Kinglets probably changed relatively little in abundance in historical times, from what we can infer about trends in forest cover. Although the original forests had a larger component of broad-leafed trees, conifers were always common enough to accommodate present numbers of this species. These birds are never extremely abundant compared to several warblers found in similar habitats, but their earlier categorization as "uncommon" in New Brunswick can hardly be explained except as applying mainly to the southwest of that province. The BBS indicated no overall change since 1966, as significant losses in the late 1970s have been mostly regained since then. The requirements of the forest industry will ensure suitable habitats for this species in the Maritimes for the future.

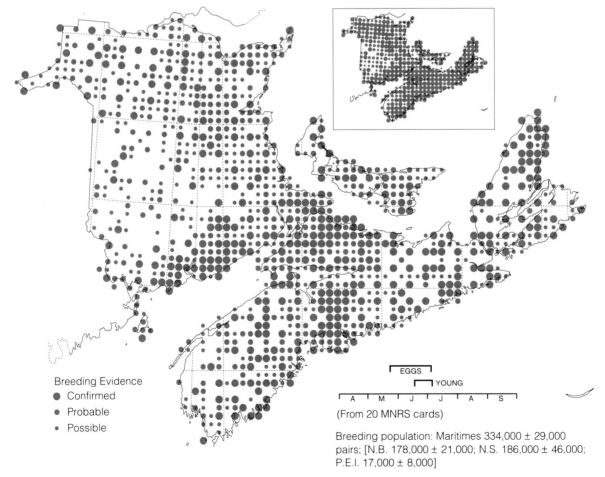

Breeding Evidence
● Confirmed
● Probable
● Possible

EGGS
YOUNG
A M J J A S
(From 20 MNRS cards)

Breeding population: Maritimes 334,000 ± 29,000 pairs; [N.B. 178,000 ± 21,000; N.S. 186,000 ± 46,000; P.E.I. 17,000 ± 8,000]

Eastern Bluebird
Merle-bleu de l'Est

Sialia sialis

This attractive bird reaches its northeastern limit in the Maritimes, whence it breeds south to Florida and Mexico, and west just into Saskatchewan. It was widely but sparsely scattered across New Brunswick and was no more general in most of Maine, so its continuous range evidently lies still farther to the southwest. Formerly, Bluebirds were more widespread in the Maritimes, including regular breeding in Nova Scotia. The arrival of Starlings in the 1920s added a major competitor for the woodpecker holes in which Bluebirds nest, as both species forage in open areas of low vegetation with scattered trees for nesting. During Atlas field work in New Brunswick, Bluebirds often were found breeding in clear-cut areas amid forests, a habitat not used by Starlings. The latter forage by walking through short grass whereas Bluebirds fly down from perches to catch insects on or near the ground, which requires less continuously low vegetation. Despite its scarcity, the blue bird with the rusty breast is so distinctive, and so attractive to all, that most of those that appeared in settled areas were reported to the Atlas. They readily use nest-boxes, and 38 of 115 squares with Bluebirds involved nests (NY,NE,ON), with newly flying young (FL) seen in 25 squares.

Bluebirds were never common in the Maritimes, as their main range is roughly coincident with the eastern broad-leafed forest, which barely reaches this far. In earlier times, the largely forested country provided less favourable habitat than resulted from efforts to clear the land for human settlement. Quite possibly the species was more common here in the late 19th and early 20th centuries than either before or since. Although existing nest-boxes are sometimes used, the few attempts to establish "bluebird trails" of nest-boxes in the Maritimes have met with no success. The Atlas data were too few to permit statistical population estimates, and.breeding was first proved in Prince Edward Island only in 1989. Projections of habitat use give no suggestion that we will see more Bluebirds in future, and the effects on forests of global warming, though likely to favour this species, will emerge only slowly. For now, we will have to enjoy those few we have.

Ref. 77

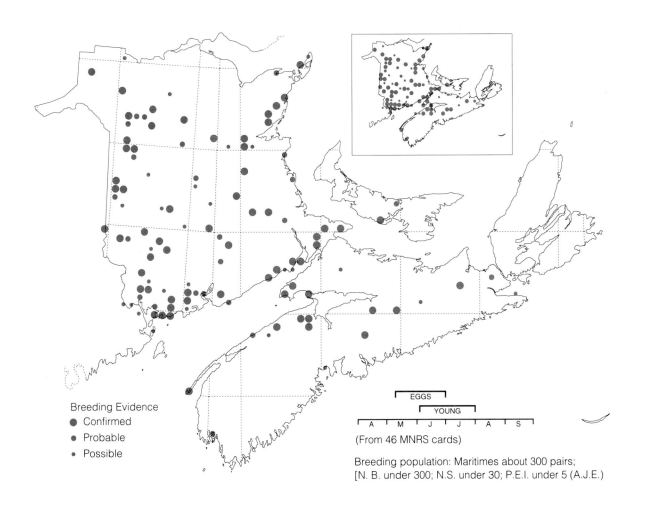

Breeding Evidence
● Confirmed
● Probable
• Possible

EGGS
YOUNG

A M J J A S

(From 46 MNRS cards)

Breeding population: Maritimes about 300 pairs;
[N. B. under 300; N.S. under 30; P.E.I. under 5 (A.J.E.)

Veery   Grive fauve

Catharus fuscescens

Our "lesser thrushes" are northern representatives of a group widely distributed in tropical America. The Veery breeds across southern Canada from interior British Columbia to southwest Newfoundland, and in the northern U.S.A., south in mountains to Arizona and Georgia. It reaches the northeast limit of its continuous distribution in the Maritimes, where it was scarce in eastern mainland Nova Scotia, and absent in much of Cape Breton Island. Veeries also were absent from the highest peaks and plateaus in north-central New Brunswick, where suitable habitat is scarce. This is the thrush of broad-leafed stands, but it occurs also in mixed forest with a substantial deciduous component. As with most thrushes, Veeries were detected usually by song, with single (H) or repeated (T) records in an area accounting for two-thirds of all squares (554 of 819) with the species. Nests, on the ground, were found in only 34 (4 per cent) of these squares.

The greater hardwood component in the primaeval forests of the Maritimes may have favoured Veeries. However, the generally more southern distribution of this bird and evidence of its recent spread within this region make it unlikely that Veeries were as or more common 400 years ago than at present. Possibly the species' range contracted southward during the "Little Ice Age" (1350–1850) and is now expanding again. Veeries are still uncommon though now widespread in Prince Edward Island, where Godfrey (1954) looked in vain for them in 1952, and they have spread eastward in Nova Scotia and increased numerically in New Brunswick during the last 40 years. Efforts by the forest industry to increase the conifer component of our forests will not improve the future prospects for the Veery. On the other hand, global warming, if it occurs on the scale often forecast, could make the hardwood forests it prefers more prevalent here, in a century or two, than at present. The long-range future for the Veery, like its past, is obscure, but it seems to be doing well here in our time.

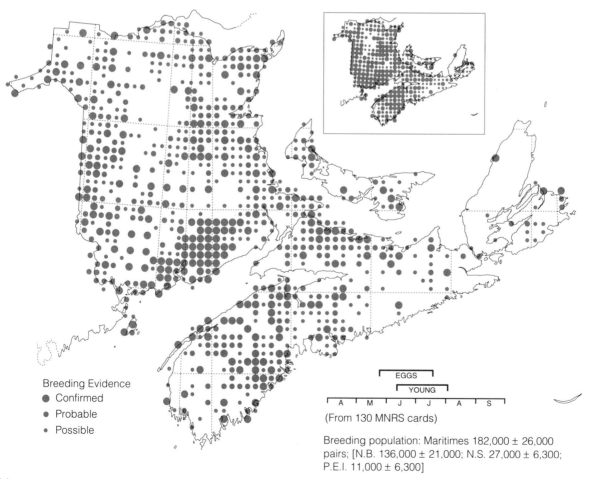

Breeding Evidence
● Confirmed
● Probable
• Possible

EGGS
YOUNG
A   M   J   J   A   S

(From 130 MNRS cards)

Breeding population: Maritimes 182,000 ± 26,000 pairs; [N.B. 136,000 ± 21,000; N.S. 27,000 ± 6,300; P.E.I. 11,000 ± 6,300]

Gray-cheeked Thrush
Grive à joues grises

Catharus minimus

This is the subarctic member of a group of small thrushes that ranges from South America to the arctic tree-line. Gray-cheeked Thrushes breed from northeast Siberia across Alaska and northern Canada to Labrador and Newfoundland, with a distinct subspecies (or species, *bicknelli*) in cool areas of the Maritimes and in New England. Here they are regular but scarce at higher elevations, mostly above 300 m, in northwestern New Brunswick and on the Cape Breton Highlands, and they also breed near sea level in the Grand Manan archipelago and in southern Cape Breton Island. A few other records came from mainland coasts of the Bay of Fundy. They formerly bred on offshore islands of Yarmouth Co., N.S., but were not found there during the Atlas period. Dense low spruces are used in the coastal habitats, but many upland records have been in dense young broadleafed or mixed stands in forested areas. As with all thrushes, the song is the usual cue to their presence. By song, our birds were all *bicknelli*, with the phrases rising towards the end; the songs of all other subspecies drop toward the end. The records, in 88 squares, included confirmed breeding in 25 squares (4 with nests), a proportion similar to more widespread thrushes. Although many atlassers were not familiar with the Gray-cheeked Thrush, the remote areas it frequents here were worked largely by experienced observers, so the records should be representative of its range.

There is little in the historical record to suggest major changes in range or abundance of Gray-cheeked Thrushes in the Maritimes since European settlement began. Human occupation of the offshore islands as fishing bases seems unlikely to have restricted their use by these birds, considered "abundant" to "common" through 1922 on Seal Island, N.S., but gone from there by 1938. The remote mountain forests they favour elsewhere have been difficult of access by forestry interests, and relatively little-logged in consequence. Future global warming will not help the species, and acid rain is thought to have damaged its conifer habitat in the mountains of New England (J.T. Marshall, letter 1988). This bird could be lost to the Maritimes as a breeding species, and the entire *bicknelli* population may be in jeopardy, as acid rain is a widespread international issue whose resolution may take decades.

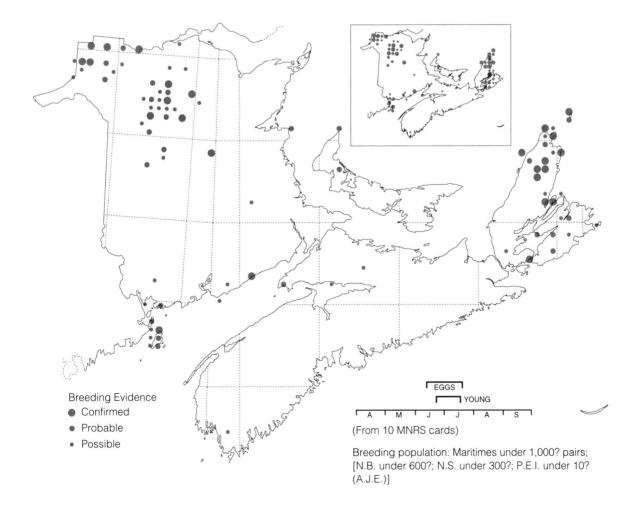

Breeding Evidence
- Confirmed
- Probable
- Possible

EGGS
YOUNG

A M J J A S

(From 10 MNRS cards)

Breeding population: Maritimes under 1,000? pairs; [N.B. under 600?; N.S. under 300?; P.E.I. under 10? (A.J.E.)]

Swainson's Thrush
Grive à dos olive

Catharus ustulatus

This is the most common small thrush of the northern forests, where it breeds from Alaska to Labrador and from northern New England to southern British Columbia, with extensions south to California and Virginia. It frequents less open woodlands than the Hermit Thrush, so it is seen less often, but census plot studies show that Swainson's is usually much the more numerous species in forested regions. These birds were found throughout the Maritimes. Swainson's Thrush breeding was confirmed slightly less frequently than Hermit Thrush (in 35 per cent vs. 36 per cent of squares with these species), possibly because the tree nests of this species were found less often (in 37 squares, 3 per cent) than the ground nests of Hermits (in 78 squares, 7 per cent). Overall, the two species were reported in almost identical numbers of squares (1094 vs. 1083), and their habitat use overlaps widely. Their songs are quite similar and may be confused at a distance; it seems likely that more Swainson's were reported as Hermit Thrushes than vice versa, but this affects their estimated numbers more than their distribution.

Swainson's Thrush may have declined slightly in absolute numbers, as well as relative to the Hermit Thrush, since the start of European settlement with its concurrent opening-up of the Maritimes forests. It is still one of our more abundant birds. Future uses of forest lands, and particularly shorter cutting cycles, seem likely to continue trends towards smaller proportions of closed woodland. The BBS suggested that Swainson's Thrush numbers here have been stable since 1966, whereas Hermit Thrushes have increased; this may reflect increasing areas of the more open woodland habitats used by the latter species. Reduction in area of closed forests (favoured by the Swainson's Thrush) through logging cannot continue indefinitely without jeopardizing the future of the forest industry here; management to ensure continued availability of forests for exploitation will also ensure future habitat for this bird. In the long term, global warming and a gradual shift to more temperate types of woodland may be of as much concern as forest management, or lack of it, for boreal forest birds.

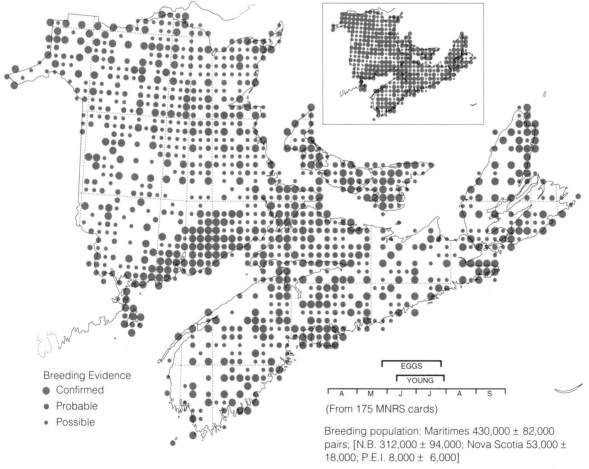

Breeding Evidence
● Confirmed
● Probable
● Possible

EGGS
YOUNG

A M J J A S

(From 175 MNRS cards)

Breeding population: Maritimes 430,000 ± 82,000 pairs; [N.B. 312,000 ± 94,000; Nova Scotia 53,000 ± 18,000; P.E.I. 8,000 ± 6,000]

Hermit Thrush Grive solitaire

Catharus guttatus

The "nightingale of the north woods" breeds all across Canada, from Alaska to Labrador and Newfoundland, and south in mountains to California and western Maryland. It was distributed all over the Maritimes, as it frequents areas of open woodland such as treed bogs, pine barrens, and cut-over areas, as well as conifer and mixed forests. The song is detectable at considerable distances, though notable for quality rather than volume, so the species is well represented in the Atlas records. Confusing it with Swainson's Thrush, which has a similar though less varied song and which also frequents a wide range of conifer and mixed woodland habitats, is easily possible. Swainson's Thrush is considerably more abundant than Hermit Thrush in forests, which are by far the most extensive habitats; the Hermit is more general near the open areas frequented by people, so it is often thought, but probably erroneously, to be the more numerous species. The ground nests of Hermit Thrush were more often found, in 78 (7 per cent) of 1083 squares with this species, than were Swainson's Thrush nests (in trees), but the overall confirmation rates of the two species (Hermit 36 per cent, Swainson's 35 per cent) were similar.

Presumably Hermit Thrushes have been common in Maritimes forests since these evolved after the glacial period. Opening of woodland by cutting expanded the availability of suitable habitats, but complete clearing for agriculture may have balanced such benefits. There is nothing to indicate that human actions have had a detrimental effect overall since settlement began in the Maritimes nearly 400 years ago. The apparent decline of Hermit Thrushes in the late 1950s (Squires 1976) was probably caused by severe winter weather in the eastern U.S.A. rather than by spraying of DDT and other toxic chemicals against forest and agricultural pests in our area. The BBS showed a significant increase since 1966, perhaps reflecting recovery from the earlier losses. Hermit Thrushes are common in all three provinces. As long as conifer forests are maintained for fibre production in the Maritimes, we are likely to be able to enjoy the song of the Hermit Thrush here.

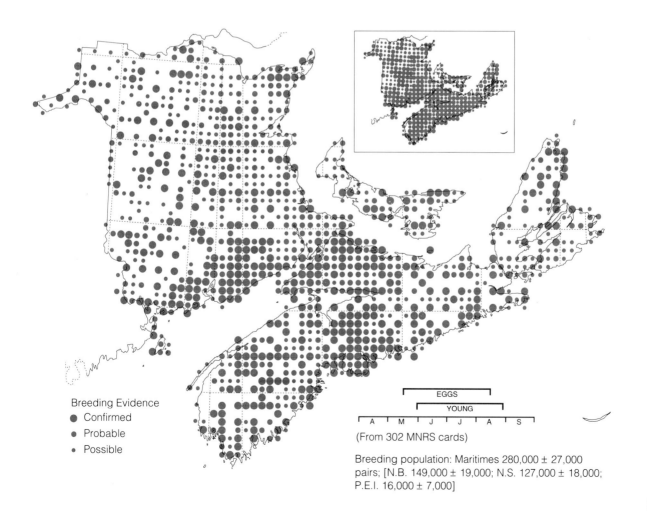

Breeding Evidence
● Confirmed
● Probable
● Possible

EGGS
YOUNG

A M J J A S

(From 302 MNRS cards)

Breeding population: Maritimes 280,000 ± 27,000 pairs; [N.B. 149,000 ± 19,000; N.S. 127,000 ± 18,000; P.E.I. 16,000 ± 7,000]

Wood Thrush
Grive des bois

Hylocichla mustelina

This large thrush is restricted as a breeder to the eastern broad-leafed forests of North America, from Texas and Florida north to Wisconsin, southern Ontario and Quebec, reaching its northeast limit in the Maritimes. Wood Thrushes were little-known here before 1950, although a few bred earlier in southwest New Brunswick, where eggs were collected in 1918. They have spread appreciably since then, as also in Maine and Ontario. In the Atlas period, they were found mostly in valleys of the southwest half of New Brunswick, with scattered records, including confirmed breeding, in southwest Nova Scotia, the areas where broad-leafed woodland is most general. Farther east and north, records were fewer and often involved lone singing birds, with no other evidence of breeding. Breeding was confirmed in only 27 (15 per cent) of 183 squares with the species, 72 per cent of squares having only singing birds (H,T), compared with 27–36 per cent confirmed breeding in the other three common woodland thrushes. Probably some Wood Thrushes were missed by less experienced observers, who may have confused their song with that of the Hermit Thrush, but here at the limit of the range many song records may have been of wandering males that failed to attract mates and breed.

Descriptions of the Maritimes forests at the time Europeans first came here suggest that Wood Thrush habitat was then as abundant as at any time since. However, that was in the middle of the Little Ice Age, and other, more obvious, southern bird species were not recorded here until recently. The apparent expansion of Wood Thrushes in our area since 1950 partly reflects that more people are able to recognize the species, especially by song, but may represent increased occurrence here recently. Present estimates of numbers are tentative, because of the many peripheral records. The future for the species is uncertain. Silviculture for fibre production discourages hardwoods in managed forests, but global warming, if it persists, will favour more southern, broad-leafed woodlands. Encouragement of mixed forests and long intervals between cutting would benefit Wood Thrushes as well as other wildlife, but this is unlikely under exploitation for maximum short-term gain.

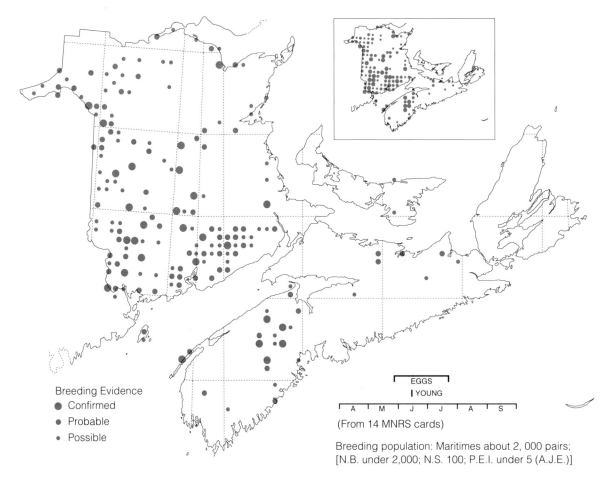

Breeding Evidence
- Confirmed
- Probable
- Possible

EGGS
YOUNG

A M J J A S

(From 14 MNRS cards)

Breeding population: Maritimes about 2, 000 pairs;
[N.B. under 2,000; N.S. 100; P.E.I. under 5 (A.J.E.)]

American Robin
Merle d'Amérique

Turdus migratorius

Our Robin should not be confused with the various distantly related birds in other continents also known by this name. It breeds throughout North America, from the tundra-edge to the Gulf of Mexico, except in treeless areas of prairie and desert regions. In the Maritimes, as in other northern parts of its range, the Robin's return in late March signals spring, even before the snow is gone, and weeks before it starts nesting. This is a bird that is familiar to all, and whose nest almost everyone recognizes, as it occurs just about everywhere except in dense continuous forest, from open woodlands and farms to coastal scrub and city gardens. The Atlas data do not over-represent the distribution, but the familiarity of the species may have exaggerated its abundance. Nevertheless, in the Atlas study, it was found in more squares (1348) than any other species, in virtually every square with appreciable field-work except a few on treeless islands or coastal barrens. It was confirmed as breeding in 1095 squares, over 100 more than any other species, 81 per cent of those with Robin data. Adults with food (AY, in 620 squares, 46 per cent) were by far the most frequent evidence of breeding, and nests (NY,NE,ON) were reported in 265 squares, more than any species except the four common swallows and (surprisingly) the Osprey.

Robins are less abundant in continuous closed-canopy forest than in the mosaic of disturbed woodland, farmland, and settlement created by European settlers since 1600, and their numbers are surely greater here than in earlier times. The BBS indicates stable numbers since 1966. Our estimates of population sizes place the Robin as the most abundant, as well as the most familiar, species in our area. As this is also one of our most adaptable birds, accepting both natural and altered habitats and even nesting on the ground in some areas, changes in land use that Robins could not tolerate are unlikely to be acceptable to people. Nuclear wastelands, and even Love Canals, provide few opportunities for wild things to survive, and even the earthworms Robins eat would not survive extremes of radiation or pollution. Robins should be around for as long as we are.

Ref. 12, 69

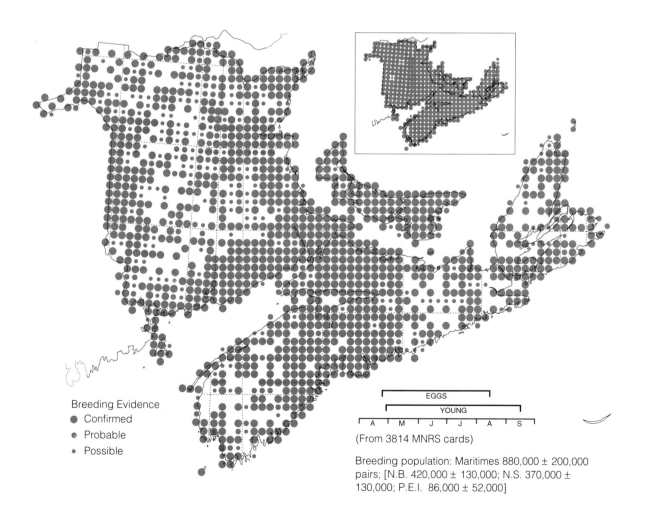

Breeding Evidence
● Confirmed
◦ Probable
· Possible

EGGS
YOUNG
A M J J A S

(From 3814 MNRS cards)

Breeding population: Maritimes 880,000 ± 200,000 pairs; [N.B. 420,000 ± 130,000; N.S. 370,000 ± 130,000; P.E.I. 86,000 ± 52,000]

Gray Catbird
Moqueur chat

Dumetella carolinensis

The Catbird is widely distributed across North America, from the southern U.S.A. to southern Canada, where it breeds from British Columbia to the Maritimes. It inhabits shrubbery in both upland and river-edge situations, mostly in areas where tree cover is of broad-leafed species. The Maritimes lies at the northeast edge of its range, and Catbirds were nearly absent here in upland areas of northern New Brunswick, in Prince Edward Island and Cape Breton Island, as well as in regions with extensive conifer forest cover. Although not an abundant bird, this is a distinctive species with a loud and striking song, so it was probably detected in most areas where it is regular. It was confirmed as breeding in 306 squares (43 per cent of 704 squares with the species), with nesting (NY,NE,ON) in 56 squares (8 per cent), quite typical percentages for these categories among birds of this size and habitat.

Probably Catbirds were originally scarcer in the Maritimes, as more continuous forest cover before European settlement would have provided fewer of the edge situations frequented by this species. There is no suggestion in the literature that it has changed markedly in numbers or range since the first summaries of bird status, around 1880, although its relatives the Northern Mockingbird (resident) and Brown Thrasher (migrant) both arrived here and have increased recently. The BBS suggests an increase in Catbirds here since 1966. For the future, shrubland resulting from old-field succession on abandoned farms will continue to decrease, but cut-over and neglected woodlands with extensive edge habitat may increase, so availability of habitat should not limit the numbers here. Global warming favours more temperate woodland types, which should improve conditions for this species.

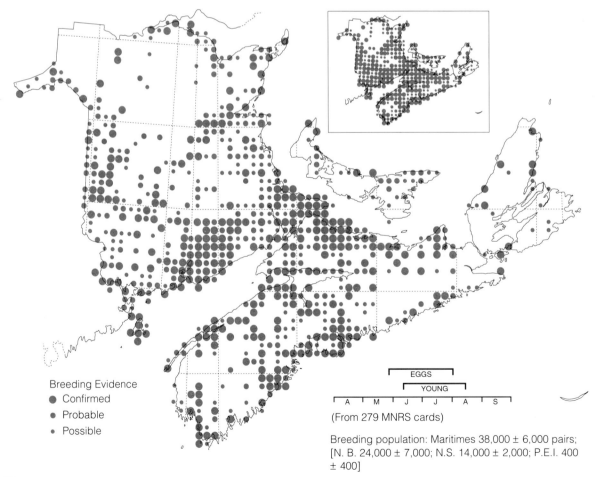

Breeding Evidence
- Confirmed
- Probable
- Possible

EGGS
YOUNG

A M J J A S

(From 279 MNRS cards)

Breeding population: Maritimes 38,000 ± 6,000 pairs; [N. B. 24,000 ± 7,000; N.S. 14,000 ± 2,000; P.E.I. 400 ± 400]

Northern Mockingbird
Moqueur polyglotte

Mimus polyglottos

The Mockingbird is one of several southern species that have spread into the
Maritimes in recent decades. It is a permanent
resident from the northern United States to
southern Mexico and the West Indies. As it does not
usually migrate to escape the rigours of winter, it colo-
nizes most successfully those areas where it can obtain
sufficient suitable food in winter, mostly at urban feeding stations. Mockingbirds do not eat grain or other
seeds, thus they rely heavily on fruits in areas where
insects and soft-bodied invertebrates are not available
in winter. The peculiarly scattered distribution here is
not continuous with the range in Maine, partly owing
to the sparse human population and scarcity of winter
feeding in both Maine and the Maritimes. All nests to
date in the MNRS were in gardens around houses, but
many Mockingbirds in the Maritimes breed in the
country, and some move into towns in winter, exceeding in number those seen there in breeding season.

There are also suggestions that part of
the Maritimes population may migrate
out of the area in winter. Over half the
reports to the Atlas were only of single (H)
or repeated (T) records of singing birds
(in 41 and 40 squares, respectively, of 139
with Mockingbirds); breeding was confirmed
in 50 squares, although only 19 squares had nests
reported.

Scattered individual Mockingbirds had been found
earlier, and one pair nested in Halifax in 1938, but the
continuing presence of the species in the Maritimes
dates only from the 1950s. They have become more
frequent each decade since, but are still uncommon
and irregular in occurrence in most areas. With such a
sparse population (only one or a few pairs in most
towns), local stocks may die out in winter, to be
replaced later by dispersing birds from other localities.
If milder winters result from the "greenhouse effect,"
improved survival may allow an increase in our
Mockingbird population in future, and development of
a migratory tradition in our birds would have a similar
effect.

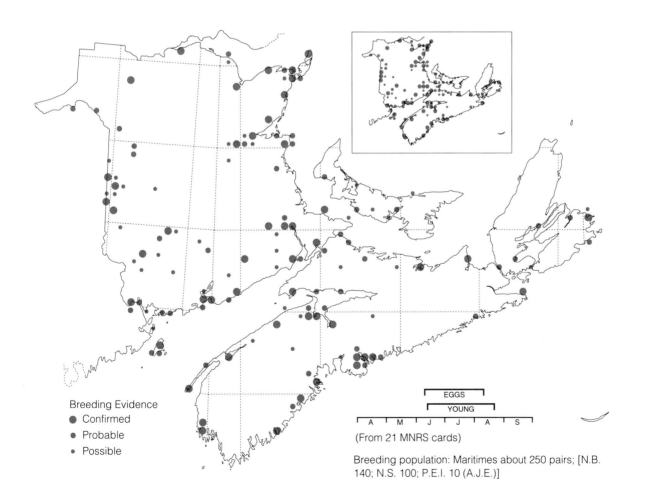

Breeding Evidence
- Confirmed
- Probable
- Possible

EGGS

YOUNG

A   M   J   J   A   S

(From 21 MNRS cards)

Breeding population: Maritimes about 250 pairs; [N.B.
140; N.S. 100; P.E.I. 10 (A.J.E.)]

Brown Thrasher
Moqueur roux

Toxostoma rufum

This bird, distinctive both in appearance and voice, is widespread across the U.S.A. east of the Rocky Mountains and breeds in parts of southern Canada from Alberta to Quebec. It frequents shrubbery, thickets, and wood-edges rather than forest, and it thus reaches its northeastern limit in our area. All records of confirmed breeding received in the Atlas were from the western half of New Brunswick, especially in the generally broad-leafed habitats of the St. John River Valley. No reports were received from Prince Edward Island or eastern Nova Scotia, and breeding has never been confirmed in those provinces. This striking bird, frequenting mostly areas near human settlement, should not often have been missed where present, and it is more likely to be over- than under-represented relative to other species. It was noted too seldom to allow generalization about types of breeding evidence, but the proportion of squares having breeding confirmed (8 out of 39) is lower than in related species (Catbird, Mockingbird). No nests were reported to the Atlas.

The thrasher is a recent settler in the Maritimes, as few were seen here at any season before 1950, and breeding in New Brunswick was not proved until 1968. Presumably this reflected a gradual spread northeastward from its earlier range, as suitable habitat has not obviously increased here in recent decades, although the climate may be marginally warmer than in previous centuries. Breeding may not occur every year in many areas where thrashers were found during the Atlas period, as losses of breeding birds while absent on migration or in winter will not be quickly replaced in such a thinly scattered population. If global warming continues, we may anticipate some increase of this more southern species in our area in future, as many inland areas of the Maritimes have suitable habitat.

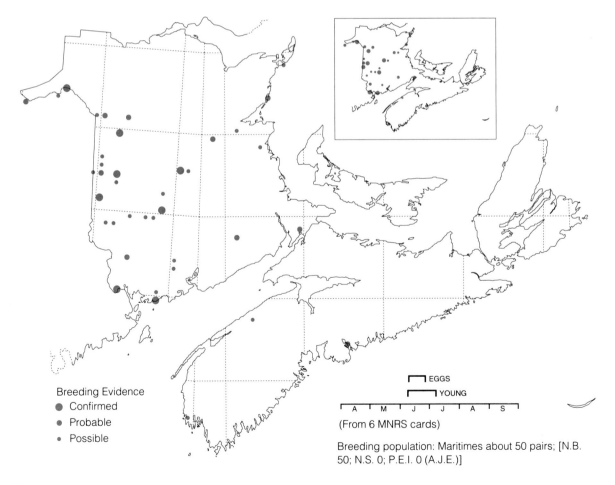

Breeding Evidence
● Confirmed
● Probable
• Possible

EGGS
YOUNG

A M J J A S

(From 6 MNRS cards)

Breeding population: Maritimes about 50 pairs; [N.B. 50; N.S. 0; P.E.I. 0 (A.J.E.)]

Cedar Waxwing
Jaseur des cèdres

Bombycilla cedrorum

This is the smaller and more southern of the two waxwings in North America; only the larger northern species occurs in Eurasia. Cedar Waxwings breed across Canada from central British Columbia to Newfoundland, and southward to California, Illinois, and Georgia. They are less erratic than their larger relatives, and they breed in the Maritimes every summer, in varying numbers. Cedar Waxwings were more than usually common here during the Atlas period, and they were noted in all regions with no obvious gaps. As they frequent forest-edges, roadsides, and gardens as well as open woodlands, they were easily detected, although their occurrence in obvious flocks is usually a sign that nesting is not yet begun or is already finished. Birds with crests seem to attract special attention from the general public, and the reports of Cedar Waxwings, in 1107 squares (ranking 17th among all species in the Atlas), were more frequent than would have been predicted. Atlas statistics showed little unusual, except overall occurrence, with breeding confirmed in 405 squares (37 per cent of those with waxwings).

However, nest-building (NB, in 80 squares) was noted more often than in most species, because waxwings nest late in the season, and combined with nests seen (NY,NE,ON, in 108 squares) accounted for more cases of confirmed breeding than even fledged young (FL, in 160 squares).

The opening-up of forests in the Maritimes after Europeans came benefited the Cedar Waxwing, which is not a bird of closed forests. There is no evidence in the literature that former numbers were appreciably greater or less than at present, but the BBS confirms the impression that these birds have increased significantly since 1980. No obvious causes for such an upswing come to mind, and it may be only temporary. Projected trends in climate and land use in the Maritimes offer no obvious threats to Cedar Waxwings, and we should be able to enjoy these handsome birds far into the future.

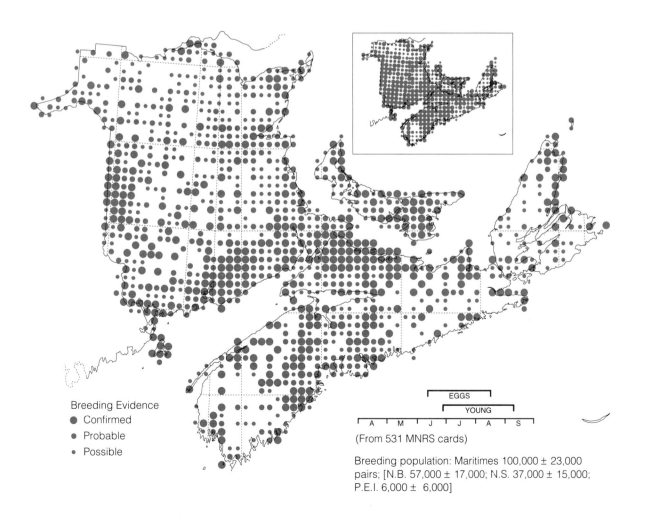

Breeding Evidence
- Confirmed
- Probable
- Possible

EGGS
YOUNG
A M J J A S

(From 531 MNRS cards)

Breeding population: Maritimes 100,000 ± 23,000 pairs; [N.B. 57,000 ± 17,000; N.S. 37,000 ± 15,000; P.E.I. 6,000 ± 6,000]

European Starling
Étourneau sansonnet

Sturnus vulgaris

Starlings were introduced into eastern North America in the late 1800s and have spread across the continent and north almost to the tree-line. They were first found in the Maritimes in 1915, but did not become regular or breed until about 1928. They are now one of our most familiar species, as they frequent towns and cities as well as farmlands, their main habitat. Access to nearby open, grassy lands for foraging restricts their breeding in forested regions, and they were absent from large areas of central and northwestern New Brunswick, with smaller gaps in other forested areas. They nest in cavities of all kinds, such as tree-holes, openings in building walls, crevices in cliffs, and rural mail-boxes, and they successfully exclude most potential competitors for such sites, to the disadvantage particularly of Eastern Bluebirds and Northern Flickers. The Starling's use of open and urban habitats, its lack of secretiveness around its well-protected and enclosed nest-sites, and its habit of flocking as soon as the young fly combine to make it the easiest of all our birds to confirm as breeding. Confirmed breeding was reported for no less than 873 (89 per cent) of the 985 squares with Starlings present, the highest proportion of any widespread species here. Fledged young (FL) and adults carrying food (AY) were almost equally frequent, and nests (NY,NE,ON) were found in 223 squares (23 per cent).

Starlings were not part of our native fauna, but by the time they arrived here much of the originally unsuitable forested country had been altered to suit their tastes. Although some Starlings migrate southwest from the Maritimes to winter, returning in March, many remain here year-round. Reduction in marginal farmland since 1900 and the gradual collapse of derelict buildings in those areas probably reduced their range somewhat; however, any related loss in numbers was more than balanced by the increased availability of waste food and other garbage, allowing more successful wintering, and we have no evidence of recent changes in numbers. Starlings are among the most numerous birds of their size in the Maritimes. There seems no reason to doubt that they will remain abundant here in the future.

Ref. 33

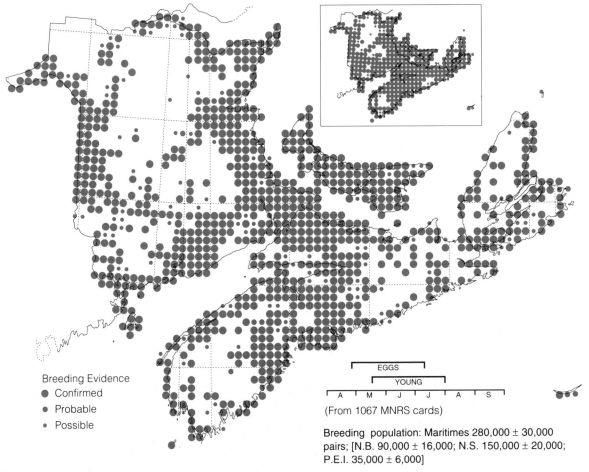

Breeding Evidence
● Confirmed
● Probable
· Possible

EGGS
YOUNG
| A | M | J | J | A | S |

(From 1067 MNRS cards)

Breeding population: Maritimes 280,000 ± 30,000 pairs; [N.B. 90,000 ± 16,000; N.S. 150,000 ± 20,000; P.E.I. 35,000 ± 6,000]

Solitary Vireo
Viréo à tête bleue

Vireo solitarius

The Solitary Vireo breeds in mixed woods and conifer edges in the boreal and montane regions of North America, across Canada from northern British Columbia just into Newfoundland, in northeast U.S.A. and south in mountains to Georgia and Central America. In many areas, it occurs in most forest types, though it is nowhere a really common species. This applies in the Maritimes too. Although recent provincial summaries termed them "uncommon" to "fairly common," Solitary Vireos were found during the Atlas project generally in all regions, though much less frequently in Prince Edward Island than elsewhere. Songs of this species may have been attributed to the more common Red-eyed Vireo by some inexperienced observers, as these species, with the scarcer Philadelphia Vireo, share a similar song pattern, and thereby this species may have been somewhat under-represented. The Solitary Vireo is much easier to see than the other vireos, but its breeding was confirmed little more often, in 300 squares, 32 per cent of 942 with the species (vs. 31 per cent for Red-eyed Vireo). Nests were noted in only 16 squares.

The history of Maritimes forests since European settlement began here gives no suggestion that Solitary Vireos were formerly much more or less common than at present, as their habitat tolerances seem rather broad. The BBS indicated a statistically significant increase here since 1966, which may reflect recovery after losses to earlier forest spraying using DDT, but it may mean only that BBS observers have become more confident in distinguishing this species by song. Forecast trends in forest stand composition suggest that Solitary Vireos should continue to do well in Maritimes forests in future, even with global warming, but the effects of deforestation in Latin American wintering areas are less easily predicted.

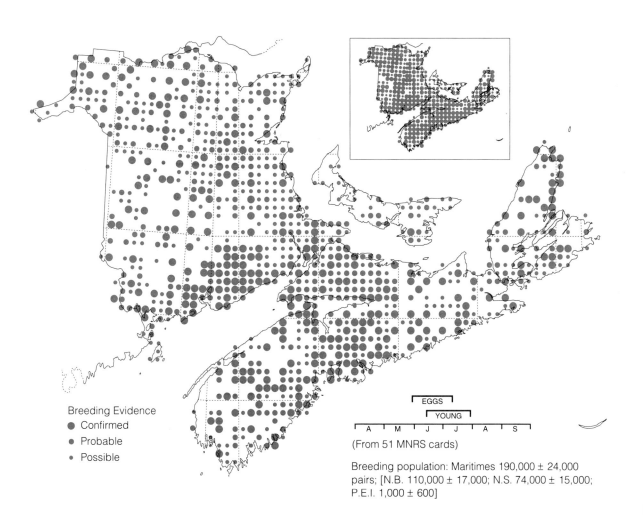

Breeding Evidence
- Confirmed
- Probable
- Possible

(From 51 MNRS cards)

Breeding population: Maritimes 190,000 ± 24,000 pairs; [N.B. 110,000 ± 17,000; N.S. 74,000 ± 15,000; P.E.I. 1,000 ± 600]

Warbling Vireo
Viréo mélodieux

Vireo gilvus

The Warbling Vireo breeds across southern Canada and far into the northwest, as well as south to northern Mexico and the Gulf coast. In the east it is largely associated with the broad-leafed forest region, which reaches its northeastern limit in western New Brunswick. Its range, as shown by the Atlas records, was largely restricted to inland valleys in New Brunswick and was thus more or less continuous with the distribution in Maine. The records in eastern Nova Scotia are perplexingly distant from the others, and in a generally cooler region. The Warbling Vireo is not distinctive in appearance, and it stays well hidden in leaves high in trees; only its song, for which "warbling" is a better description than for most "warblers," is easily detectable and recognizable, though it can be mistaken for a Purple Finch song. Most records in the Atlas were of singing birds, detected once (H, 49 squares) or more often (T, 30 squares). Only one nest was found in the 102 squares with this species, near Tide Head, N.B.

(1989), the third ever reported in the Maritimes; the earlier ones were at Hammond River in 1972 and at St. Leonard in 1981, roughly spanning the mapped range in New Brunswick. Other confirmed breeding reports were based on adults carrying food (in 14 squares).

As this species is obscure and difficult to detect except by song, the historical record in the Maritimes is scanty. Although former mature broad-leafed and mixed forests were partly replaced by conifers or cleared land in recent centuries, the present numbers make it unlikely that the species was formerly more abundant here. The Maritimes were and are peripheral to its main range, and the Warbling Vireo was probably always scarce here. Breeding had not previously been reported in Nova Scotia. The Atlas data provide little basis for estimation of numbers, but leave little doubt that it is more common than we can show. Although the "greenhouse effect" is projected to provide a warmer climate here in future, it is difficult to translate this into an obviously improved environment for Warbling Vireos. These birds are likely to remain scarce in the Maritimes.

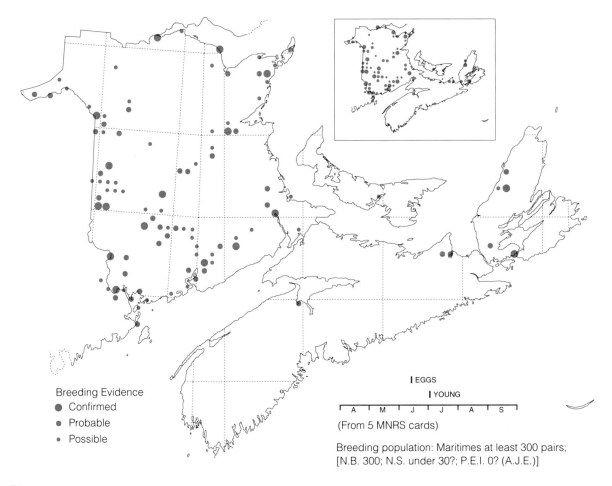

Breeding Evidence
- Confirmed
- Probable
- Possible

EGGS

YOUNG

| A | M | J | J | A | S |

(From 5 MNRS cards)

Breeding population: Maritimes at least 300 pairs; [N.B. 300; N.S. under 30?; P.E.I. 0? (A.J.E.)]

Philadelphia Vireo
Viréo de Philadelphie

Vireo philadelphicus

This is a northern vireo, breeding in the boreal forest from northeast British Columbia to the Maritimes and perhaps southwest Newfoundland, and south just into the northernmost states—North Dakota to Maine. Like other vireos, these birds are more often heard than seen, and their songs are very similar to those of Red-eyed Vireos; many that are heard are assumed to be the more common species. Like the Red-eyed Vireo, this species is found mainly in broad-leafed trees, in pure or mixed woods, but it sings and forages more often in young stands and in the sub-canopy than does the larger vireo. In the Maritimes, Philadelphia Vireos were widespread in the northern highlands of New Brunswick, with scattered records farther south, where others were probably missed. Breeding has never been proved in Nova Scotia (where only 11 records were accepted in 1986–90), but 2 confirmed records in Prince Edward Island were the first in that province. The species is easily missed, so its absence in many areas meant partly that it was not expected; it would have been detected, if at all, only if looked for intensively. Over two-thirds of the records were of single or repeated (H,T) sightings or calling birds, but breeding was confirmed in 35 squares (22 per cent of 163 with Philadelphia Vireos), although only two nests were found.

Changes in habitat since the start of European settlement in the Maritimes seem unlikely to have had major effects on the distribution or numbers of Philadelphia Vireos here, and there are no data to suggest either increases or decreases. Reports of the species are more frequent in recent years as more people learn to distinguish it, but this increase is not more than in proportion to the numbers of observers, so "no change" is the most plausible assessment. In the future, more intensive forestry seems unlikely to work against this bird, which uses young poplar stands as well as older forests. However, as with other northern species, global warming may tend, over the long term, to reduce opportunities for Philadelphia Vireos here, at the southern limit of their range.

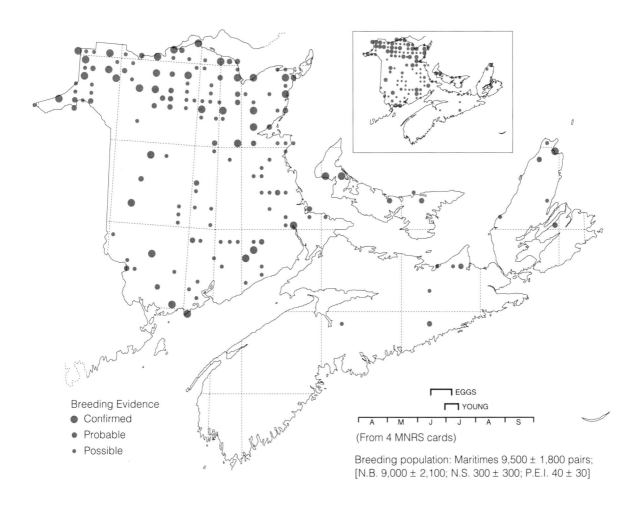

Breeding Evidence
- Confirmed
- Probable
- Possible

EGGS
YOUNG

A M J J A S

(From 4 MNRS cards)

Breeding population: Maritimes 9,500 ± 1,800 pairs;
[N.B. 9,000 ± 2,100; N.S. 300 ± 300; P.E.I. 40 ± 30]

157

Red-eyed Vireo
Viréo aux yeux rouges

Vireo olivaceus

The Red-eyed Vireo breeds across Canada from British Columbia and southern Mackenzie to Newfoundland, and south through the U.S.A. to Oregon, Texas, and Florida, with closely related forms in Central and South America. In the Maritimes, it was found in all regions, wherever broad-leafed trees dominate the canopy, whether as climax hardwoods (sugar maple, yellow birch, beech) or in succession (aspen, white birch) before conifers. This is the bird that sings all day in the shade trees in towns and cities and roadsides, as well as in the forests, so in 59 per cent of 1132 squares with the species the breeding evidence was of singing birds (H,T) only. Breeding was confirmed in 31 per cent of squares with the species, with nests seen in 42 squares (4 per cent). The song of the Philadelphia Vireo is very similar to that of the Red-eyed Vireo, and the song of Solitary Vireo may also be confused with them; some novices have even confused vireo songs with those of Robins. The Red-eyed Vireo is so common and widespread that such confusion should not have altered its representation on the Atlas maps.

More extensive coverage by broad-leafed trees in the primaeval forests of the Maritimes may have supported larger numbers of Red-eyed Vireos, but their distribution can hardly have been much broader than at present. They are almost everywhere now! The BBS showed a significant increase since 1966, apparently in part a recovery from earlier losses to DDT spraying in New Brunswick forests. Their present numbers make them our most numerous vireo, and one of the most numerous small passerines in broad-leafed woodlands. Although forest management in the Maritimes is, and presumably will continue to be, focused on fibre production from conifers, there seems no likelihood of broad-leafed trees being eliminated permanently over large areas. The tolerant Red-eyed Vireo makes use of scattered hardwood trees, including old poplars or birches remaining above growing conifer stands, so it is likely to find abundant opportunities to continue as a common bird in the Maritimes.

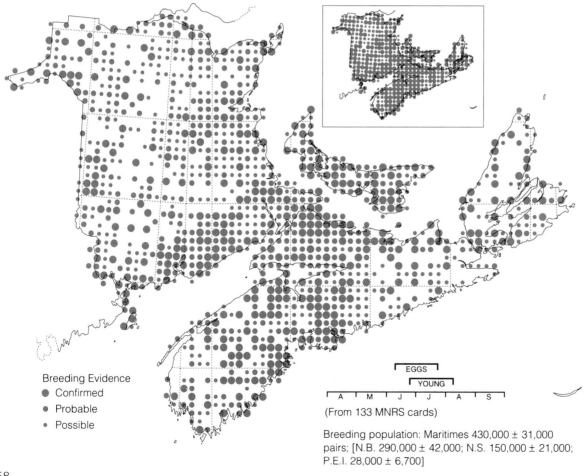

Breeding Evidence
● Confirmed
● Probable
• Possible

EGGS
YOUNG
A M J J A S

(From 133 MNRS cards)

Breeding population: Maritimes 430,000 ± 31,000 pairs; [N.B. 290,000 ± 42,000; N.S. 150,000 ± 21,000; P.E.I. 28,000 ± 6,700]

Tennessee Warbler
Paruline obscure

Vermivora peregrina

This is a northern warbler, breeding across the boreal forests of Canada from southeast Alaska to southern Labrador, south into the northernmost United States; it occurs in Tennessee only as a migrant. These birds were found all over the Maritimes, but with only sparse records, especially of confirmed breeding, in southwestern Nova Scotia, where it probably reaches the southern limit of its range, as it does in southwest Maine. The apparent gap in northwest Cape Breton Island may have been produced by late-season coverage or by recent clear-cutting, as this species was noted there frequently in 1960–65. The Tennessee is a forest bird, one of several warblers that increase enormously during spruce budworm outbreaks. It is associated with areas of small broad-leafed trees and shrubs within conifer forests, but it feeds on budworms mostly in coniferous trees. As it is the only "budworm-following" warbler with a loud and distinctive song, it is easily detected by those attuned to songs, but it was probably under-rep-

resented in records by other observers as it lacks bright colours and conspicuous patterns. Over 60 per cent (577 of 940 squares with the species) of the records were apparently of singing birds only (H or T). Nests were reported in only 11 squares, although breeding was confirmed in 305 squares (32 per cent).

Presumably Tennessee Warblers were in the Maritimes before European settlement, although they were missed by early naturalists. Budworm outbreaks in the extensive spruce/fir stands, which developed in New Brunswick after the original hardwood-dominated forest was opened up or burned, resulted in wide variations in abundance of this species, which continue to this day. The Atlas project spanned a relatively low period in the generally epidemic levels of budworm that have prevailed since forest spraying began in New Brunswick in 1952, but Tennessee Warblers were common as well as widespread. With Maritimes forest management dedicated to perpetuating conifers for fibre production, habitat for Tennessee Warblers should remain abundant in future.

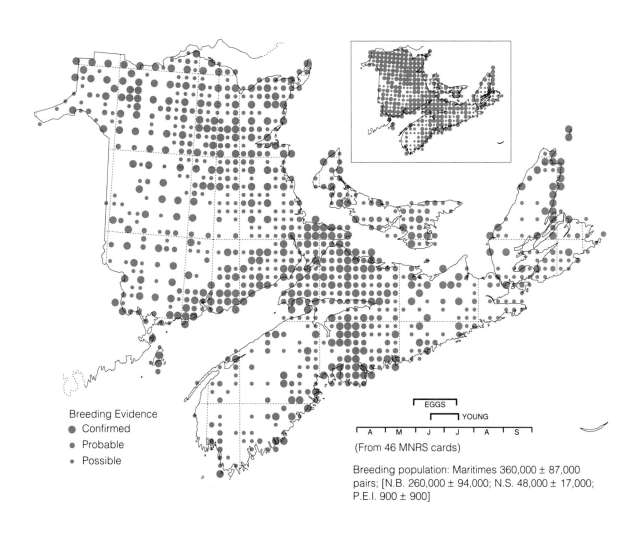

Breeding Evidence
- Confirmed
- Probable
- Possible

EGGS
YOUNG

A M J J A S

(From 46 MNRS cards)

Breeding population: Maritimes 360,000 ± 87,000 pairs; [N.B. 260,000 ± 94,000; N.S. 48,000 ± 17,000; P.E.I. 900 ± 900]

Nashville Warbler
Paruline à joues grises

Vermivora ruficapilla

The Nashville Warbler inhabits open conifer woodlands and brushy habitats in the eastern boreal forests of Canada, from the Maritimes to Saskatchewan, and south in the mountains to West Virginia, with a distinct subspecies in dry shrublands of the western intermountain country from British Columbia to California. In the Maritimes, these birds were found generally in New Brunswick, but often sparsely elsewhere. They were absent from much of northern Cape Breton Island and were scarce also in southwest Nova Scotia. The songs of Nashville and Tennessee Warblers, though superficially similar, are unlikely to have been confused during atlassing, when efforts were made to see singing birds for additional breeding evidence. Possibly the rather poorly defined habitat preferences of this species, apart from its use of larches, hindered specific searches in areas where it was missed on a first survey.

Atlas statistics were quite typical for warblers, with breeding confirmed in 31 per cent (261) of 837 squares with the species, but only one nest reported, among the fewest for widespread species in this family.

As Nashville Warblers are now near their northeast limit in our area, and as they frequent relatively open woodlands in preference to closed forests, the Maritimes probably presented less favourable conditions for this species in the colder climate of the pre-settlement era. Warming during the last century and forest cutting in the past 200 years may have allowed some increase over primaeval numbers, but there has been no change according to the BBS in recent decades. No obvious threats are evident in the future, either from land use here, pollution, or global warming. We can say little about the effects of deforestation in tropical America, but a species like the Nashville Warbler, which uses shrublands, is less likely to be threatened there than birds of closed forest habitats.

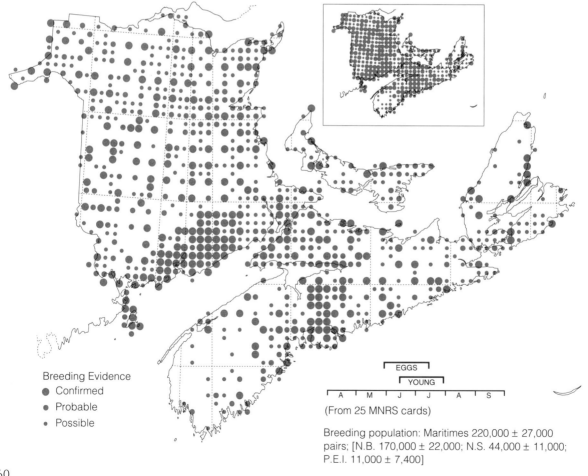

Breeding Evidence
- Confirmed
- Probable
- Possible

EGGS
YOUNG

A M J J A S

(From 25 MNRS cards)

Breeding population: Maritimes 220,000 ± 27,000 pairs; [N.B. 170,000 ± 22,000; N.S. 44,000 ± 11,000; P.E.I. 11,000 ± 7,400]

Northern Parula Warbler
Paruline à collier

Parula americana

The Parula is one warbler with which the Maritimes have been especially blessed. This is a bird of eastern forests, ranging from the Gulf Coast of the U.S.A. to southern Canada, scarcely penetrating the boreal region farther west. It is common only in the southeast (notably Louisiana, Florida, and South Carolina) and in the northeast (Maine, New Brunswick, and Nova Scotia). Elsewhere it is scarcer, as are the arboreal plants, Spanish moss in the southeast, beard-lichens in the northeast, that it uses as nest-sites. In the Maritimes, it was generally distributed, scarcer in northwest New Brunswick, western Prince Edward Island, and the Cape Breton Highlands, where the conifer habitat is mostly low or second-growth, as this is a bird of mature forests. The song is distinctive, and the bird, when seen, is hard to confuse with others, so Parulas should have been adequately represented. Atlas statistics were typical of most warblers, with breeding confirmed in 356 squares (35 per cent of 1022 squares

with Parulas), but nests were found in only 19 squares (2 per cent).

The mature, closed forests of the pre-settlement era in the Maritimes, with larger proportions of older trees than at present, provided favourable conditions for this species. Replacement of mature forests with second-growth was a deterioration, but the piecemeal cutting of forests in previous centuries should not have greatly reduced Parulas on more than a local scale. The widespread clear-cutting of recent decades may pose threats to species associated with mature, even if not necessarily old-growth, forests. The BBS suggested that Parula numbers increased substantially since 1966, as likely recovering after losses to spraying of DDT against spruce budworm in New Brunswick in 1952–67 as reflecting habitat changes. Projected trends in forest management, with shorter cutting rotations, are unlikely to let trees stand long enough to develop heavy growths of beard-lichens. However, the future of the Maritimes' economy depends on maintenance of large areas under conifer forest cover, so the Parula should find places here to its taste in the future.

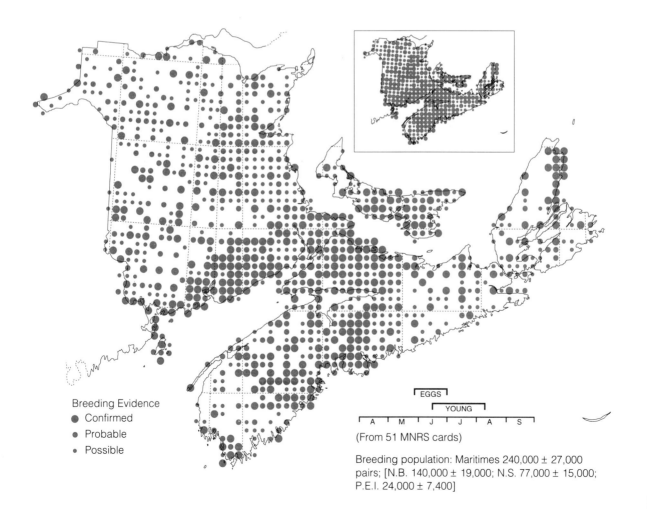

Breeding Evidence
- ● Confirmed
- ● Probable
- ● Possible

EGGS
YOUNG

A M J J A S

(From 51 MNRS cards)

Breeding population: Maritimes 240,000 ± 27,000 pairs; [N.B. 140,000 ± 19,000; N.S. 77,000 ± 15,000; P.E.I. 24,000 ± 7,400]

Yellow Warbler
Paruline jaune

Dendroica petechia

This is one of the most widely distributed warblers, breeding north almost to the tree-line across Canada and in Alaska and south to the West Indies and northern South America. It is also the most familiar warbler to many people, as it often nests in town and suburban gardens in ornamental shrubbery. Away from human settlements, it frequents shrubs and bushes in old-field situations and especially shrubbery along streams and marshes. Yellow Warblers thus are birds of edges and disturbed habitats and are much scarcer in the main forested parts of the Maritimes, particularly the uplands of northern New Brunswick. Their use of settled areas and the ease with which these all-yellow birds may be recognized means that in general they are over-represented in the Atlas relative to other warblers. Their nests, presumably most found being in gardens, were reported more often than those of any other warbler species, in 92 of 983 squares with Yellow Warblers, but by far the most frequent confirmation of breeding was adults carrying food (AY, in 330 squares).

Like other species that use habitats altered by humans, the Yellow Warbler has increased greatly since Europeans first settled in the Maritimes. It was probably a relatively scarce bird here 400 years ago, as it still is across the generally forested parts of Canada. Its present breeding population ranks it no higher than 8th in numbers among our warblers. Its former pre-eminence as *the* suburban warbler may now be rivalled by the American Redstart, which nests higher in the shrubs and trees and thus is less affected by disturbance from power mowers and household pets. There is no reason to view these as serious threats to the future of Yellow Warblers in the Maritimes, and neither land-use patterns nor climatic change seem likely to alter its status here.

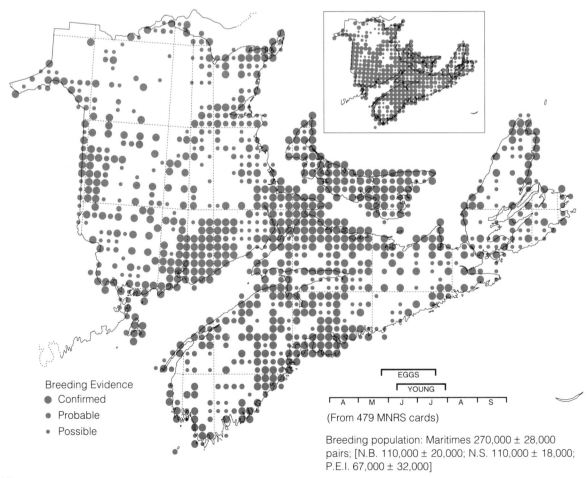

Breeding Evidence
- Confirmed
- Probable
- Possible

(From 479 MNRS cards)

Breeding population: Maritimes 270,000 ± 28,000 pairs; [N.B. 110,000 ± 20,000; N.S. 110,000 ± 18,000; P.E.I. 67,000 ± 32,000]

Chestnut-sided Warbler
Paruline à flancs marron

Dendroica pensylvanica

This bird is one of the many warblers of forest regions in northeastern America. It breeds mainly in southern Canada, west to Manitoba and locally farther, and south to New Jersey and Nebraska. The Maritimes includes the northeast limits of its range. In the Atlas project here, it was found less often in the highlands of north-central and northwest New Brunswick, and in Prince Edward Island; it was nearly absent on Cape Breton Island and in Guysborough Co., N.S. Although it regularly sings from trees, its nests are in low shrubs or raspberry canes, and it breeds in disturbed woodland and forest-edge situations rather than in closed forest. Adults carrying food (AY; in 29 per cent of 749 squares with this species) were by far the most frequent confirmation of breeding; nests were reported in only 12 squares. Although the song is at least average among warblers in volume and distinctiveness of pattern, it can be confused with other more common species in similar habitats, such as Yellow Warbler and American

Redstart, so Chestnut-sided Warblers may be somewhat under-represented. However, their use of roadsides and other disturbed areas means that they are encountered frequently, relative to some other warblers.

This species benefited greatly by the opening-up of the primaeval forests in the Maritimes, as it did elsewhere. It was formerly a scarce bird, which Audubon saw only once (!), but it has become a common if hardly abundant species over the past 150 years. The availability of early successional growth in cut-over areas has helped it in the past, and may do so in future. The BBS gives no indication of changed numbers here since 1966. The demands of the forest industry for conifer wood and the use of herbicides to suppress early growth of hardwood saplings and low shrubbery and to expedite regeneration of commercial tree species, do not favour this species in managed forests. The Chestnut-sided Warbler may become scarcer in future, but there should be ample shrubland in other edge situations for it to remain widely distributed here.

Ref. 49

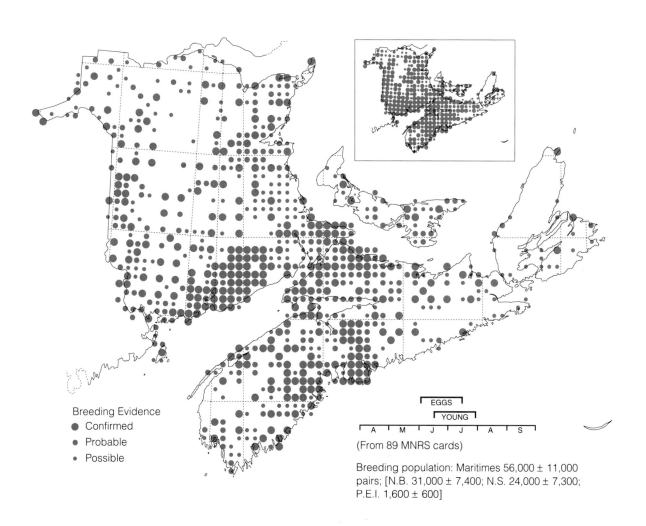

Breeding Evidence
- Confirmed
- Probable
- Possible

EGGS
YOUNG

A M J J A S

(From 89 MNRS cards)

Breeding population: Maritimes 56,000 ± 11,000 pairs; [N.B. 31,000 ± 7,400; N.S. 24,000 ± 7,300; P.E.I. 1,600 ± 600]

Magnolia Warbler
Paruline à tête cendrée

Dendroica magnolia

This is one of the most common warblers of the eastern boreal forest in North America, its density diminishing westward to northern British Columbia and southern Mackenzie and southward in mountains to Virginia. Magnolia Warblers were found throughout the Maritimes, in conifer and mixed forests of all stages from scattered spruces in old fields through to mature stands. They are most characteristic of younger and more open woodlands and of habitats with balsam fir rather than pure spruce stands. Among the warblers associated mainly with conifers, the Magnolia Warbler is familiar to more people than any except the Yellow-rumped, as it is neither secretive nor obscurely patterned, and it often sings and forages low in trees where it is easily seen. It was probably recorded roughly in proportion to its occurrence. Breeding was confirmed in 594 (53 per cent) of the 1127 squares with the species, although nests were noted in only 25 squares (2 per cent), both about average for common warblers in conifer habitats.

Changes in the forest cover of the Maritimes, caused by cutting and clearing since 1600, have somewhat increased the opportunities for Magnolia Warblers. As they frequent both mature and successional habitats, they probably were always common here. Their numbers seem not to change much in response to food availability during spruce budworm outbreaks, but they continue to make use of the habitat even with the changes that follow budworm damage to forests. The BBS showed no significant trends in Magnolia Warbler populations since 1966. The continued use of conifer woodland in the forest industry guarantees abundant habitat for Magnolia Warblers for generations to come. Concern over loss of woodland habitats in Latin America, where these and other boreal warblers winter, has not focused on this species as being under threat.

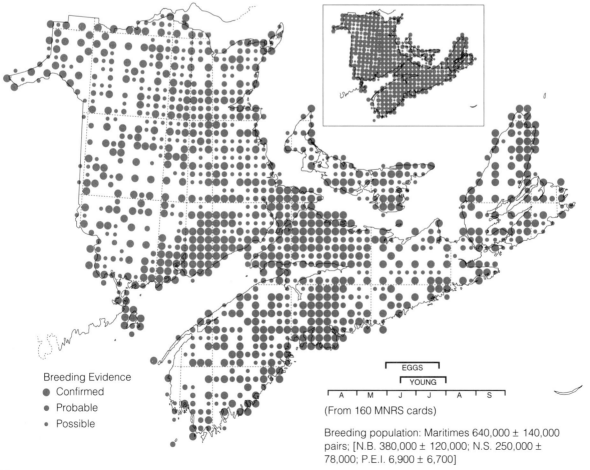

Breeding Evidence
- ● Confirmed
- ● Probable
- · Possible

EGGS
YOUNG

A M J J A S

(From 160 MNRS cards)

Breeding population: Maritimes 640,000 ± 140,000 pairs; [N.B. 380,000 ± 120,000; N.S. 250,000 ± 78,000; P.E.I. 6,900 ± 6,700]

Cape May Warbler
Paruline tigrée

Dendroica tigrina

This well-marked but little-seen warbler is found in summer in the northern conifer forests, ranging from the Maritimes across Ontario, and more sparsely west to Alberta and Mackenzie, but south only to Minnesota and Maine. It does not reach the subarctic fringes of the boreal forest and is absent from Newfoundland and Ungava. The Cape May is one of several warblers that attain high densities during spruce budworm outbreaks, but this bird is more usual in mature spruces than in balsam fir stands. The mapped range in the Maritimes is patchy, because this species is often missed. Its weak and undistinctive voice attracts little attention, and its secretive activity in the tip-tops of tall spruces is unfamiliar to many observers and thus undetected. The scarcity of records in the hardwood regions of the St. John River Valley and southwest Nova Scotia is likely real, reflecting a scarcity of suitable habitat; some other gaps mean only that the species was not detected, or was absent in the year of survey. Song and sight records (H,T) accounted for 71 per cent of the 422 squares with Cape May Warblers, and breeding was confirmed in only 94 squares (22 per cent), with only two reports of nests, which are usually high in trees.

The past history of this reclusive species cannot easily be reconstructed. The greater extent of broad-leafed forest before European settlement worked against it, but conifer stands more often attained maturity then. The uncontrolled forest fires that often accompanied early clearing and lumbering led, after intervals of 40–60 years, to regeneration of even-aged stands of spruce and fir in much of New Brunswick, setting the stage for spruce budworm outbreaks and Cape May Warbler incursions. The vast Miramichi burn of 1871 led, indirectly, to the budworm infestation of 1915–18 that allowed Philipp and Bowdish (1919) and others to find over 40 Cape May Warbler nests near Tabusintac. Estimates of its populations are biased by poor detectability and by local budworm numbers. The future for this species is similarly obscure, with lumbering based on fibre production from conifers encouraging suitable habitat, if it is allowed to remain long enough before it is cut. We may not be able to detect and measure future trends in its numbers, but we do not foresee its disappearance here.

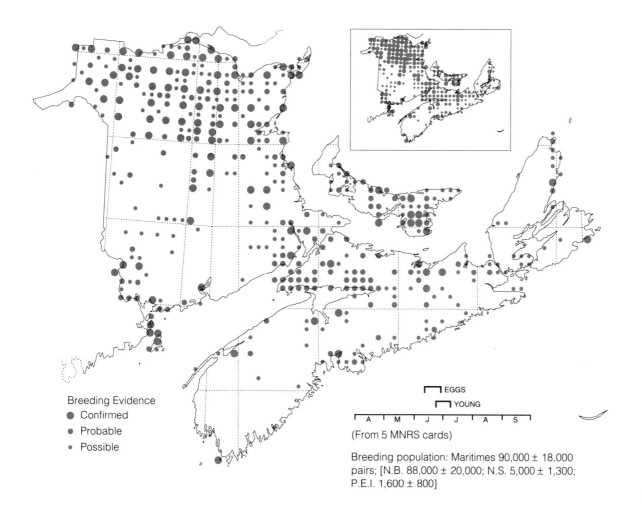

Breeding Evidence
● Confirmed
● Probable
● Possible

EGGS
YOUNG

A M J J A S

(From 5 MNRS cards)

Breeding population: Maritimes 90,000 ± 18,000 pairs; [N.B. 88,000 ± 20,000; N.S. 5,000 ± 1,300; P.E.I. 1,600 ± 800]

Black-throated Blue Warbler
Paruline bleue à gorge noire

Dendroica caerulescens

This is one of the most distinctive warblers, and also one of the most restricted in overall range. It breeds only in southeast Canada, from Nova Scotia to western Ontario, and in the northeastern U.S.A., from Minnesota to Pennsylvania, and in mountains to Georgia. The Black-throated Blue is a warbler of broad-leafed woodland and is nowhere abundant in the Maritimes. It was found generally but sparsely in New Brunswick and in Nova Scotia west of a line south from Truro. In eastern and northern Nova Scotia and in Prince Edward Island it was much scarcer, perhaps in response to fewer and less extensive broad-leafed stands. Its virtual absence in northern Cape Breton Island during the Atlas period was odd, as the species was noted regularly in summer there, in the Margaree Valley and near Baddeck, in 1954 and the 1960s. The Atlas records were mostly of singing birds, single (H) and repeated (T) detections accounting for records in

69 per cent of 400 squares with these birds. Breeding was confirmed in only 22 per cent of the squares with records, and only one nest was reported. As mature broad-leafed forests were more general in the Maritimes in 1600 than at present, Black-throated Blue Warblers might have been more common here in the past.

However, this is not a common warbler, although it is more regular farther south in New England where hardwood forests are much more general, so present scarcity here does not point to a decline from former numbers. There is simply no evidence for trends in overall populations here, in the short or long term. Large-scale lumbering here is focused on fibre production from conifers, so the medium-aged hardwood stands that harbour this species are under no immediate threat, except perhaps near the one mill in New Brunswick that uses hardwood for pulp. There seem to be few reasons to expect appreciable changes in numbers of these attractive birds in the years to come.

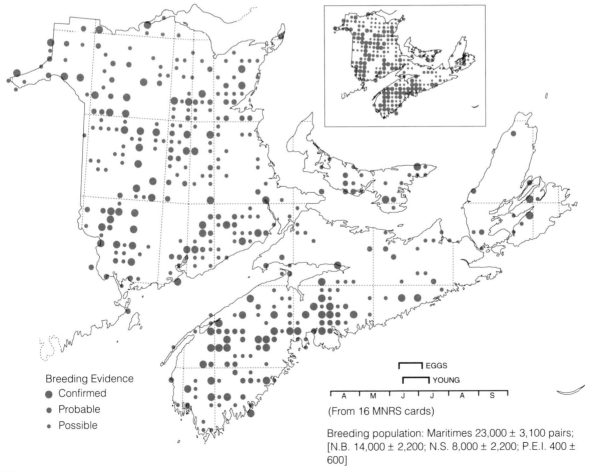

Breeding Evidence
● Confirmed
● Probable
· Possible

EGGS
YOUNG

| A | M | J | J | A | S |

(From 16 MNRS cards)

Breeding population: Maritimes 23,000 ± 3,100 pairs; [N.B. 14,000 ± 2,200; N.S. 8,000 ± 2,200; P.E.I. 400 ± 600]

Yellow-rumped Warbler
Paruline à croupion jaune

Dendroica coronata

The Yellow-rump is the earliest warbler to return here in spring and the last to leave in fall. A few even stay in the Maritimes for the winter, in places where they find bayberry (wax myrtle) fruits for food close to dense spruce cover. This species includes the western Audubon's race, which has a yellow throat patch, as well as the eastern white-throated form. These birds breed north to the arctic tree-line across Canada and Alaska, south to the northern states and in the mountains to northern Mexico and to Pennsylvania. In the Maritimes they were found throughout the forested areas, which occur in most squares. They are tolerant of open and disturbed woodlands as well as of continuous forest, provided coniferous trees are present. Because of their prolonged season here, more people are familiar with them than with many more common forest warblers. The yellow rump patch makes them easy to identify, fortunately, as their various songs are often confused with other species, especially later in the season. Most records of confirmed breeding were of adults carrying food (AY) or of fledged young (FL), in 33 per cent and 20 per cent, respectively, of 1140 squares with these birds. Their nests were seldom found, in only 30 squares.

This is one forest bird whose numbers may not have changed much since European settlement began. Increased secondary succession of conifers may well have balanced out losses caused by clearing, the species being quite flexible in its habitat requirements. These birds pass through in far larger numbers in spring and fall than are apparent in summer, as these generalists evidently breed in various forest habitats at relatively low densities compared to many other warbler species. Our estimates of breeding populations confirm that the species is one of our more numerous warblers, but certainly not the most abundant. As long as wood-fibre production from conifer forests remains a major industry in the Maritimes, suitable habitat for this species will not restrict its numbers, and no other threats are obvious.

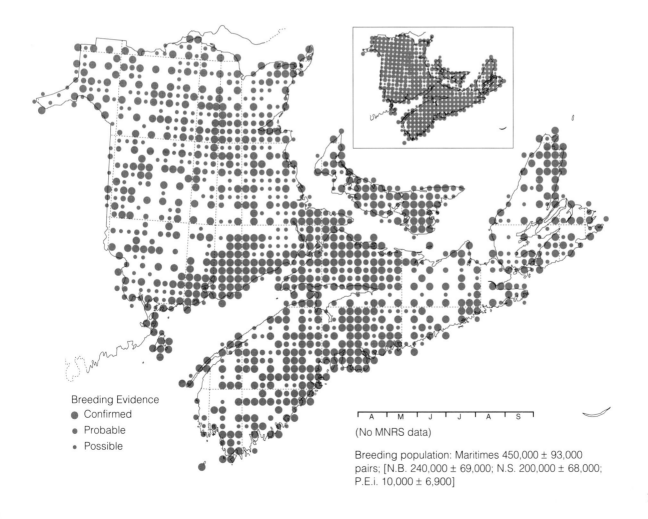

Breeding Evidence
- Confirmed
- Probable
- Possible

(No MNRS data)

Breeding population: Maritimes 450,000 ± 93,000 pairs; [N.B. 240,000 ± 69,000; N.S. 200,000 ± 68,000; P.E.i. 10,000 ± 6,900]

Black-throated Green Warbler
Paruline verte à gorge noire

Dendroica virens

This conspicuous warbler breeds in the eastern boreal and temperate mixed forests of North America,
with densities diminishing westward to northern
Alberta and south in the mountains to Georgia. The
Black-throated Green Warbler is most characteristic of
conifer forests, where it selects the richest stands available in a region, whether these feature hemlocks, balsam firs, or spruces, alone or with hardwoods admixed.
In the Maritimes, it was general throughout, the obvious
gaps in Cape Breton Highlands and western Prince
Edward Island being areas where conifers are mostly
stunted or second-growth. Both the appearance and
song of these birds are distinctive and easily learned,
making this one of the few forest warblers familiar to
every bird student. Few 10 x 10 km squares lacked it
completely, but in any one square its numbers were
tens rather than hundreds. Statistics for this bird were
quite typical of warblers, with breeding confirmed in 41

per cent of 966 squares with the species,
mostly by seeing adults carrying food
for young (AY, in 243 squares, 25 per
cent), and nests in only 15 squares.
Minimal detections (H,T) were obtained in
only 54 per cent of squares, indicating that the bird
is easily seen, allowing detection of other breeding
evidence, as well as easily identified.

The original mature mixed forests of the Maritimes
were suitable for Black-throated Green Warblers, but so
were the more broken fir and spruce forests that succeeded them. We cannot assume that European settlement markedly affected numbers of this species,
upward or downward. These birds increase in response
to spruce budworm outbreaks, but not as much as several other warblers characteristic of conifers. There has
been no population change since 1966, according to
the BBS. Breeding estimates confirm it as a common
but not abundant bird. Projected trends in use of forest
lands in the Maritimes give no obvious cause for concern, and this species should continue to brighten our
woods and our lives in future.

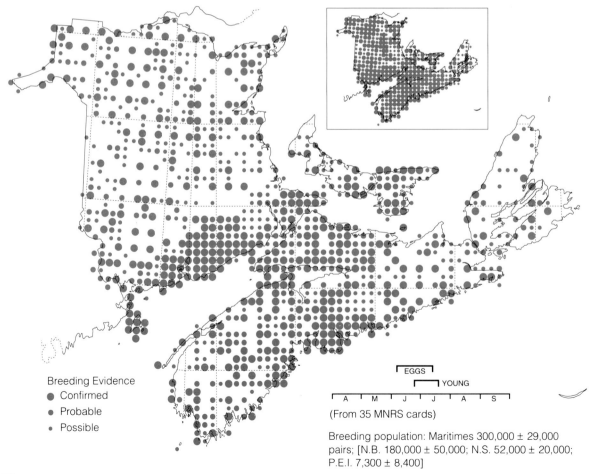

Breeding Evidence
- Confirmed
- Probable
- Possible

EGGS
YOUNG
A M J J A S
(From 35 MNRS cards)

Breeding population: Maritimes 300,000 ± 29,000
pairs; [N.B. 180,000 ± 50,000; N.S. 52,000 ± 20,000;
P.E.I. 7,300 ± 8,400]

Blackburnian Warbler
Paruline à gorge orangée

Dendroica fusca

The Blackburnian, with Cape May and Bay-breasted, is another of the thin-voiced warblers of the tree-tops in Canadian conifer forests. An eastern species, it breeds from the Gulf of St. Lawrence to the upper Great Lakes, with extensions west to central Alberta and south in mountain forests to Georgia. In the Maritimes, it was found widely, in almost every square in well-worked regions. Considering that its song is high-pitched and faint, and identified mainly by experienced observers, gaps on the map were probably caused as often by failure to detect it as by its absence. This species often increases during spruce budworm outbreaks, but no widespread increase in budworm occurred to account for these birds being so frequent. Blackburnian Warblers evidently are much less "uncommon" than suggested by earlier summaries. Atlas statistics were typical of other warblers, with breeding confirmed in 35 per cent of squares with these birds (289 of 819), although only one nest was reported, fewer than for most other common warblers.

The Blackburnian Warbler reaches highest densities in mature hemlock stands and generally frequents the most fertile conifer habitats available in a region. As those forest types would have been exploited early in the settlement of the Maritimes, these birds may now be less numerous here than they were 400 years ago. The BBS suggests an increase during the last 25 years, which seems more likely to be a response to the phasing-out of DDT in forest spraying than to increased availability of suitable habitat. With forest cutting on shorter rotational cycles, and subsequent planting of commercial woodlands to black spruce and jack pine—trees characteristic of poor habitats—the prospects for Blackburnian Warblers look rather unpromising. These birds survived the many changes in eastern forests during and since the glacial epoch, so they may be more adaptable to different forest types than we often assume. Deforestation in its wintering areas may be more of a threat than habitat changes in the Maritimes.

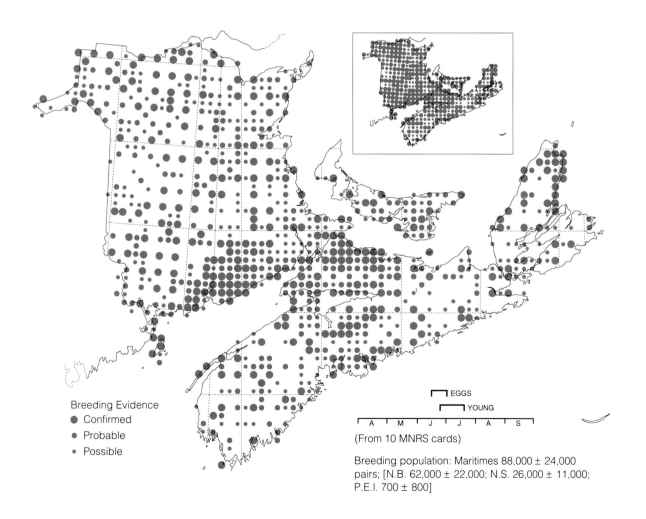

Breeding Evidence
- Confirmed
- Probable
- Possible

EGGS
YOUNG

| A | M | J | J | A | S |

(From 10 MNRS cards)

Breeding population: Maritimes 88,000 ± 24,000 pairs; [N.B. 62,000 ± 22,000; N.S. 26,000 ± 11,000; P.E.I. 700 ± 800]

Pine Warbler
Paruline des pins

Dendroica pinus

This warbler frequents only pine woods, including open stands and small groves, in breeding season. It thus occurs locally throughout the U.S.A. east of the prairies, and from southeast Manitoba to south Quebec. It was widespread across southern Maine, so the 17 records received by the Atlas in New Brunswick, the first near St. Stephen (FL) in 1987 and the easternmost near Minto, evidently represent an extension of the known range. The species is not very distinctive, in appearance or song, to observers unfamiliar with it, and it was probably missed in other suitable locations in south-west New Brunswick. Breeding was confirmed in 6 squares, including 3 with newly fledged young fed by adults (FL), and one nest was found.

Breeding by Pine Warblers in the Maritimes had not been demonstrated earlier, although the possibility of its doing so had been recognized long ago. Earlier breeding could easily have been missed, but it seems unlikely that the species bred here in substantial numbers in the past without having been detected at some time. Breeding numbers of Pine Warblers here may increase in future if global warming becomes general. Maturing of pine woods, planted more widely after 1945, also should increase availability of habitat for this species.

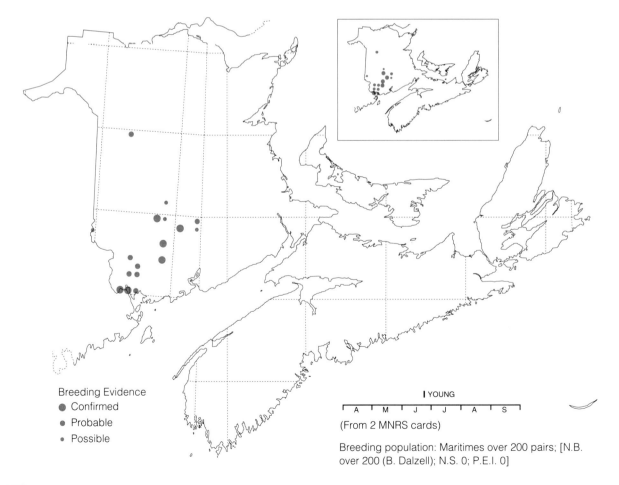

Breeding Evidence
- Confirmed
- Probable
- Possible

I YOUNG

| A | M | J | J | A | S |

(From 2 MNRS cards)

Breeding population: Maritimes over 200 pairs; [N.B. over 200 (B. Dalzell); N.S. 0; P.E.I. 0]

Palm Warbler
Paruline à couronne rousse

Dendroica palmarum

The Palm Warbler breeds across the boreal regions of Canada, from Newfoundland to the Rockies, and south into the northernmost states. It is found mainly in areas with muskegs and other bogs, where scattered low conifers are interspersed with open sedges or low shrub cover, but it also occurs, less regularly, in similar cover types in drier areas. Its distribution pattern in the Maritimes, as shown by the Atlas (with frequent occurrence on the Atlantic slope of Nova Scotia and through much of eastern and northeast New Brunswick, but sparser in southwest New Brunswick and eastern Prince Edward Island), was unique. This corresponds well with avoidance of the upland regions, which is plausible for a species associated with bogs. Earlier accounts called the Palm Warbler "fairly common" in Nova Scotia and "uncommon" in New Brunswick, which agree with the Atlas findings, but Godfrey (1954) found it commonly in Prince Edward Island in 1952. Many people are not familiar with Palm

Warblers except as migrants, and their breeding habitats are likely to have been neglected, being difficult to walk in and with little variety of birds. This is not a conspicuous or well-known bird, so we may assume that it was under-represented by our data. About half the records were of confirmed breeding, mostly by seeing adults carrying food (AY) or fledged young (FL) (in 28 per cent and 18 per cent, respectively, of 434 squares with Palm Warblers), so it was not a difficult species to observe where it was found. Nests were found in 9 squares.

There is no suggestion that this species' numbers have changed substantially in historic times, as its habitats have seldom been converted to other uses, and other pressures are not apparent. With its main habitats, bogs, being relatively limited both in extent and in productivity, Palm Warblers are unlikely to have been very abundant at any time. Our data mark it as one of our scarcer regular warblers. There are no obvious reasons to expect this to change, for better or for worse, in future.

Ref. 152

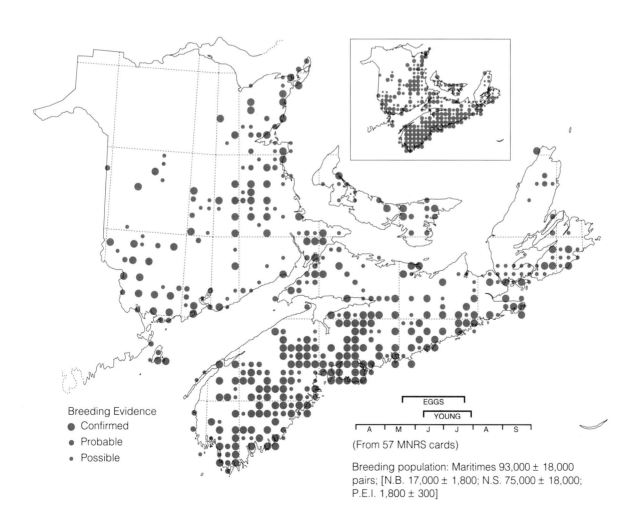

Breeding Evidence
- Confirmed
- Probable
- Possible

EGGS
YOUNG

A M J J A S

(From 57 MNRS cards)

Breeding population: Maritimes 93,000 ± 18,000 pairs; [N.B. 17,000 ± 1,800; N.S. 75,000 ± 18,000; P.E.I. 1,800 ± 300]

Bay-breasted Warbler
Paruline à poitrine baie

Dendroica castanea

The Bay-breasted is one of the less widespread warblers, breeding in a narrow band across the closed boreal forests from northeast British Columbia to western Newfoundland, and south just into the U.S.A. Although during migrations and while foraging it is often seen in mixed stands, this bird nests only in conifers, reaching highest densities in balsam fir forest infested with spruce budworm. Given the prevalence of this insect here, the mapped distribution of the species is an under-representation in many areas, though perhaps not in areas with much broadleafed forest. The song of the Bay-breasted Warbler is soft and undistinctive, and probably was often missed, especially from roadsides, by some observers. Nests were hardly ever seen, in 2 squares only, but adults carrying food (AY) respond to "spishing" readily so breeding was often confirmed (in 207, 32 per cent, of 652 squares with the species).

Broad-leafed trees probably formed a larger proportion of the Maritimes forests, but relatively more of the conifers present would have been mature, when Europeans first settled in the Maritimes. However, spruce budworm epidemics were probably less frequent, or less widespread, in the distant past than in the last century, and Bay-breasted Warblers may have been less common, but this is only speculation. They can hardly have been more common than during the budworm outbreaks of 1917–19 and the 1950s–60s in New Brunswick, when densities of over 4 singing males/ha were estimated on some small census plots (< 10 ha) in areas of high budworm infestation. The average density over a 10 x 10 km Atlas square would never exceed 1/ha, as suitable habitat is never continuous over such a large area. The under-representation of this species makes our estimates of present numbers minimal. In future, the trend to shorter forest-cutting cycles will reduce the area of relatively mature forest, and planting of black spruce and especially of jack pine, which are more resistant to budworm, instead of balsam fir will provide less attractive habitat. Bay-breasted Warblers will not disappear in the Maritimes, but their numbers will continue to fluctuate, and they are unlikely to be as common in the future as in the recent past.

Ref. 51

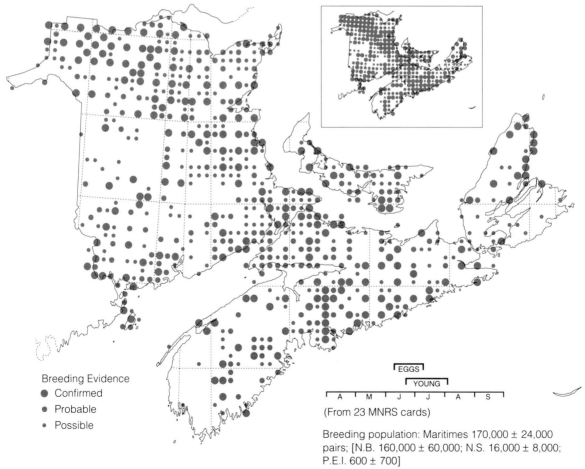

Breeding Evidence
- Confirmed
- Probable
- Possible

EGGS
YOUNG
A M J J A S

(From 23 MNRS cards)

Breeding population: Maritimes 170,000 ± 24,000 pairs; [N.B. 160,000 ± 60,000; N.S. 16,000 ± 8,000; P.E.I. 600 ± 700]

Blackpoll Warbler
Paruline rayée

Dendroica striata

This is among the northernmost of the warblers, breeding through the open boreal and subarctic forests of Canada and Alaska to the tree-line. It also extends southward along the Atlantic coasts where the cold waters of the Labrador and Nova Scotia currents simulate subarctic conditions. The occurrence of Blackpoll Warblers at high elevations, and locally at high densities, in northwest New Brunswick and on the Cape Breton Highlands represents the southern edge of the continental range, whereas the coastal birds around Grand Manan Island, the Cape Chignecto area, and the outer coasts of Nova Scotia, reflect the maritime influence. As in the main range, they breed here mainly in cool, damp spruce forests. Blackpolls are usually late migrants, despite a few appearing in early to mid-May, and in some years considerable numbers are still pass-

ing through in mid-June; some mapped records (H,T) in the southern Maritimes were probably of stragglers rather than breeding birds. The species is easily detected, despite having a faint and very high-pitched song. With a very few exceptions, all confirmations (FL or AY) were from coastal or northern highland areas. Only one nest was found in the 232 squares with Blackpoll Warblers during the Atlas period.

The status of the species here seems unlikely to have changed much in historic times, as these remote and inhospitable habitats were among those least affected by settlement. Despite the insatiable demands on conifer forests by the pulp and paper industry, the remoteness of the Blackpoll's habitat here, particularly in the coastal forest, and the generally small size of trees in it should ensure that they continue to enliven those sombre shades.

Ref. 30

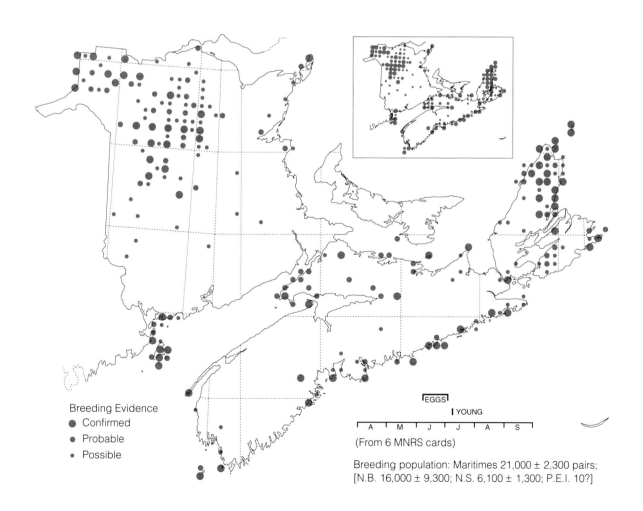

Breeding Evidence
● Confirmed
● Probable
● Possible

EGGS

┃ YOUNG

A   M   J   J   A   S

(From 6 MNRS cards)

Breeding population: Maritimes 21,000 ± 2,300 pairs;
[N.B. 16,000 ± 9,300; N.S. 6,100 ± 1,300; P.E.I. 10?]

Black-and-white Warbler
Paruline noir et blanc

Mniotilta varia

The Black-and-White is a warbler of broad-leafed and mixed woodlands in the eastern U.S.A. and southern Canada. It breeds from Texas and Georgia north to Newfoundland, and westward to northeast British Columbia and southern Mackenzie, but nowhere is it among the most common warblers. In the Maritimes, it was found throughout, in varying frequency, perhaps least common in the northwest and north-central highlands of New Brunswick where broad-leafed trees are scarcer. The breeding evidence in different regions suggested that it was detected and confirmed especially by more intensive efforts, and was detected only rarely in areas covered mainly by rapid "square-bashing." This is a little surprising, as the appearance and song of the Black-and-White Warbler are both distinctive, even if not very conspicuous. The Atlas statistics did not point to marked under representation, with breeding confirmed in 452 squares (44 per cent of 1019 with the species), and song detections (H,T) in 486 squares (48 per cent).

However, confirmed breeding was mostly concentrated in five regions, and the Atlas records elsewhere were much scarcer, so this must be among the more "difficult" of the more common warblers for Atlas work. The nests, placed on the ground among tree-roots, were found quite often, for a warbler, although only 13 were reported to the Atlas.

Descriptions of the forests of the Maritimes when the area was first discovered by Europeans suggest that habitat for Black-and-White Warblers was then as abundant as or more so than at present. These birds are not especially numerous now, and both human actions and ecological succession in their wake are working to reduce the hardwood component of our forests on which these warblers depend. Although silviculture is directed towards increasing conifers for fibre production, the effects of global warming may tend to push forest succession in the reverse direction. Diversifying forests, for multiple use and to minimize their vulnerability to insect pests, also should help this species. There are few direct and obvious threats to its well-being at present, in eastern Canada, but its wintering areas may be less secure.

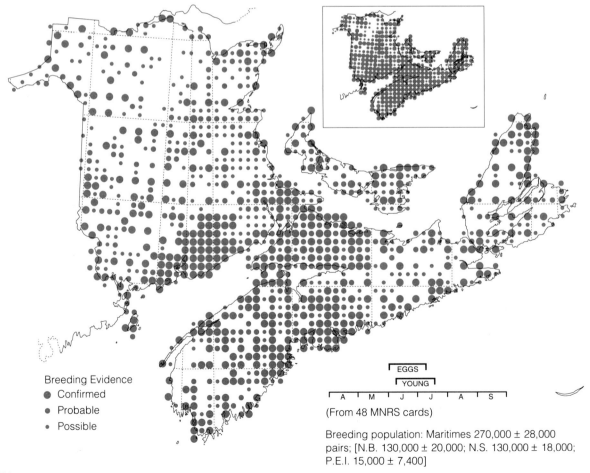

Breeding Evidence
● Confirmed
● Probable
● Possible

EGGS
YOUNG

| A | M | J | J | A | S |

(From 48 MNRS cards)

Breeding population: Maritimes 270,000 ± 28,000 pairs; [N.B. 130,000 ± 20,000; N.S. 130,000 ± 18,000; P.E.I. 15,000 ± 7,400]

American Redstart
Paruline flamboyante

Setophaga ruticilla

This active, striking little bird breeds across Canada, from the Yukon Territory to Labrador, and south through the U.S.A. to Oregon, Louisiana, and Georgia. It frequents both mixed and deciduous woodlands, with alders and other deciduous understory trees providing enough of the broad-leafed element even in mainly conifer areas. In addition to being an abundant bird of forest edges and understory, it also frequents open woodland, tree-rows, tall shrubland, and even shade trees and garden shrubbery in towns and cities. In contrast to the foliage-gleaning habits of most warblers, the Redstart is a warbler adapted to flycatching, darting out to seize insects flying past. Its nests are relatively low (2–5 m) in small trees, often in the first fork of the trunk. With its bright colours, constant activity, and accessible habitats, its detection almost everywhere is easily understood. In the Maritimes the Redstart was our most widely reported warbler, found in 1230 squares in all regions. Breeding was confirmed in 689 squares (56 per cent), a higher proportion than for most warblers, exceeded only by the equally familiar Yellowthroat (62 per cent), Yellow (58 per cent), and Yellow-rumped (57 per cent). Redstart nests, in 63 squares (5 per cent), were found more often than those of any warbler except Yellow Warblers (9 per cent).

Opening-up of the formerly more continuous forests of the Maritimes since European settlement benefited this species, which probably was already abundant and widespread here. Subsequent changes in land use left many opportunities for the tolerant Redstart, and even recent extensive clear-cutting and more intensive agriculture have had little effect on its population. The BBS suggests an overall increase of Redstarts in the Maritimes since 1966, which may be interpreted as a recovery from effects of earlier DDT spraying rather than a result of recently improved habitat conditions here. It is certainly one of our most abundant warblers. Future threats to its numbers are more likely to arise in wintering or migration areas than in the Maritimes.

Refs. 107, 108

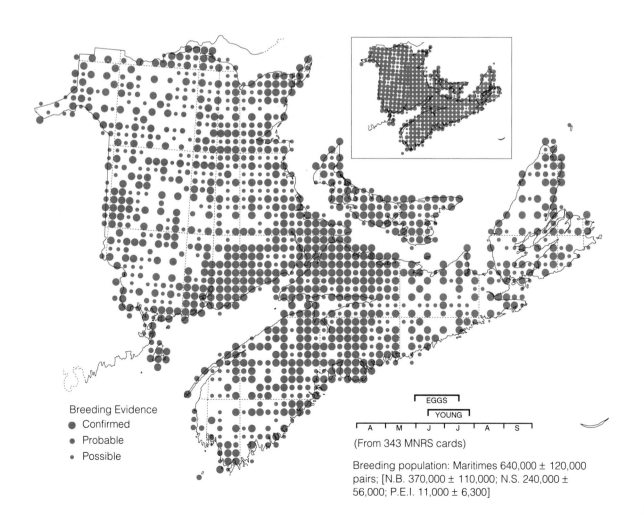

Breeding Evidence
- Confirmed
- Probable
- Possible

EGGS
YOUNG

A   M   J   J   A   S

(From 343 MNRS cards)

Breeding population: Maritimes 640,000 ± 120,000 pairs; [N.B. 370,000 ± 110,000; N.S. 240,000 ± 56,000; P.E.I. 11,000 ± 6,300]

Ovenbird
Paruline couronnée

Seiurus aurocapillus

The Ovenbird is a warbler of the eastern temperate and boreal forests, breeding from Oklahoma and Georgia to southern Canada, where it ranges from Newfoundland west to northeast British Columbia. This is a bird of broad-leafed and mixed woods without a dense understory. Its nests are roofed "oven-like" structures on the ground, and it forages on or near the forest floor. In the Maritimes, Ovenbirds were found almost everywhere, with obvious gaps mainly near the Atlantic coast of Nova Scotia, where broad-leafed trees are scarce, especially in Guysborough Co. and on Cape Breton Island. Although conifers dominate our forests, few stands of any great extent have less than 10 per cent broad-leafed trees, and this is often sufficient to provide a niche for Ovenbirds. This species is more often heard than seen, and its loud song is familiar to most people; such detections (H,T) occurred in 60 per cent of 1035 squares with Ovenbirds, and breeding was confirmed in another 321 squares (31 per cent). Nests were reported in 42 squares (4 per cent, about average for warblers).

The primaeval forests of the Maritimes, more mature than at present and with a larger hardwood component, included much good Ovenbird habitat. Probably this species remained common throughout, although recent summaries for Nova Scotia and Prince Edward Island termed it only "fairly common," vs. "common" in New Brunswick. The BBS suggests that Ovenbirds have increased in the Maritimes since 1966, which may be only a recovery from losses owing to forest spraying with DDT during 1952–67 in New Brunswick. Recent efforts in silviculture to suppress, with herbicides, hardwood regeneration among planted conifers affect mainly the earliest stages of forest succession. The Ovenbird does not occupy woodlands until a full canopy has developed and shade-intolerant undergrowth has declined, 20 years later. Forest management here is not likely to favour this species, but is unlikely to have much adverse effect. Ovenbirds should be around our woods for years to come.

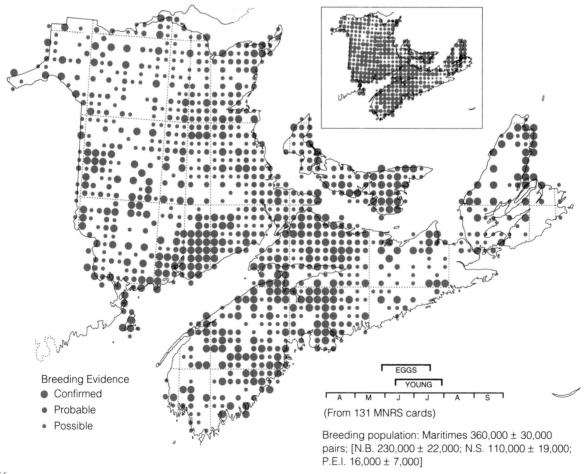

Breeding Evidence
● Confirmed
● Probable
● Possible

(From 131 MNRS cards)

Breeding population: Maritimes 360,000 ± 30,000 pairs; [N.B. 230,000 ± 22,000; N.S. 110,000 ± 19,000; P.E.I. 16,000 ± 7,000]

Northern Waterthrush
Paruline des ruisseaux

Seiurus noveboracensis

This large warbler breeds across Canada except west of the British Columbia Coast Mountains, and from near the arctic tree-line south into the northernmost United States. It is most often found in damp mixed woodlands, including alder and cedar swamps and along streams and rivers, where it walks on the ground and along fallen logs by the water. As access to these habitats is often difficult, and the birds are not conspicuous or confiding, only the loud song betraying their presence, the species was probably somewhat under-represented by the Atlas data. In the Maritimes, Northern Waterthrushes were widely distributed, but the Atlas record in Nova Scotia was strangely patchy, even in well-worked regions, suggesting that some observers were not familiar with the species. Its scarcity in southwest Nova Scotia and virtual absence from most of Prince Edward Island, where habitat is present although breeding was not known before the Atlas period, may only reflect its more northern distribution, but this cannot explain all the other gaps. Confirmation of breeding, in 20 per cent of 649 squares (the lowest proportion for any warbler), most often involved adults carrying food (AY), or fledged young (FL), but nearly half the records (310 squares) were of singing birds only (H).

The historical record gives no easy clues to likely changes in numbers or range of the Northern Waterthrush in the Maritimes since the start of European settlement. Its poorly drained habitats would not have been likely candidates either for clearing as farmland or as timber production sites; probably many of them have changed rather little. The BBS indicates no sustained or significant change in the Maritimes since 1966. The future outlook for the species includes no obvious threats here, as global warming seems unlikely to restrict habitats for waterthrushes.

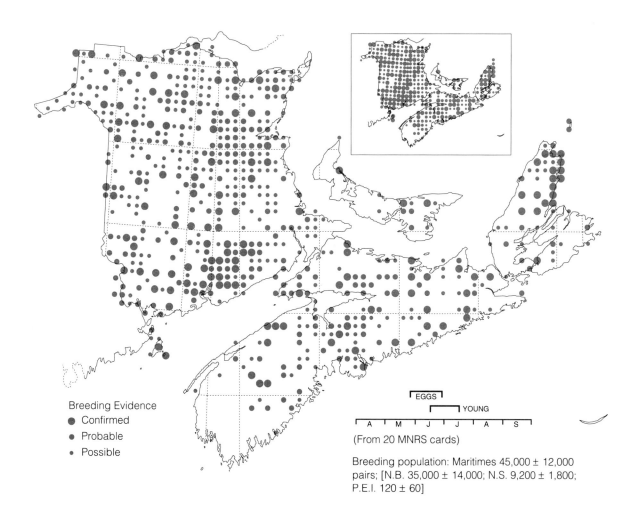

Breeding Evidence
- Confirmed
- Probable
- Possible

EGGS
YOUNG
A M J J A S
(From 20 MNRS cards)

Breeding population: Maritimes 45,000 ± 12,000 pairs; [N.B. 35,000 ± 14,000; N.S. 9,200 ± 1,800; P.E.I. 120 ± 60]

Mourning Warbler
Paruline triste

Oporornis philadelphia

Like all our warblers, this species is restricted to North America, where it breeds in boreal and cooler temperate regions west to the Rockies and south in the Appalachians to Virginia. The closely related MacGillivray's Warbler replaces it in and beyond the Rocky Mountains. In the Maritimes, they were widespread, but conspicuously absent from southwest Nova Scotia, the coastal lowlands of northern Nova Scotia and western Prince Edward Island, and a broad band across the interior lowlands of New Brunswick, a pattern similar to that of Lincoln's Sparrow. The habitat of Mourning Warblers is dense deciduous shrubbery, in openings within forest or along woodland edges; shrubbery and brush piles within clear-cut areas are regularly used. The birds are confirmed skulkers, but their loud distinctive songs betray their presence; once detected, they were readily confirmed (in 210 squares, 36 per cent of 589 with the species). Notwithstanding their dis-

tinctive voice and appearance, Mourning Warblers are unfamiliar to many birders, and our data may under-estimate their occurrence. They are late migrants in spring, with few appearing before the last week of May.

Clearing and fragmentation of the formerly continuous forests of the Maritimes increased the habitats suitable for Mourning Warblers in the past. Even within the last 30 years, these birds seem to have become more frequent in the southern part of our area, perhaps in response to salvage cutting after budworm damage to formerly continuous forests, though this is still one of our less common warblers. It is unlikely to lack habitat either in the Maritimes or in its tropical wintering grounds. Herbicide suppression of hardwood shrubbery in replanted clear-cuts may limit its occurrence in intensively managed forest lands here, but there is no shortage of neglected cut-over areas in the Maritimes. Neither global warming nor acid rain are likely to affect the successional habitats it uses throughout its range.

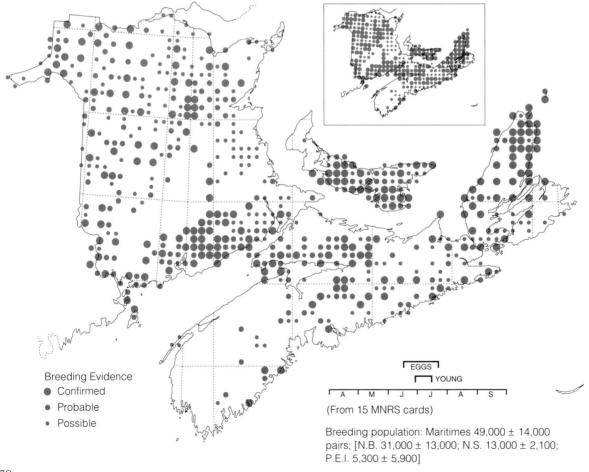

Breeding Evidence
- ● Confirmed
- ● Probable
- · Possible

EGGS
YOUNG
A M J J A S
(From 15 MNRS cards)

Breeding population: Maritimes 49,000 ± 14,000 pairs; [N.B. 31,000 ± 13,000; N.S. 13,000 ± 2,100; P.E.I. 5,300 ± 5,900]

Common Yellowthroat
Paruline masquée

Geothlypis trichas

This bird of shrubby habitats is one of our most familiar warblers. It breeds across Canada from southeast Alaska to Newfoundland, and south through the U.S.A. to Florida and Mexico. In the Maritimes, the most widespread warbler except American Redstart, Yellowthroats were found wherever suitable brushy sites occur — along roadsides, field edges, forest cut-overs, around marshes and along streams, in forested regions as well as farming areas. With its distinctive appearance and easily recognized song, and its use of relatively accessible habitats, this also was one of the best represented species in the warbler family. Adults carrying food (AY) were readily detected, giving Common Yellowthroats the highest proportion of confirmed breeding records (62 per cent) of any warbler, although nests (NY,NE,ON) were detected in only 39 (3 per cent) of 1212 squares.

Yellowthroats were much scarcer before Europeans came to the Maritimes, as the open areas whose edges they frequent were fewer and less extensive. Subsequent clearing of forests gave them the opportunity to expand. Probably Yellowthroats were originally associated only with marsh-edge shrubs, as they still are in the west, but from Ontario to the Atlantic the eastern subspecies also frequents upland shrubbery of many kinds. Although this adaptation is correlated with the duration of white settlement, it is most unlikely that the eastern subspecies to which it is restricted evolved this recently. Yellowthroats have benefited by human alterations of the natural habitats, in the Maritimes and elsewhere. They are one of the most numerous warblers here, essentially unchanged in numbers since 1966, according to the BBS. With their versatile use of altered habitats, there seems good reason to assume that these birds will continue to thrive in our region, as global warming should pose no threats to them.

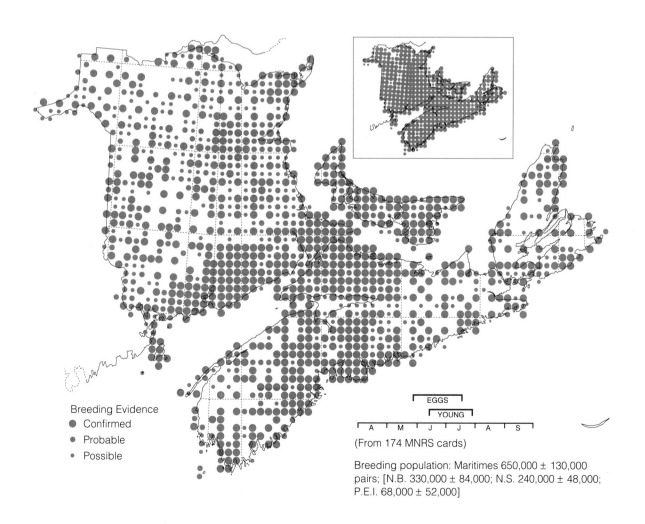

Breeding Evidence
● Confirmed
● Probable
· Possible

EGGS
YOUNG

| A | M | J | J | A | S |

(From 174 MNRS cards)

Breeding population: Maritimes 650,000 ± 130,000 pairs; [N.B. 330,000 ± 84,000; N.S. 240,000 ± 48,000; P.E.I. 68,000 ± 52,000]

Wilson's Warbler
Paruline à calotte noire

Wilsonia pusilla

This is a northern species, breeding across the boreal forests from Newfoundland and Labrador west to Alaska, and south through the mountains to California and New Mexico, but not extending south of central Ontario, north Maine, and the Maritimes in the east. The range of Wilson's Warbler in our area included most of New Brunswick except the hardwood regions of the lower St. John valley, but it thinned out in southeast New Brunswick. In Nova Scotia groups of confirmed breeding records were obtained in several widely spaced areas, but the species was not at all general there, and it was scarcely found in Prince Edward Island. Southern New Brunswick and Nova Scotia records were mostly in cooler areas near the Fundy and Atlantic coasts, but enough, including some of confirmed breeding, were in warmer areas to weaken this tentative pattern. The Wilson's Warbler song is not very distinctive, and the bird, except for the male's black cap, would be correctly described as a "yellow war-bler," so it was probably under-represented in the Atlas records. Among 409 squares with the species, breeding was confirmed in 102 (25 per cent, well below the average for war-blers), and 64 per cent of squares had only singing birds (H,T). This adds up to a quite incomplete understanding of the species' status in much of our area here.

As a bird of shrubland and early forest succession, Wilson's Warbler seems likely to have been scarcer in former times when more of the Maritimes was clothed in old-growth mixed forest. Subsequent cutting and clearing provided many opportunities for this bird, but its present patchy distribution suggests that availability of habitat was not its main limitation here. It is more regular and numerous in northern New Brunswick than elsewhere in the Maritimes, which is at the southern edge of its range. The species should persist in future in successional habitats provided by forest harvesting in New Brunswick, but no confident predictions can be made for the other provinces, and global warming will act against this as well as other northern species.

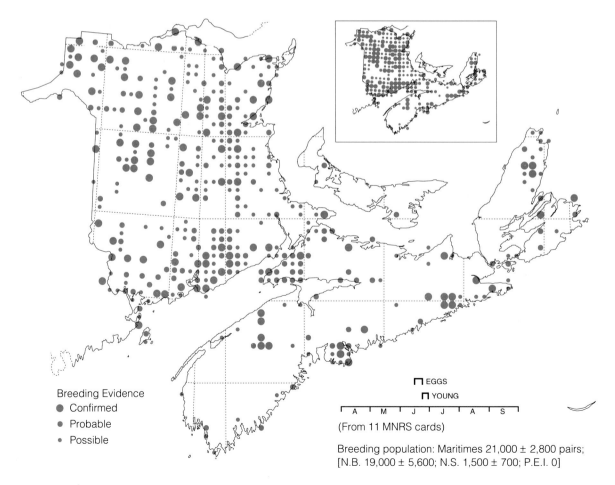

Breeding Evidence
- ● Confirmed
- ● Probable
- · Possible

⊓ EGGS
⊓ YOUNG

A   M   J   J   A   S

(From 11 MNRS cards)

Breeding population: Maritimes 21,000 ± 2,800 pairs; [N.B. 19,000 ± 5,600; N.S. 1,500 ± 700; P.E.I. 0]

Canada Warbler
Paruline du Canada

Wilsonia canadensis

This is one of the scarcer warblers in most areas. Its breeding range is mainly in the eastern mixed forests from Michigan and New Jersey to the Maritimes and central Ontario, with extensions west through the southern boreal region to northeast British Columbia, and south in the mountains to Georgia. The Maritimes, with Anticosti Island, Quebec, form the northeast limit of the range. Canada Warblers were found only sparsely on Cape Breton Island and in extreme southwest Nova Scotia, but they were widespread elsewhere in our area. They are less predictable from habitat than most warblers; they are usually found in dense understory vegetation of mature to mid-aged mixed forests, most closely associated with broadleafed trees and shrubs, but with conifers usually present too. This bird is less familiar to most people than other warblers, but its plumage pattern and song are quite distinctive, and it is not a difficult bird to observe once detected. With breeding confirmed in 32 per cent of squares with the species (249 of 780), and with sub-

stantiation mainly by song (H,T) in only 53 per cent of squares, these birds were represented as well as were other, more common warblers.

There is nothing in the historical record to suggest that Canada Warblers increased or decreased markedly since the first Europeans came to the Maritimes. Our difficulty in defining explicitly its habitat niche makes inference about its past status nearly impossible. Probably these birds were always one of the less common warblers in our forests, as they are elsewhere in their range, and they are still relatively scarce and inconspicuous. The BBS indicated no significant change since 1966. No obvious threats to their future well-being have been identified. Management practices aimed at reducing the broadleafed component in our forests seem likely to have a negative effect on Canada Warbler numbers, if pursued intensively over wide areas. Changes predicted in consequence of global warming would work in the opposite direction. Thus, we foresee that this species should persist in the Maritimes, but this is more of a hunch than a firmly based prediction.

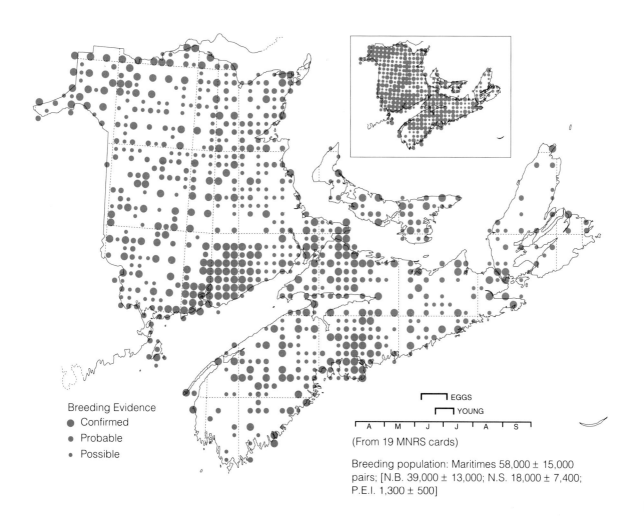

Breeding Evidence
- Confirmed
- Probable
- Possible

EGGS
YOUNG

A   M   J   J   A   S
(From 19 MNRS cards)

Breeding population: Maritimes 58,000 ± 15,000 pairs; [N.B. 39,000 ± 13,000; N.S. 18,000 ± 7,400; P.E.I. 1,300 ± 500]

Scarlet Tanager
Tangara écarlate

Piranga olivacea

This colourful bird is the only northeastern representative of a tropical subfamily with about 200 species in South America. It breeds in southern Canada from New Brunswick to Manitoba, and through the eastern U.S.A. south to Oklahoma and Georgia, its range thus coinciding with the eastern broad-leafed forest biome. Scarlet Tanagers were widespread but rather scarce in New Brunswick, their absence in much of the northern highlands, the Fundy coastal belt, and the eastern lowlands reflecting scarcity of extensive mature hardwoods there. The mapped range in Nova Scotia is very patchy, not paralleling closely the relative intensity of Atlas field work, but its near-absence on the Atlantic slope and Cape Breton Island is probably real. In northwestern Nova Scotia, where coverage was intensive, it was surprising to find so few records, as several tanagers were noted there in the 1970s. Single (H) or repeated (T) sightings accounted for 66 per cent of 216 squares with records of Scarlet Tanagers. Breeding was confirmed only in 34 squares (16 per cent), with only one nest reported.

The greater prevalence of mature trees and broad-leafed species in the forests before European settlement began around 1600 may have favoured this species, but the cooler climate associated with the Little Ice Age before the mid-1800s would have worked against it. As the Scarlet Tanager is so conspicuous that it would seldom have been missed where present, it probably was always a marginal species in the Maritimes. The more frequent records in Nova Scotia in recent decades parallel the increase in field effort and reporting, and may not indicate a real increase there. The continuing regrowth of forests in Prince Edward Island perhaps may see this species establish itself there soon. Any tendency towards forests of a more temperate nature, caused by global warming, would favour Scarlet Tanagers and other southern species, but efforts to increase production of conifers for fibre production exert pressure in the opposite direction. We are unlikely to see any major change in total numbers of tanagers here in the near future.

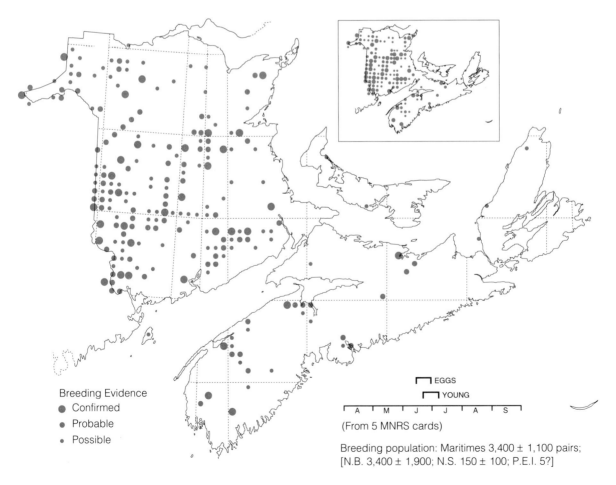

Breeding Evidence
- Confirmed
- Probable
- Possible

EGGS
YOUNG

| A | M | J | J | A | S |

(From 5 MNRS cards)

Breeding population: Maritimes 3,400 ± 1,100 pairs;
[N.B. 3,400 ± 1,900; N.S. 150 ± 100; P.E.I. 5?]

Northern Cardinal
Cardinal rouge

Cardinalis cardinalis

The Cardinal is an American bird, and a permanent resident from California, Texas, Wisconsin, southern parts of Ontario, Quebec, and Maine, southward to Central America, the Gulf Coast, and Florida. The recently established population in the Maritimes was restricted to a few suburban areas in the southwest of New Brunswick and Nova Scotia. As elsewhere, our Cardinals frequent shrubbery, thickets, and wood-edges, close to the garden feeding stations on which they depend for winter survival, so these conspicuous birds are unlikely to have been missed, where present. Although recent winter sightings at feeders have been more widespread, the Atlas received records from only 18 squares, with breeding confirmed in 8 of these. Fledged young, usually brought to feeders in late summer, were the usual evidence, and only one nest was reported.

Cardinals have been extending their range to the north and northeast for decades, as winter feeding opportunities increased survival in these marginal populations. They reached southern Ontario as nesting birds by 1901, but breeding in Maine was not detected until 1969. The Maine atlas (1978–83) had no confirmed records in the sparsely settled areas east and north of Bangor. A few pairs have bred in the Maritimes, in Yarmouth and Pubnico, N.S., since 1975, and in Saint John, N.B., since 1980, but not yet in Prince Edward Island. Confirmed records in the Atlas extended little farther north and east than these original stations. These areas are 150–300 km from the continuous range in south-central Maine, and Cardinals may not yet have a stable population in our region, as the numbers in any area must be very small. Their continued presence depends on enough birds surviving each winter at feeders to maintain local stocks. Such small populations are vulnerable to extinction, but designation as "endangered" or "threatened" would be inappropriate, as they exist here solely because of human actions. Cardinals are here because people want birds around in winter.

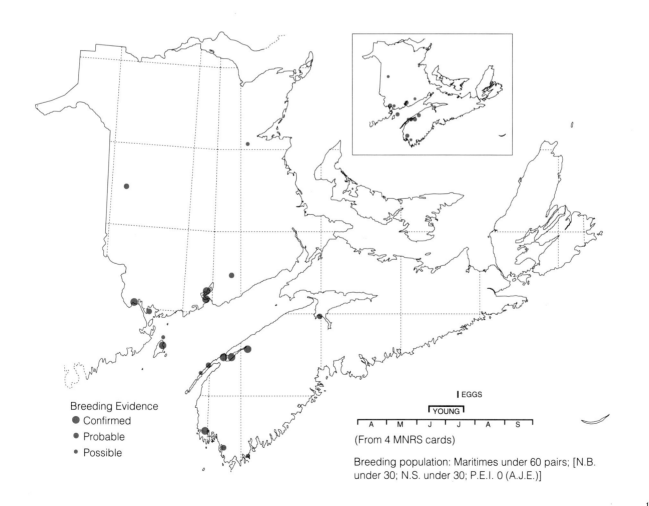

Breeding Evidence
● Confirmed
● Probable
● Possible

EGGS
YOUNG

A M J J A S

(From 4 MNRS cards)

Breeding population: Maritimes under 60 pairs; [N.B. under 30; N.S. under 30; P.E.I. 0 (A.J.E.)]

Rose-breasted Grosbeak
Cardinal à poitrine rose

Pheucticus ludovicianus

The Rose-breasted Grosbeak is a bird of mixed and broad-leafed woodland. It ranges through the temperate forests east of the prairies from Georgia and Kansas to southern Canada and across the southern boreal region from Nova Scotia to north-east British Columbia. In the Maritimes, it was previously described as uncommon or rare, but in the Atlas period it was found in all regions, though evidently scarcer in Nova Scotia on the Atlantic slope and on Cape Breton Island. Its song resembles that of the ubiquitous Robin, which may have led to its being passed over during many earlier surveys. Over half of the Atlas reports (467 of 901 squares) were of single (H) or repeated (T) records, usually of singing birds. Nests were found in only 30 squares (3 per cent), breeding being most often confirmed, as with most species, by sightings of adults carrying food (AY, 141 squares) or of newly fledged young (FL, 121 squares).

The forests of the Maritimes before European settlement were more mature and with a larger component of broad-leafed trees than in the early 20th century, when continuous conifer forests were widespread. This suggests that Rose-breasted Grosbeaks may have been more common in the distant past compared to recent times. With the break-up of the solid conifer forests by spruce budworm damage and widespread clear-cut logging since 1970, these birds became much more general in some areas, including the New Brunswick – Nova Scotia border region, where only one record had been obtained by extensive earlier work (Boyer 1972). The present abundance thus may be very recent in some areas. The BBS indicated a significant increase in 1966–78, because people at first mistook their songs for those of Robins. Similar trends were seen in Quebec and Ontario, but the trend has levelled out since then. Although the forest industry will continue to depend on softwoods for fibre production, it seems likely that the continuous conifer cover of the past, which proved so vulnerable to budworm outbreaks, will not be encouraged. More varied forests would include adequate habitat for the Rose-breasted Grosbeak, which is nowhere really abundant.

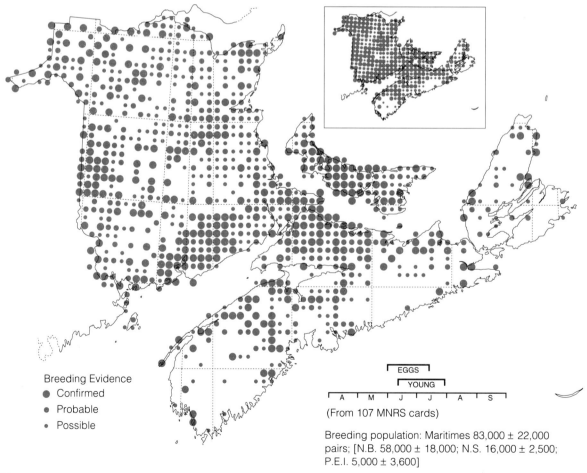

Breeding Evidence
- Confirmed
- Probable
- Possible

EGGS

YOUNG

A M J J A S

(From 107 MNRS cards)

Breeding population: Maritimes 83,000 ± 22,000 pairs; [N.B. 58,000 ± 18,000; N.S. 16,000 ± 2,500; P.E.I. 5,000 ± 3,600]

Indigo Bunting
Passerin indigo

Passerina cyanea

The Indigo Bunting breeds through much of the eastern temperate broad-leafed forest region, from southern Manitoba to New Brunswick and south to Texas and Florida. This is a bird of forest-edges, thickets, and shrubbery rather than woodland. Most Maritimes records, both before and during the Atlas project, have been in the southwest third of New Brunswick, with a few records in the Miramichi region and in southwest Nova Scotia. The latter included the first confirmed breeding in Nova Scotia (FL, 1990). Given the frequency with which this species turns up in April, in Nova Scotia even more than New Brunswick, many records beyond the main range are more likely to be lingering vagrants than pioneering breeders; however, one male Indigo Bunting was found singing in the same spot north of Chatham, N.B. (the northernmost record mapped), both in 1989 and 1990.

With only six confirmed breeding records, generalization on types of evidence would be premature; 31 of 43 squares with the species had sightings (H,T) only. Only one nest, in St. Stephen, N.B., (1988) was found during the Atlas period, and only three nests had been found previously (all in New Brunswick).

Indigo Buntings were not found breeding in New Brunswick until 1977. As they bred regularly in Maine a century ago, their presence in our area evidently is recent rather than delayed detection of a scarce species. The opening-up of the forests over the past 400 years created more suitable edge habitat than existed here in pre-settlement times, and this, with some warming of the climate, may explain the species' spread into our area. Any further tendency towards climatic warming is likely to improve the opportunities for Indigo Buntings in our region. They are scarcely established here at present.

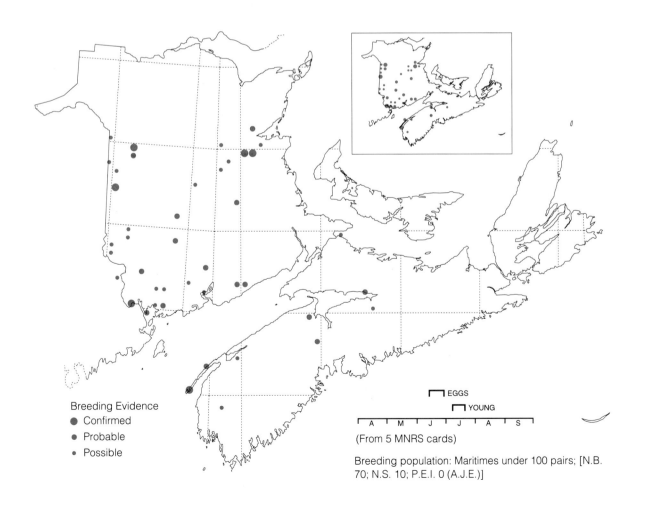

Breeding Evidence
● Confirmed
● Probable
● Possible

EGGS
YOUNG

A M J J A S
(From 5 MNRS cards)

Breeding population: Maritimes under 100 pairs; [N.B. 70; N.S. 10; P.E.I. 0 (A.J.E.)]

Chipping Sparrow
Bruant familier

Spizella passerina

The Chipping Sparrow breeds across Canada well south of the tree-line from Alaska to Newfoundland, and south through the U.S.A. to Florida and Nicaragua. In the east it is best known as a bird of edges and gardens, but it is common also in open woodlands, especially in western Canada where White-throated Sparrows are scarce or lacking. In the Maritimes, Chipping Sparrows were found in most areas, the most obvious gaps being in southwest Nova Scotia and on the Cape Breton Highlands, the latter perhaps verging on the subarctic shrubland habitat where Tree Sparrows replace this species farther north. Chippies were probably detected in most squares where they occur, as they frequent accessible areas and are not shy. Their song may be confused, at a distance or unthinkingly, with the less incisive trill of the Junco;

the latter overlaps in habitat but is more of a woodland bird. The Atlas statistics were quite typical among the sparrows, with breeding confirmed in 608 squares (61 per cent of 993 with the species), most often by seeing adults carrying food (AY, in 269 squares) or flying young (FL, in 247 squares). Nests (NY,NE,ON) were seen in 65 squares (7 per cent).

The closed forests that clothed the Maritimes before European settlement presumably contained enough openings to accommodate some Chipping Sparrows. The opening-up of the forests since then provided more habitats for this species, whose numbers increased subsequently. The BBS showed no change since 1966, which supports the impression that the vast increase in clear-cut forest areas has provided few attractive opportunities for Chipping Sparrows. Projected trends suggest that this species will have little difficulty in maintaining itself in the Maritimes for the foreseeable future.

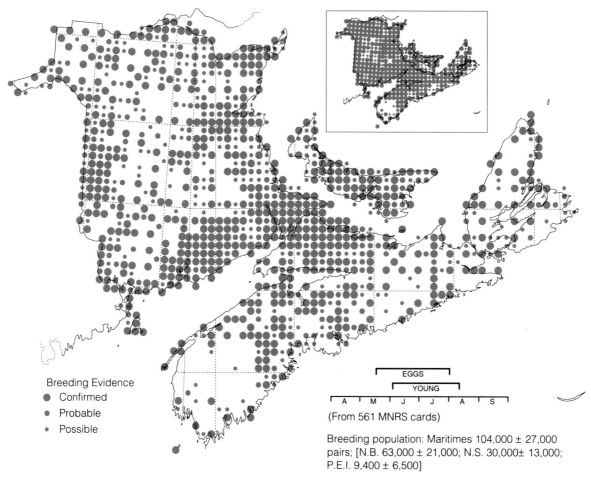

Breeding Evidence
- Confirmed
- Probable
- Possible

EGGS
YOUNG

A M J J A S

(From 561 MNRS cards)

Breeding population: Maritimes 104,000 ± 27,000 pairs; [N.B. 63,000 ± 21,000; N.S. 30,000± 13,000; P.E.I. 9,400 ± 6,500]

Vesper Sparrow
Bruant vespéral

Pooecetes gramineus

The Vesper Sparrow is distributed across North America, from the Maritimes to British Columbia and from California to North Carolina. Its main range comprises the prairie grasslands of the west, but it was scattered thinly across the more open areas of the Maritimes. It is characteristic of areas with short grass or low shrubs, such as pastures, blueberry fields, and clearings, where scattered trees and taller shrubs are often used as song posts. The clear, plaintive song reveals the presence of the Vesper Sparrow long before the white outer tail-feathers, which set it apart from the other grassland sparrows in the east, are seen. It is so scarce in our area now that many observers do not recognize the species by song. Confirmed breeding, in 28 of 94 squares with Vesper Sparrows, comprised about one-third of the records obtained in the Maritimes Atlas, but nests were found in only 2 squares.

Vesper Sparrows must have been scarce in the Maritimes before European settlement, when the clear-ing of forests made available an abundance of stump pastures and other poor farmland in which the species prospered for a century or more. It has dwindled markedly over the last 50 years in the Maritimes, and also west across south-ern Quebec and Ontario, as a result of changes in land use, with farm-land becoming either intensively used or abandoned and reverting to forest. The relatively few records obtained during the Atlas field-work were often in blueberry fields, but showed no obvious pattern otherwise. Estimates of numbers here are little more than guesswork, but this is our scarcest regularly breeding sparrow. It seems unlikely that this present scarcity will change much, unless the species adapts to use clear-cut areas during the two to five years after cutting before regeneration makes them unsuitable. Global warming, unless accompanied by reduced rainfall, is unlikely to make habitat suitable for Vesper Sparrows more available. Given their scarcity and restricted habitat, Vesper Sparrows may deserve attention as approaching "threatened" status in our region, though they are still abundant in the prairies.

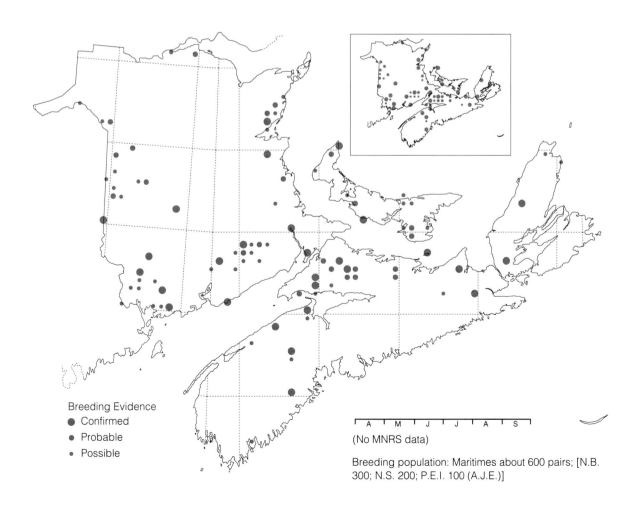

Breeding Evidence
- Confirmed
- Probable
- Possible

A M J J A S
(No MNRS data)

Breeding population: Maritimes about 600 pairs; [N.B. 300; N.S. 200; P.E.I. 100 (A.J.E.)]

Savannah Sparrow
Bruant des prés

Passerculus sandwichensis

This ubiquitous grassland sparrow is found from the tundra of northern Canada to Central America, though in the east south only to Maryland. It breeds only in open vegetated areas, but uses all kinds of low grass and sedge habitats, including meadows, prairies, bogs, salt marshes, and tundra, nesting on the ground well-concealed among the vegetation. In the Maritimes, it was absent from the forested hinterlands, including most clear-cut areas, but was found wherever grasslands persist in all regions, even on islands offshore. The Savannah Sparrow is not a conspicuous bird in appearance or song, though the latter, once learned, is the easiest clue to its presence. The Ipswich Sparrow, a conspicuously large, pale form, was treated as a separate species until 1957. It is almost confined to Sable Island, Nova Scotia, in breeding season with a population of about 1,000 pairs. Data collected for the Atlas undoubtedly under-represent its occurrence in some regions, but the distribution pattern is obvious. It was confirmed as breeding in more than half (488 of 875; 56 per cent) of the squares where it was found, with newly flying young (FL) the most frequent evidence (280 squares), and nests found in 64 squares.

Savannah Sparrows in the Maritimes before European settlement were restricted to the grassy edges between forest and salt marsh or river bank, to the low-sedge areas of bogs, and to the wind-swept coastal barrens. They have increased greatly over the last 350 years, following clearing of forest cover from much of the land. They decreased, in parallel with the subsequent reduction in farmland area, during the past half-century, as confirmed by the BBS, which showed a significant decline since 1966. The decrease in breeding habitat is likely to continue at a slower rate in future, but is unlikely to make Savannah Sparrows a rare bird here. Continued loss of wintering habitat in the eastern United States may ultimately set limits to their numbers here.

Refs. 26, 140, 153

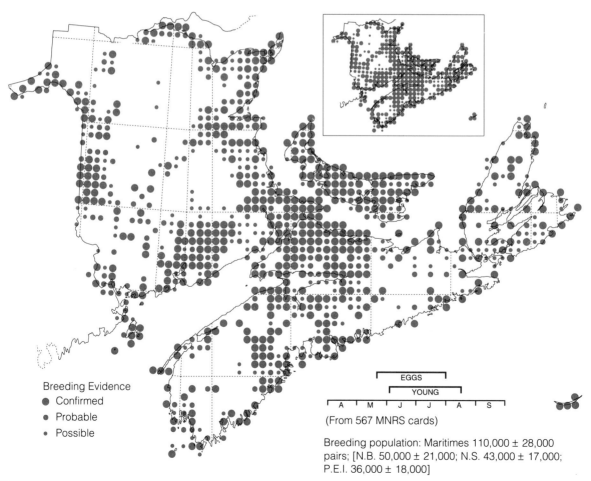

Breeding Evidence
- Confirmed
- Probable
- Possible

EGGS
YOUNG
A M J J A S

(From 567 MNRS cards)

Breeding population: Maritimes 110,000 ± 28,000 pairs; [N.B. 50,000 ± 21,000; N.S. 43,000 ± 17,000; P.E.I. 36,000 ± 18,000]

Sharp-tailed Sparrow
Bruant à queue aiguë

Ammodramus caudacutus

The Sharp-tailed Sparrow is associated with salt marshes and other saline habitats with little regard for north-south zonation. In the east, it breeds around salt marshes from the St. Lawrence estuary in Quebec to North Carolina; other subspecies are found around Hudson and James bays, and around lakes and ponds from Minnesota northwest to Great Slave Lake. In the northeast, little groups of these birds are found also well inland along the St. John and other rivers, in areas with no present tidal influence but to which the sea extended during post-glacial marine submergence. Records far inland are suspect unless well documented, as this is *the* salt-marsh sparrow of the Maritimes; the subspecies of the north Atlantic coast *(subvirgatus)* was described from specimens collected near Hillsborough, N.B. Sharp-tailed Sparrows are not familiar to many observers, nor are they very distinctive, and they were probably under-represented in the Atlas. The records were fairly evenly divided among confirmed, probable,

and possible breeding, with adults carrying food (AY) and newly flying young (FL) being the most frequent confirmations, in 13 per cent and 12 per cent, respectively, of 202 squares with the species. Nests were found in only 5 squares.

The loss of salt marshes to dyking and draining during the last 300 years reduced available habitat for Sharp-tailed Sparrows by one-half or more. The remnant populations that persisted inside the dykes on former salt marsh have mostly been squeezed out as draining and cultivation of these areas progressed. Tufts (1962) emphasized the importance of undisturbed rank grassland (a scarce habitat on managed dykelands), for nesting by this species, as with related sparrow species. The Atlas data allow only rough estimates of populations. The impression persists that this bird has declined over the last 30 years, here as well as in Quebec, where a status report for COSEWIC concluded that the species did yet not warrant designation in any category. Future threats to it here come from competing uses of salt marshes, including flooding them for waterfowl, as well as draining and in-filling as in the past.

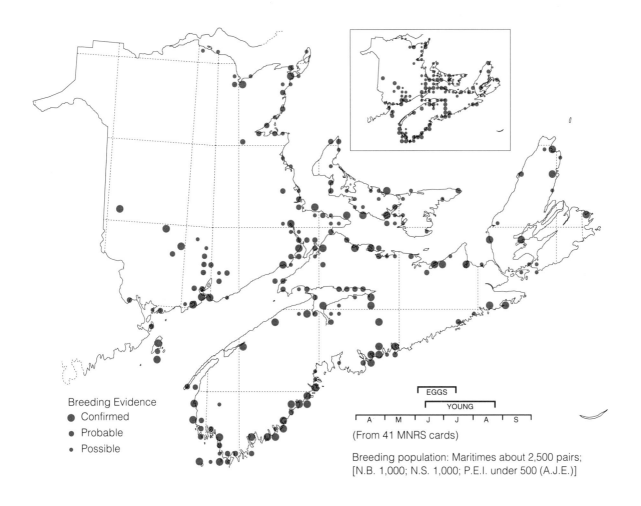

Breeding Evidence
- Confirmed
- Probable
- Possible

EGGS
YOUNG
A M J J A S

(From 41 MNRS cards)

Breeding population: Maritimes about 2,500 pairs; [N.B. 1,000; N.S. 1,000; P.E.I. under 500 (A.J.E.)]

Fox Sparrow
Bruant fauve

Passerella iliaca

This large sparrow breeds northward from our area to the arctic tree-line, and west across the continent to Alaska and British Columbia, with other forms occurring south through the western mountains. Fox Sparrows breed in northern New Brunswick, and on Cape Breton Island and outer islands and headlands of the Nova Scotia Atlantic coast. These are generally higher or cooler areas, but they also occurred in summer on the warm coastal plain around the Baie des Chaleurs, N.B. The habitat is often dense damp shrubbery of alders and other small broad-leafed trees, as in its inland range elsewhere, but they were often found in regenerating areas in northern New Brunswick. On Nova Scotia's outer coasts, Fox Sparrows frequented stunted spruces and shrubby bogs as in Newfoundland and Labrador. The loud, musical song is the easiest clue to its presence, as the inland breeding haunts are often difficult to penetrate. Rather few atlassers encountered this species, and only about 25 per cent of the 175 squares with Fox Sparrows—vs.

over 50 per cent in most common sparrows—had breeding confirmed, usually adults carrying food (AY) or fledged young (FL).

In the Maritimes, Fox Sparrows as breeding birds were detected rather recently; the one confirmed breeding record obtained in Prince Edward Island was the first known there, and the species became established on Cape Breton Island only from 1965. Fox Sparrow distribution in the Maritimes has not been greatly influenced by human actions, although use of regenerating areas and other partly open habitats suggests a net benefit. Our region is peripheral to the main range, and most records were in habitats not drastically altered from a natural state. The remote areas frequented by most of our Fox Sparrows were less thoroughly sampled than much of the Maritimes, and the Atlas records probably under-represent the species. Projected land-use patterns seem unlikely to threaten their future here, but gradual warming under the "greenhouse effect" might, in a century or so, push back the southern range limits beyond our area.

Ref. 32

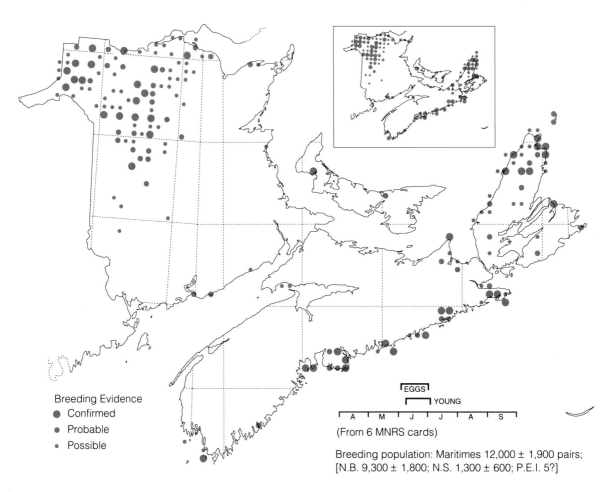

Breeding Evidence
● Confirmed
● Probable
● Possible

EGGS
YOUNG
A M J J A S
(From 6 MNRS cards)

Breeding population: Maritimes 12,000 ± 1,900 pairs;
[N.B. 9,300 ± 1,800; N.S. 1,300 ± 600; P.E.I. 5?]

Song Sparrow
Bruant chanteur

Melospiza melodia

This cheerful songster is one of our most familiar birds, though less distinctive than the Robin or swallows. They return early in spring, even before the snow is gone, and a few half-hardy individuals survive through most winters with us. Song Sparrows breed across Canada, in brushy edges and waterside shrubbery, from northern British Columbia to western Newfoundland, and south to California, Iowa, and North Carolina. They were found nearly everywhere in the Maritimes, but more sparsely where forests are more continuous, especially in the central and northwest highlands of New Brunswick. Probably no area in the Maritimes except coastal sandbars, such as Sable Island, completely lacks suitable places for this species. Song Sparrows are conspicuous if not showy, with a vigorous and pleasing song, and they frequent places where people live, so they were probably detected more frequently relative to their occurrence than most species. Breeding was confirmed in 791 squares (66 per cent of 1194 with Song Sparrows), only exceeded by Robin,

Barn and Tree Swallows, Starling, and Common Grackle; all of these others were found at nests far more often and more easily than were Song Sparrows, whose nests were noted in only 95 squares (8 per cent).

Presumably the more continuous forests of the Maritimes prior to European settlement offered fewer opportunities for Song Sparrows than now exist, although these birds were doubtless here, and common, even then. The lengthy process of clearing forests and establishing settlements and farms must have favoured these birds as much as any others, until they are now among our most common birds. It is difficult to imagine future land-use practices here, or climatic changes, that could seriously threaten this bird without also putting an end to many other living things. Nevertheless, the BBS shows a statistically significant decline in its numbers here since 1966, as well as in other species that winter in agricultural areas in the eastern U.S.A. These birds depend on other areas besides those in which they breed. Song Sparrows are here to stay, but even such a common species is not immune to the habitat changes caused by human demands for "development."

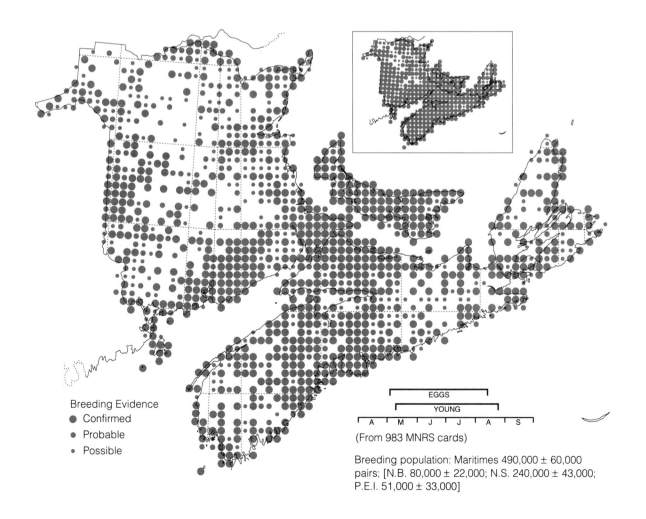

Breeding Evidence
- Confirmed
- Probable
- Possible

EGGS
YOUNG
A M J J A S

(From 983 MNRS cards)

Breeding population: Maritimes 490,000 ± 60,000 pairs; [N.B. 80,000 ± 22,000; N.S. 240,000 ± 43,000; P.E.I. 51,000 ± 33,000]

Lincoln's Sparrow
Bruant de Lincoln

Melospiza lincolnii

This secretive but melodious sparrow is wide-spread, breeding from Alaska to Labrador across the boreal regions of Canada, and south in mountains to California in the west, but only reaching to central Ontario and central Maine in the east. In the Maritimes it was almost absent from Nova Scotia southwest of a line from Windsor to Chester (like the Mourning Warbler), and it was scarce throughout the hardwood-dominated areas of the St. John River Valley in New Brunswick, and in western Prince Edward Island. The greatest abundance of Lincoln's Sparrows was probably found on Cape Breton Island. It frequents areas with shrubs and small trees scattered amid grasses or sedges, thus including bogs and old fields as well as alder swales. A shy bird, it is easily detected by song, but rather few people know it well. The relatively high proportion of squares with confirmed breeding (284 out of 669, 42 per cent) suggests that it was found mostly in the course of intensive field work by experienced observers. Nests (only in 5 squares, 1 per cent) are hard to find everywhere in its range.

As this species is not a forest bird, Lincoln's Sparrow was scarcer when Europeans first came to the Maritimes than at present. The species was not described until 1833, when Audubon found it on the North Shore of the Gulf of St. Lawrence. Cutting forests and clearing land for farms and settlements in the Maritimes created an abundance of brushy edges, but the combination of small trees amid open areas may not have become abundant until marginal farmlands were being abandoned, mostly in this century. These birds probably increased from primaeval numbers, but less than other, more adaptable species such as Song and White-throated Sparrows, though their numbers in natural habitats would have been unchanged. Recently, increasing areas of forest clear-cuts in early succession-al stages have provided much suitable habitat, so their numbers now may be as high as ever in the past. Land-use by the forest industry promises well for Lincoln's Sparrows in future, but few other trends allow predictions bearing on this species.

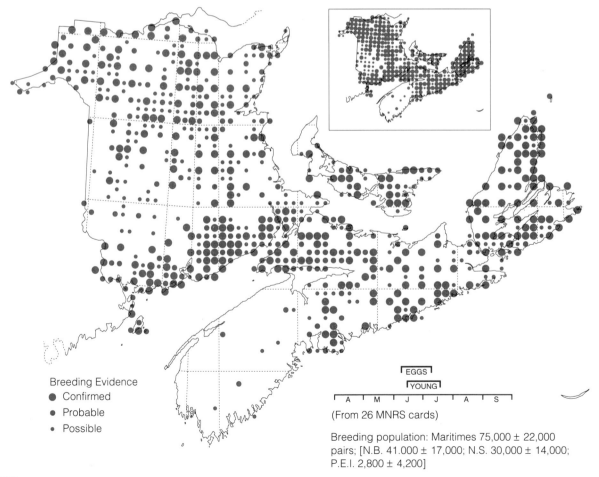

Breeding Evidence
- Confirmed
- Probable
- Possible

EGGS
YOUNG

A M J J A S

(From 26 MNRS cards)

Breeding population: Maritimes 75,000 ± 22,000 pairs; [N.B. 41.000 ± 17,000; N.S. 30,000 ± 14,000; P.E.I. 2,800 ± 4,200]

Swamp Sparrow
Bruant des marais

Melospiza georgiana

This is the breeding sparrow of wetlands with tall or shrubby vegetation throughout eastern boreal and temperate North America, from northeast British Columbia to Newfoundland and south to Delaware and Nebraska. Earlier accounts in the Maritimes termed it uncommon in New Brunswick, fairly common in Nova Scotia, and rather common in Prince Edward Island. However, the Atlas records showed Swamp Sparrows in every square in the well-worked regions, but with many gaps where effort was less intensive, which suggests that the species was general but often unrecognized. However, the largest gaps were in upland parts of northern New Brunswick, where steep terrain limits the amount of standing waters. Wetlands with suitable habitats, though in most squares, are often few in number and difficult of access, and the bird is not striking in appearance or in song, though distinctive once learned, so its status may be under-represented by the Atlas records. The statistical data, however, are in no way unusual, with breeding confirmed in 313 squares (41 per cent of 760 with Swamp Sparrows), mostly by seeing adults carrying food (AY, 24 per cent) or newly fledged young (FL, 14 per cent), whereas nests (NY,NE,ON) were found in only 18 squares (2 per cent), and detections by song (H,T) in 366 squares (48 per cent).

Drainage of large wetlands and filling of smaller ones since the start of European settlement reduced available habitat for this species somewhat, although Swamp Sparrows also use shrubby growth along small creeks and ditches that have been less altered. Earlier statements on its abundance are suspect if, then as now, the species often passed unrecognized, but some decline in the past is plausible. The BBS suggests no change since 1966, but the more extensive wetlands are not well sampled by these roadside surveys. With wetland restoration accelerating under the North American Waterfowl Management Plan, we may expect to see some reversal of earlier habitat losses in the more accessible areas, and perhaps some increase in these birds, in future.

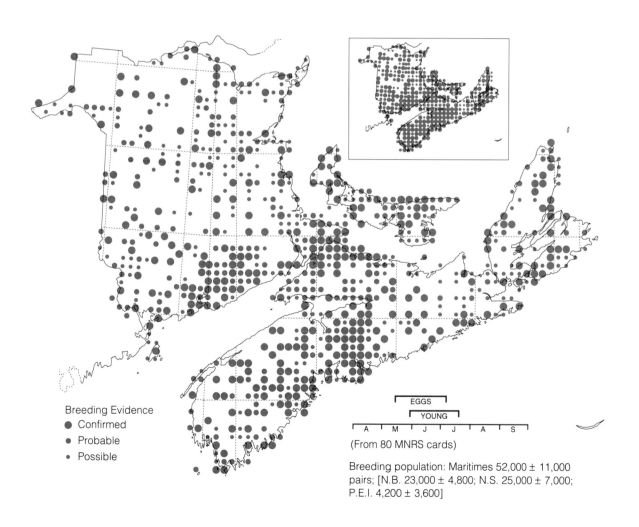

Breeding Evidence
- Confirmed
- Probable
- Possible

EGGS
YOUNG

A M J J A S

(From 80 MNRS cards)

Breeding population: Maritimes 52,000 ± 11,000 pairs; [N.B. 23,000 ± 4,800; N.S. 25,000 ± 7,000; P.E.I. 4,200 ± 3,600]

White-throated Sparrow
Bruant à gorge blanche

Zonotrichia albicollis

This is an abundant woodland bird in eastern Canada and northeast U.S.A., but it is much scarcer and less general in its westward extension to northeast British Columbia and south Mackenzie. The main range extends from near the arctic tree-line south to the eastern temperate forest zone. White-throated Sparrows breed in forest-edge and disturbed woodland, including both conifer and mixed stands, their nests being on the ground, usually under branches or other vegetation. They were found just about everywhere in the Maritimes that field work was done during the Atlas period, including the southwest New Brunswick and southwest Nova Scotia areas where hardwoods are more prevalent. As almost everyone recognizes this species by song, the Atlas data should represent its distribution here adequately. Whitethroats were confirmed as breeding in 749 squares (59 per cent of 1264 with the species), although nests (NY,NE,ON) were found in only 79 squares (6 per cent).

Probably White-throated Sparrows were already common in the Maritimes when Europeans first came here. Subsequent clearing and cutting in the formerly more continuous forests created much more suitable habitat for this species, which is now one of our most abundant woodland birds. MNRS data indicated that Whitethroat nests were found each year more frequently in New Brunswick than in Nova Scotia, and BBS data also showed that the species is markedly more numerous in New Brunswick. Estimates of breeding numbers from the Atlas data confirm higher densities in New Brunswick, allowing for its 30 per cent larger area than Nova Scotia, and are of the same order as earlier, independent estimates (Erskine, 1980). The BBS indicated a significant decline in this species since 1966, which does not correlate with known environmental influences on the species in the Maritimes, unless sub-lethal effects of forest pesticides are responsible. Despite this decline, there is no reason to believe that White-throated Sparrows are in any danger of becoming scarce birds here in the future.

Ref. 18

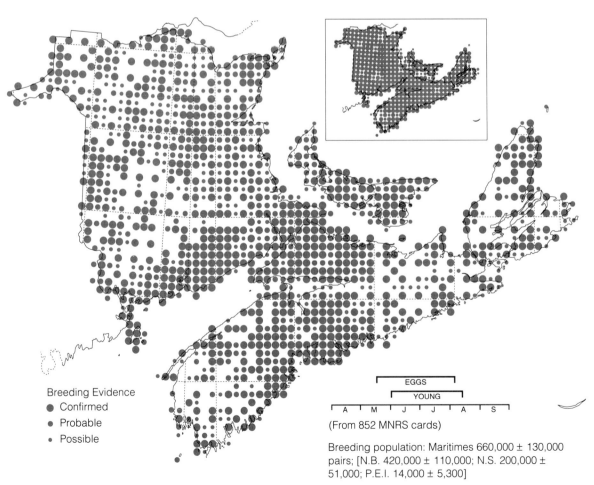

Breeding Evidence
● Confirmed
● Probable
● Possible

EGGS
YOUNG

A M J J A S

(From 852 MNRS cards)

Breeding population: Maritimes 660,000 ± 130,000 pairs; [N.B. 420,000 ± 110,000; N.S. 200,000 ± 51,000; P.E.I. 14,000 ± 5,300]

194

Dark-eyed Junco
Junco ardoisé

Junco hyemalis

This species in one or another of its forms breeds in wooded regions all across Canada and Alaska north to the tree-line, and in mountains south to Georgia and to the Mexican border, the only place where it meets Yellow-eyed Juncos. Juncos are birds of forest openings and edges. Although they often sing from the top of a tree, usually a conifer, they are more often seen on or close to the ground, and their nests are placed on the ground. Juncos were detected in the Maritimes almost everywhere that was visited, as virtually all squares include some suitably open woodland or coastal scrub. Confirmation of breeding, in 706 squares (61 per cent of 1149 with these birds), was most often by sightings of newly flying young (FL), as the white outer tail-feathers make even young birds of this species obvious and easily identified. Nests too were readily found, in 106 squares (9 per cent).

The older and more continuous forests that clothed the Maritimes when Europeans first came here contained suitable niches for many Juncos, though perhaps fewer than at present. The tree-cutting and partial clearing that accompanied settlement and that has continued to this day provided increased edges and openings where this species proliferated. The numbers evidently levelled off later, as the BBS suggested no change in the last quarter-century. The Junco is not a remarkably abundant bird, despite its wide distribution. Nest records have consistently indicated Juncos to be more common in Nova Scotia than New Brunswick, the reverse of the situation for White-throated Sparrows, and the Atlas extrapolations support that impression. Shorter harvest cycles in conifer forest management in the Maritimes should provide lots of breeding habitat suitable for Juncos in the foreseeable future, but wintering habitat, mostly in settled areas of eastern North America, may eventually limit their numbers here.

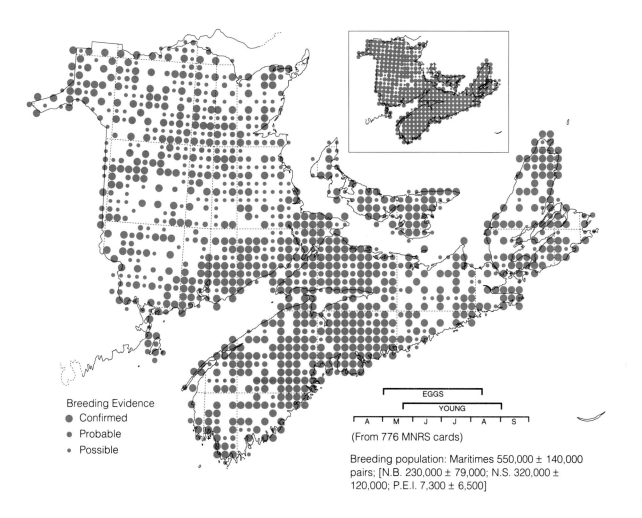

Breeding Evidence
- Confirmed
- Probable
- Possible

EGGS
YOUNG
A M J J A S

(From 776 MNRS cards)

Breeding population: Maritimes 550,000 ± 140,000 pairs; [N.B. 230,000 ± 79,000; N.S. 320,000 ± 120,000; P.E.I. 7,300 ± 6,500]

Bobolink
Goglu

Dolichonyx oryzivorus

The Bobolink, sometimes called "skunk-blackbird" from the male's "reversed" plumage pattern (white above and black below), is the blackbird of lush meadows in the east. It is the only member of its family in Canada that does not rely on trees and bushes for song-posts, as its advertising song is delivered in flight. The nesting area is not defended as an exclusive territory, and the mating system involves more than one female with each male. These birds breed across Canada and the United States from Nova Scotia and New Jersey west to California and British Columbia, but their centre of abundance is east of the Mississippi. Their distribution in the Maritimes, expectably in a largely forested region, is patchy; they were not found in large areas of north and central New Brunswick, nor in parts of south-west and eastern mainland Nova Scotia, nor in the Cape Breton Highlands. In Prince Edward Island they were found in most squares, although the species was of rare and sporadic occurrence there until the 1960s. The striking appearance and distinctive bubbling song make this an easily detected species, and the open habitats it frequents allow easy confirmation of breeding. Adults carrying food (AY) were the most common evidence of breeding (in 34 per cent of 790 squares), but nests (NY,NE,ON) were quite often found (45 squares).

Before European settlement and the opening-up of the forests, Bobolinks in the Maritimes were restricted to beaver-meadows and other areas of lush grass, mostly along river flood-plains. They have undoubtedly increased greatly here over the past four centuries, and in Prince Edward Island they have gone from rare to common within the last 25 years. Their present numbers rank them third among the blackbirds in the Maritimes. The frequently damp climate here prevents early cutting of hay-fields, a practice that affects the species' nesting in other parts of its range. Reduction of land area used for hay crops, accompanying a trend towards rearing of beef cattle in feed-lots, may gradually reduce the available habitat for Bobolinks here, but they should be around for years to come.

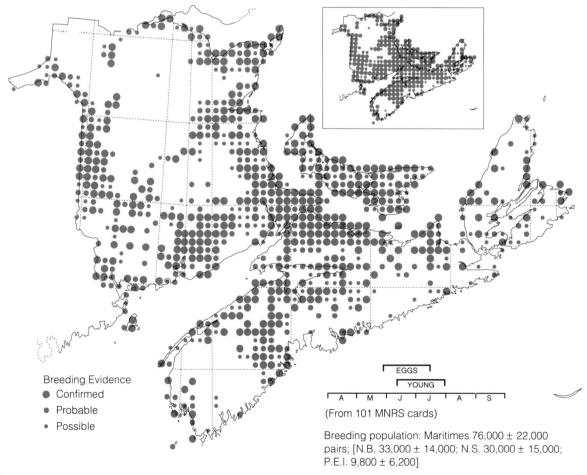

Breeding Evidence
● Confirmed
● Probable
● Possible

EGGS
YOUNG
A   M   J   J   A   S

(From 101 MNRS cards)

Breeding population: Maritimes 76,000 ± 22,000 pairs; [N.B. 33,000 ± 14,000; N.S. 30,000 ± 15,000; P.E.I. 9,800 ± 6,200]

Red-winged Blackbird
Carouge à épaulettes

Agelaius phoeniceus

The Redwing breeds across Canada and the U.S.A., and south through Mexico to Costa Rica, a widespread northern representative of a subfamily centred in tropical America. This bird nests in freshwater marshes with cattails, bulrushes, and low shrubs, which were its original habitat, in all parts of its range; although it also breeds in upland fields in some areas, it rarely if ever does so in the Maritimes. Redwings were found here in all regions, but their numbers varied greatly with the availability of suitable habitats. Wetlands in farming areas provided nesting and foraging habitats in close proximity and supported higher Redwing densities than more forested areas and rugged terrain where open lands were scarce. Many squares in upland forests, e.g., in northern New Brunswick, had fewer than 10 pairs each, whereas some squares in the St. John River Valley, the New Brunswick–Nova Scotia border area, and in Prince Edward Island supported over 1,000 pairs. Male Redwings in fertile marshes may defend "harems" comprising two to six females, but in the less fertile habitats typical of the Maritimes a defendable Redwing territory often supports only one female and brood. Redwing breeding was easily confirmed (in 61 per cent of 998 squares with the species), and nests (NY,NE,ON) were seen in 142 squares (14 per cent). Our records should provide a representative picture for this easily recognized bird.

The historical record leaves doubt as to whether Red-winged Blackbirds bred in the Maritimes before European settlement began. If present, they were certainly uncommon, in part because open habitats for foraging, at all seasons, were scarce. This species increased greatly in the Maritimes in the 20th century, colonizing the Atlantic slope of Nova Scotia mainly after 1960, although still incompletely. Recent development (by DUC and others) of impounded marshes for waterfowl production has increased habitat locally, replacing some of that lost through drainage of other marshes. Red-winged Blackbird numbers here are probably higher now than they have ever been. With major efforts going into wetland conservation to increase waterfowl populations, Redwings are likely also to maintain or increase their numbers in the coming decades.

Ref. 12

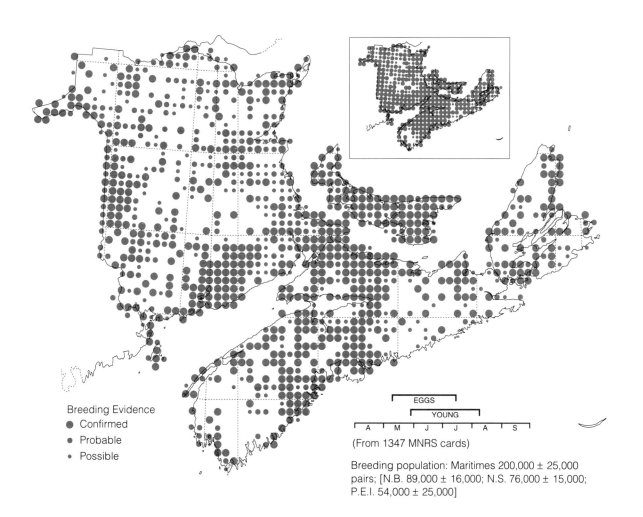

Breeding Evidence
- Confirmed
- Probable
- Possible

EGGS
YOUNG
A M J J A S

(From 1347 MNRS cards)

Breeding population: Maritimes 200,000 ± 25,000 pairs; [N.B. 89,000 ± 16,000; N.S. 76,000 ± 15,000; P.E.I. 54,000 ± 25,000]

Eastern Meadowlark
Sturnelle des prés

Sturnella magna

Meadowlarks, as their name implies, are songbirds of grasslands. This species breeds from Brazil and Colombia through Central America and the central and eastern U.S.A. into southern Ontario and Quebec. Although breeding occurs regularly in open areas of southern Maine and in the St. Lawrence Valley to east of Quebec City, Meadowlarks have always been scarce breeders in New Brunswick and have largely been vagrants elsewhere in the Maritimes. The Atlas confirmed that picture, with only scattered breeding found in the St. John and Kennebecasis/Petitcodiac valleys, N.B., and in the Annapolis Valley, N.S. Hantsport (1986) was the east-ernmost confirmed location, although the first breeding in Nova Scotia, at Windsor in 1945, was a few kilome-tres farther southeast. Single birds wander in spring and especially in autumn, and the records of "possible breeding" (H) in northeast New Brunswick and Cape Breton Island were probably of vagrants. Breeding was confirmed in 10 of 29 squares with the species, only one nest being reported during the Atlas period, at Hantsport.

Presumably scarcity of open lands, in New England and Quebec as well as in our area, restricted the eastward spread of Meadowlarks before European settlement of the Maritimes, but this does not explain fully their continuing scarci-ty here. Though present in New Brunswick since the first bird compilations in the 19th century, this was and remains a very scarce bird here. Atlas records were too few for extrapolated estimates. As this is a striking and conspicuous bird with a far-car-rying song, frequenting open habitats where it can be detected from a distance, the few records should repre-sent its status here quite accurately. Meadowlarks are vulnerable to any kind of alteration of their few breed-ing areas here, and both this species and its western counterpart seem to be declining all across Canada (BBS). This is a species that easily could be lost as a breeding bird in the Maritimes, and its status should be monitored regularly.

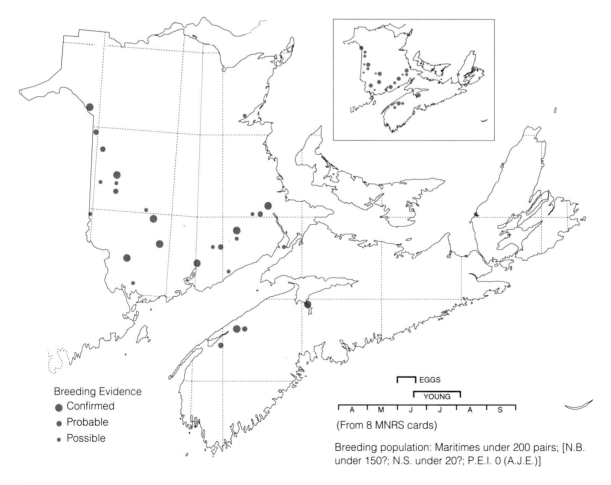

Breeding Evidence
● Confirmed
● Probable
● Possible

EGGS
YOUNG

| A | M | J | J | A | S |

(From 8 MNRS cards)

Breeding population: Maritimes under 200 pairs; [N.B. under 150?; N.S. under 20?; P.E.I. 0 (A.J.E.)]

Rusty Blackbird
Quiscale rouilleux

Euphagus carolinus

This is the northernmost of the icterine black-birds, a group originating in tropical America. The Rusty breeds across the boreal regions of Canada from Alaska to Labrador, north almost to the tree-line, and reaching south across the United States border only into northern New York and New England. In the Maritimes, it was found in all regions, but nowhere commonly, being absent in many well-worked areas. It frequents cool habitats in spruce bogs, swamps, and damp alder swales, but on Cape Breton Island (a generally cooler region where Common Grackles approach their range limit), it also use drier sites such as pasture-edges. Because of their restricted habitats, which are often difficult of access, the Rusty Blackbird was probably under-represented, and it was confirmed as breeding much less regularly than other blackbirds, in only 169 (38 per cent) of 446 squares with the species. Nests were seldom found, in only 9 squares (2 per cent, vs.

11–14 per cent for other blackbirds), which accounts partly for the low rate of confirmation.

Probably the wetland habitats where Rusty Blackbirds breed, although remote and difficult to farm, have been reduced since the start of European settlement along with other wetlands. The species was termed "uncommon" in most previous provincial summaries, and it may have become even scarcer in recent decades. Clear-cutting of forests around wooded swamps favours Grackles, which depend on more open areas for foraging, and may lead to replacement of the Rusty by the larger blackbird. The BBS, though working from meagre samples, suggests a significant decline since 1966. Forecasts of global warming may imply further declines in future for this as well as other species in our area that are near the southern limits of their ranges here. As long as the cold ocean keeps parts of our region cool, Rusty Blackbirds are unlikely to disappear completely, but we may find them scarcer birds than in the past.

Ref. 34

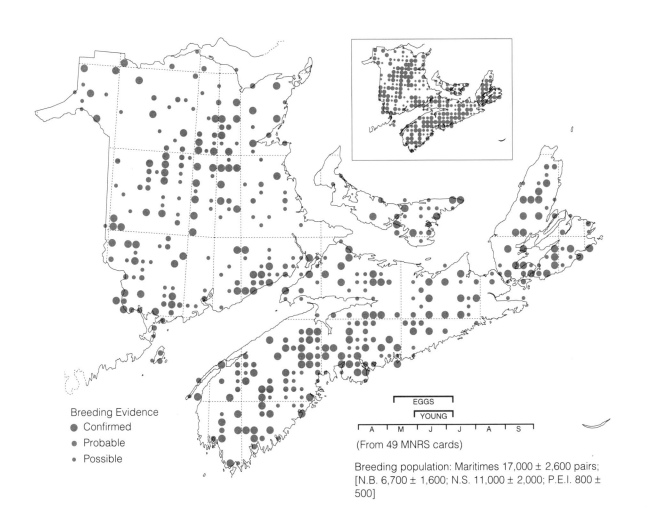

Breeding Evidence
- Confirmed
- Probable
- Possible

EGGS
YOUNG

A M J J A S

(From 49 MNRS cards)

Breeding population: Maritimes 17,000 ± 2,600 pairs; [N.B. 6,700 ± 1,600; N.S. 11,000 ± 2,000; P.E.I. 800 ± 500]

Common Grackle
Quiscale bronzé

Quiscalus quiscula

The Grackle is an eastern American species that breeds from the Atlantic to the Rockies and from the boreal forest to the Gulf of Mexico. Its habitat is mainly agricultural areas and other open habitats where there are trees, bushes, or buildings for nesting, but it also nests, in much lower densities, in alders and other shrubby areas in forested regions. Grackles are large, conspicuous birds that frequent open habitats, so they were easily detected and confirmed. In the Atlas, they were found in all regions, and often in every square. In forested regions, they were scarcer, with gaps in northwest New Brunswick, in the Cape Breton Highlands and in southwest Nova Scotia. Feeding of young (AY or FL), in or out of the nest, was the most frequent confirmation, but nests were reported in only 11 per cent of 1111 squares where the species was found. Grackles feed mainly on insects obtained from the ground in spring and summer, but they also make use of other animal foods such as eggs and nestlings of birds, small fish, and small mammals; waste grain is the major food at other seasons.

Evidently the Grackle was a scarce and local bird in the Maritimes before European settlement and clearing of the forests. Provincial bird summaries agree that it did not become widespread and common here until after 1900. Abandonment of marginal farmland after 1930 and its reversion through shrubland to forest caused a temporary upsurge in numbers; densities in Prince Edward Island in the late 1960s (from BBS indices) were the highest northeast of New York, but the BBS indicated a significant decline in the Maritimes since then. As a bird that uses urban and farmland habitats at higher densities than natural habitats, our data probably over-estimate its numbers. It has undoubtedly benefited from human settlement here and will continue to thrive. Recent land-use patterns, including Christmas tree plantations and conifer planting in forest clear-cut areas, ensure an adequate supply of suitable habitat for Grackles.

Ref. 34

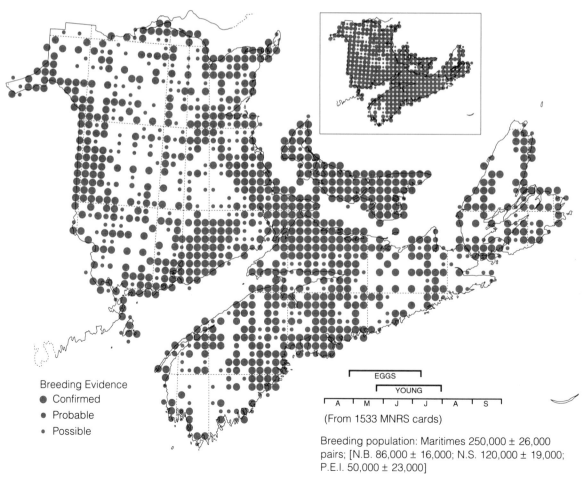

Breeding Evidence
- ● Confirmed
- ● Probable
- ● Possible

EGGS

YOUNG

A M J J A S

(From 1533 MNRS cards)

Breeding population: Maritimes 250,000 ± 26,000 pairs; [N.B. 86,000 ± 16,000; N.S. 120,000 ± 19,000; P.E.I. 50,000 ± 23,000]

Brown-headed Cowbird
Vacher à tête brune

Molothrus ater

This is the northernmost of a South American group of small blackbirds, not all of which are parasitic. It breeds throughout the U.S.A. and southern Canada, mainly in settled areas, with highest densities in the prairies. It is quite widespread in the Maritimes, but was virtually absent in the forested regions of northern and central New Brunswick and eastern Nova Scotia. Farming areas in southern New Brunswick, central Nova Scotia, and central Prince Edward Island had more continuous Cowbird distribution than elsewhere. This, our only regularly parasitic bird, lays its eggs in nests of other species, the latter rearing the young Cowbird, which often crowds out the young of the host/victim. Cowbird eggs have been recorded in nests of 37 species here (MNRS), and over 250 species have been parasitized across its range. Species most frequently parasitized in the Maritimes, relative to the numbers of their nests found, were Veery, Solitary and Red-eyed Vireos; Chestnut-sided, Magnolia, Yellow-rumped, and Black-and-White Warblers; and American Redstart. Song Sparrows and Dark-eyed Juncos were commonly victimized in Prince Edward Island where most warblers are scarcer. Atlas records of Cowbirds, in 683 squares overall, were mostly of single birds (H, 289 squares) or pairs (P, 128 squares), and nests (mostly NE) were seen in only 32 (18 per cent) of 175 squares with breeding confirmed. Newly fledged young (FL, in 143 squares) fed by hosts were the most frequent confirmation of breeding.

Cowbirds came late to the Maritimes as they spread east from the original prairie range. Open habitats were scarce here until well into the 19th century, but Cowbirds were uncommon until the 1950s, when feeding stations and cattle feed-lots allowed more birds to winter here. The area of agricultural land has decreased since 1930, with marginal farms reverting to forest. The BBS suggests a marked decline in Cowbirds since 1966 in the Maritimes, and also across eastern Canada, so the decrease here may not have a local cause. Given the impact of Cowbirds on other small birds, no one is much perturbed by a decrease in this species. The mix of habitats and availability of suitable hosts ensure that Cowbirds will still be here in the future.

Ref. 49

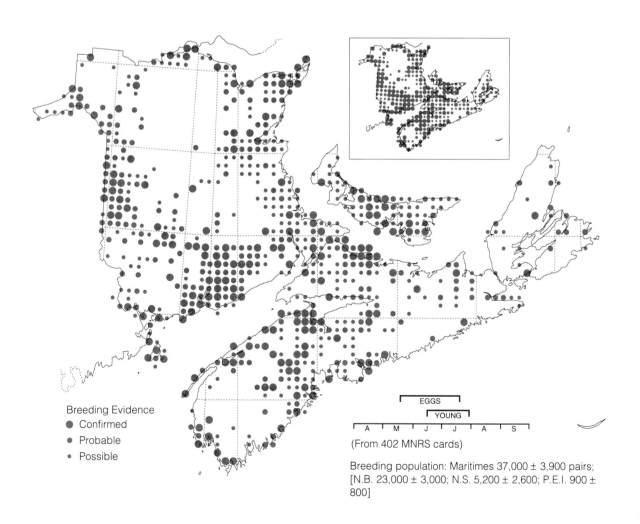

Breeding Evidence
● Confirmed
● Probable
● Possible

EGGS
YOUNG
A M J J A S

(From 402 MNRS cards)

Breeding population: Maritimes 37,000 ± 3,900 pairs; [N.B. 23,000 ± 3,000; N.S. 5,200 ± 2,600; P.E.I. 900 ± 800]

Northern Oriole
Oriole du Nord

Icterus galbula

This showy black-and-orange bird breeds across southern Canada and south through the U.S.A. (except the southeast) to northern Mexico. The western birds (subspecies *bullockii*) were formerly considered a separate species. Orioles generally frequent open woodlands, forest edges, and scattered trees in farmlands and towns where tall shade trees, in the east especially elms, provide suitable nest-sites. In the Maritimes, elms occur mostly along rivers and in towns, and these birds were found breeding mainly in the major valley systems: the St. John, Miramichi, and Petitcodiac in New Brunswick, and the Annapolis-Cornwallis in Nova Scotia. None were found on Cape Breton Island, and relatively few in northern New Brunswick, indicating that the range limit is reached there. Orioles are conspicuous and their nests are distinctive, so the data should represent the status of this species acceptably. Breeding was confirmed in 50 per cent of 237 squares with orioles. Although active nests (NY,NE,ON) were reported in only 35 squares, another 25 squares had breeding confirmed by sightings of old nests (UN) after the leaves fell from the trees.

Floodplain forests have been disturbed by lumbering in the Maritimes, as elsewhere, since European settlement began, but elm is not valued either for construction or firewood so it was not sought out for harvest. The widespread planting of elms for shade in residential areas probably balanced losses of these preferred nest-trees elsewhere. The Northern Oriole has increased in Nova Scotia in this century, perhaps when planted elms in towns and farmlands there matured in that period. However, Dutch elm disease began to reduce numbers of elm trees in the Maritimes since about 1960, although there is no evidence that this has limited the range of orioles yet. Our estimates of breeding populations make this one of our scarcer well-established breeding birds. The loss of elms may in the long run force use of other shade trees, but this seems unlikely to pose a serious threat to the species continuing to breed here, especially if global warming provides a generally milder climate.

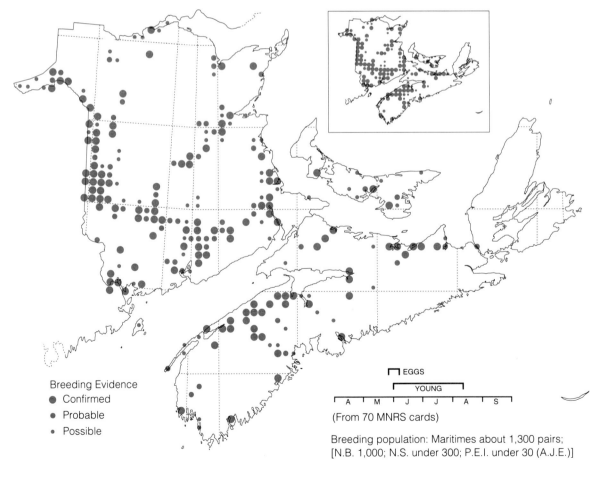

Breeding Evidence
● Confirmed
• Probable
· Possible

EGGS
YOUNG
A M J J A S
(From 70 MNRS cards)

Breeding population: Maritimes about 1,300 pairs;
[N.B. 1,000; N.S. under 300; P.E.I. under 30 (A.J.E.)]

Pine Grosbeak
Dur-bec des pins

Pinicola enucleator

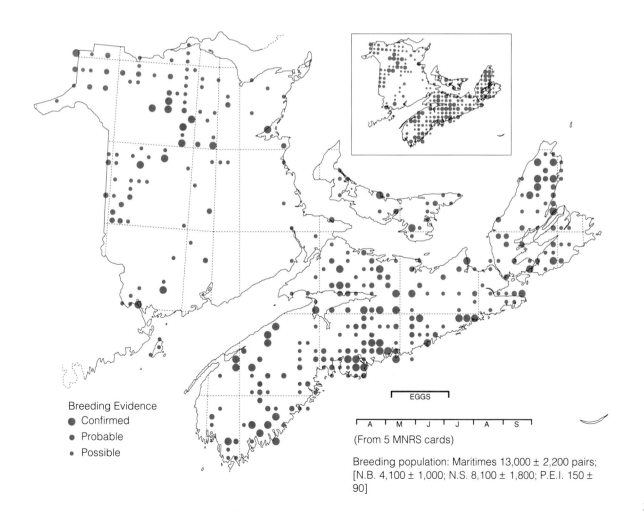

The Pine Grosbeak breeds in the boreal and subarctic forests of America, from Alaska to Labrador, and southward in the western mountains to California and New Mexico, but only to northern New England in the east. It also breeds in Eurasia from northern Scandinavia to Kamschatka, Siberia, and northern Japan. In the Maritimes, it approaches the southern limit of its range. Pine Grosbeaks were found quite generally in Nova Scotia, but were scarce in Prince Edward Island and virtually absent from large areas of New Brunswick, except the northwest and north-central highlands and, oddly, the central part of the St. John River Valley. All but the last-noted areas reflect avoidance of warmer, hardwood-dominated regions. Breeding was seldom confirmed, in only 61 (18 per cent) of 339 squares with Pine Grosbeaks, and sightings without further breeding evidence (H,T) occurred in nearly three-quarters of the squares with these birds. Although such birds might have been only wandering in search of suitable feeding areas as they do at other seasons, those records were generally intermingled with confirmations of breeding, so summer sightings point to secretive nesting. Only 2 squares had nests reported.

Pine Grosbeaks were formerly more regular and numerous in the Maritimes, especially in winter, as they were in the northeast United States (Bent, 1968), but they seem to have declined in the last 70 years. Possibly the "Little Ice Age" (1350–1850) had made conditions here more generally suitable for this northern bird up into this century. It seems unlikely that changes caused by human action in the forests here since 1600 would have greatly restricted their use by Pine Grosbeaks. Like other finches, and especially the Evening Grosbeak, this bird was often encountered in areas of spruce budworm infestation during the 1960s in New Brunswick. As this is a rather scarce breeding bird in the east, especially in New Brunswick, a trend to global warming is likely to make its status here even more precarious. Restorative action is unlikely to be feasible, but monitoring may be appropriate.

Breeding Evidence
- Confirmed
- Probable
- Possible

EGGS

A M J J A S

(From 5 MNRS cards)

Breeding population: Maritimes 13,000 ± 2,200 pairs; [N.B. 4,100 ± 1,000; N.S. 8,100 ± 1,800; P.E.I. 150 ± 90]

Purple Finch
Roselin pourpré

Carpodacus purpureus

A colourful bird with a loud and pleasing song, the Purple Finch ranges across southern Canada from Newfoundland to northern British Columbia, with southward extensions west of the coast mountains to California, and in the east to Wisconsin and Pennsylvania. Although they usually nest in conifers, they frequent open mixed woodland and well-treed gardens as well as spruce/fir forests. With such catholic tastes, it was not surprising that this bird was found all over the Maritimes, with no sizable gaps in well-worked regions. The male Purple Finch is virtually unmistakable, other species similar in appearance (House Finch, crossbills) or song (Warbling Vireo, Pine Grosbeak) being generally much scarcer; even the female is, for a stripey bird, quite distinctive, so the Atlas data should represent the species' occurrence adequately. The adults are conspicuous, especially when the male is giving forth a passionate torrent of song, but the Purple Finch was not especially easy to confirm as breeding (in 392 of 1091 squares, 36 per cent, about average). Newly flying young (FL, in 243 squares) were noted much more often than adults carrying food (AY, in 111 squares). This pattern (more FL than AY) was shared with other cardueline finches, but differs from sparrows (Emberizinae) and warblers (Parulinae) in which adults carrying food were more often seen. Nests of Purple Finches were noted in only 16 squares.

Attempted reconstruction of past habitats in the Maritimes gives no suggestion that Purple Finches were formerly much more or less numerous than at present. They seem likely to have been common here throughout the period since European settlement began, and they remain so to this day, although the BBS suggests a non-significant decrease since 1966. Predictable trends in land use and climate pose no obvious threats to their well-being, so Purple Finches should continue to brighten our lives well into the future.

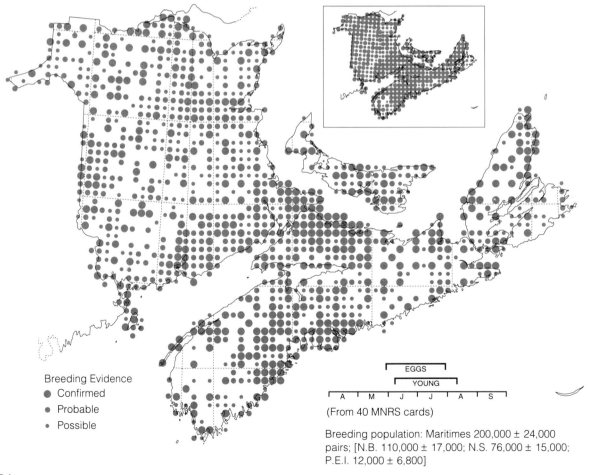

Breeding Evidence
● Confirmed
● Probable
● Possible

EGGS
YOUNG

A M J J A S

(From 40 MNRS cards)

Breeding population: Maritimes 200,000 ± 24,000 pairs; [N.B. 110,000 ± 17,000; N.S. 76,000 ± 15,000; P.E.I. 12,000 ± 6,800]

House Finch
Roselin familier

Carpodacus mexicanus

The House Finch is native to western North America, from British Columbia (first recorded 1937) to Mexico and east to Colorado and Texas. From unauthorized release of birds held illegally in pet-shops in New York City in 1940, it has spread ever more widely, reaching North Dakota, Missouri, and Mississippi by 1989. Breeding in Canada was first proved in southern Ontario in 1978, the same year the species was first detected in the Maritimes. House Finches are usually associated with human settlement, in towns and villages, often nesting in shrubs or vines in gardens. Breeding in the Maritimes was first proved in 1987, with newly flying young seen in St. Stephen, N.B. All records to date are from urban centres, well-spaced across southern New Brunswick and southwestern Nova Scotia, with isolated clusters near Antigonish, N.S., and Charlottetown, P.E.I. Owing to

male House Finches being mistaken for Purple Finches and female House Finches being drab inconspicuous birds, the species is probably more widespread than is recognized at present. Breeding was confirmed in 8 of 41 squares with House Finches, all involving sightings of newly fledged young except for one nest, in a hanging flower-pot in a garden at Yarmouth (1988).

House Finches are the latest alien species to sweep across North America as a result of unplanned actions by people. Their recent spread through eastern North America, now about to merge with the original western range, has taken them in many areas from an exciting novelty to a commonplace bird within a decade. The pattern in other northern areas suggests that House Finches will establish themselves widely in the Maritimes as well, but the harsher climate, with the lower frequency of human settlement and winter bird feeders here, will restrict their potential numbers in our area. Our estimates of present populations probably will be outdated before they are published.

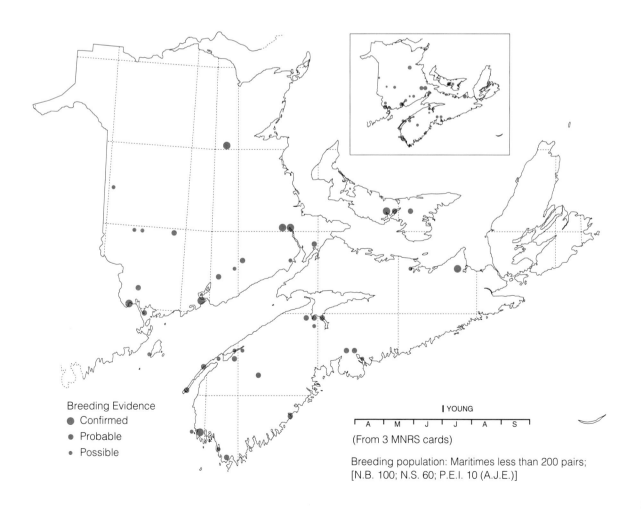

Breeding Evidence
- ● Confirmed
- ● Probable
- • Possible

YOUNG

A M J J A S

(From 3 MNRS cards)

Breeding population: Maritimes less than 200 pairs;
[N.B. 100; N.S. 60; P.E.I. 10 (A.J.E.)]

Red Crossbill
Bec-croisé rouge

Loxia curvirostra

The crossbills are finches specialized for feeding on conifer seeds, which become available in quantity irregularly in both time and place. Thus the breeding range of this species is variable, by year and season, but moves back and forth across North America in the southern boreal, transitional, and western montane forest zones, as well as in Eurasia. Red Crossbills were well known as breeding birds in the Maritimes in the distant past, with virtual absences in some years, but there has been no major irruption here recently. The Atlas records were too sparse to provide any clear pattern. The more scattered occurrences in New Brunswick could have resulted from surveys in many areas being of brief duration and almost entirely in summer, whereas crossbills often breed in January through April. No completed nests were reported to the Atlas, and most confirmed breeding, in 25 squares (16 per cent of 159 with Red Crossbills) involved fledged young (FL). The latter is expectable in finches, which carry food for their young in the crop, thus not visible, so few AY records occur. This may also mean that most field work missed the peak of the nesting season.

Red Crossbills occurred regularly in the Maritimes and throughout the northeast from before European settlement until the 20th century. The virtual disappearance of the regional subspecies in the eastern U.S.A. after 1912 has been correlated with very extensive logging of their food trees—white pine and eastern hemlock—all across their range. Their virtual absence from Nova Scotia for 40 years after 1922, with no nests found after 1913, and the history of lumbering in the Maritimes, fit well with that correlation. The return of breeding Red Crossbills here in the 1960s and 1970s may represent some recovery of the local subspecies, as the food trees mature again, or overflow breeding by another race, perhaps the larger, darker form from Newfoundland. The low numbers and sparse distribution found during the Atlas period suggest the latter explanation may be more plausible; Red Crossbills were breeding both in Cape Breton Island and in western Newfoundland in July-August 1968. Our population estimates are based on very small samples, and thus are little more than guesses. Proposals for forest management in the future do not offer much hope for recovery of former food sources, but recurring immigrants may maintain the present sparse population.

Refs. 25, 146

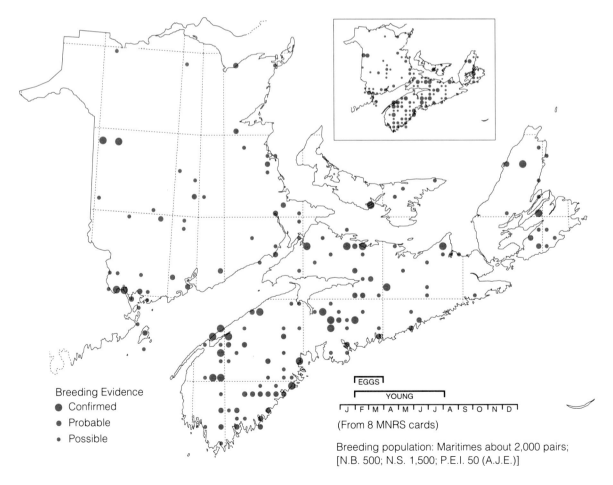

Breeding Evidence
- Confirmed
- Probable
- Possible

EGGS
YOUNG
J F M A M J J A S O N D

(From 8 MNRS cards)

Breeding population: Maritimes about 2,000 pairs;
[N.B. 500; N.S. 1,500; P.E.I. 50 (A.J.E.)]

White-winged Crossbill
Bec-croisé à ailes blanches

Loxia leucoptera

These crossbills breed in the boreal forests across northern America and Eurasia. Breeding in any area is erratic, depending on seed crops of spruce, fir, and larch. Crossbills feed their young on partially digested conifer seeds; as they do not require insect food for breeding, they may be found nesting in all months of the year, the main breeding efforts in the Maritimes being in January to April and July to October. During the Atlas study, the spruce cone crop in the summer and autumn of 1988 was the heaviest for many years. White-winged Crossbills appeared in large numbers and bred over most of the conifer regions of the Maritimes. Their scarcity in the hardwood-dominated regions of southern New Brunswick, southwestern and central Nova Scotia, and Prince Edward Island was striking; the map summarizes all 1986–90 records, but is dominated by those from late summer 1988. Breeding evidence was largely the detection of singing birds on territory (H,P,T; in 416 squares, 78 per cent of 536 squares with these birds). Relatively few people were still atlassing in late August and in the fall when young were being fed in or out of the nest; FL were noted in 100 squares, but only 3 squares had nests reported.

When the White-winged Crossbills arrived in force in 1988 (7 July in Region 13), their sweet trilling could be heard in the tree-tops literally everywhere in the forests and would hardly have been missed by observers then in the field. Gaps in the mapped distribution probably represent gaps in coverage in that period, or unsuitable habitat, rather than absence of the species. Our extrapolated estimates of populations apply mainly to 1988. These birds were virtually absent from large areas in the other years. Modern forestry practices, including short growth cycles, clear-cutting, and re-planting, are unlikely to inhibit the future occurrence of crossbills in the Maritimes; cutting before conifers grow to sufficient age to bear cones and set seed is uneconomical under almost any imaginable regime.

Ref. 8

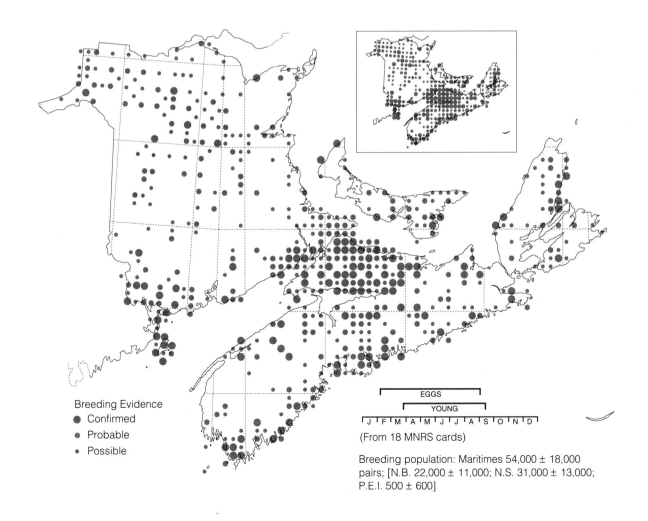

Breeding Evidence
- Confirmed
- Probable
- Possible

EGGS
YOUNG
J F M A M J J A S O N D

(From 18 MNRS cards)

Breeding population: Maritimes 54,000 ± 18,000 pairs; [N.B. 22,000 ± 11,000; N.S. 31,000 ± 13,000; P.E.I. 500 ± 600]

Pine Siskin
Chardonneret des pins

Carduelis pinus

Pine Siskins breed in conifer forests from Alaska to southern Labrador, southward through the western mountains to Central America and in eastern woodlands to Kansas and Pennsylvania. Their numbers in an area vary greatly between seasons and between years, although some appear quite regularly in the Maritimes in most years. During the Atlas, Siskins were found in all regions, but rather unevenly. They were seldom found in the hardwood-dominated regions of southern New Brunswick and southwest Nova Scotia, nor in the low second-growth mixed woodlands of eastern New Brunswick and western Prince Edward Island, scarcity of mature conifers being a feature common to all these areas. Although easily detected by their calls, Siskins are most conspicuous when moving around in flocks, mainly before they have settled to breed in an area or after the young have fledged. The Atlas data may be reasonably representative; they are as likely to have included some flocked wandering birds as to have missed unrecognized breeding birds, so these errors may cancel each other. Breeding of Siskins was confirmed about as often as with other finches (in 24 per cent of 812 squares with these birds), but two-thirds of all records were of single (H) or repeated (T) sightings or birds heard calling. Only 7 squares had nests reported.

Presumably Pine Siskins moved around between suitable places in the primaeval forests of the Maritimes, as they do now. Early forest cutting was focused on timber trees, including white pines, and on general clearing, whereas more recent lumbering took mainly balsam fir and spruce for pulpwood. Reduction in conifers need not have reduced Siskin numbers, as these birds feed on buds and seeds of successional tree species such as birches and alders as well as of conifers, and also on insects. Siskins probably are doing as well in modern forests as in those of the past, and the BBS shows no change in their population here since 1966. There are no obvious reasons to think that they will not continue as a common if erratic breeding bird in the Maritimes.

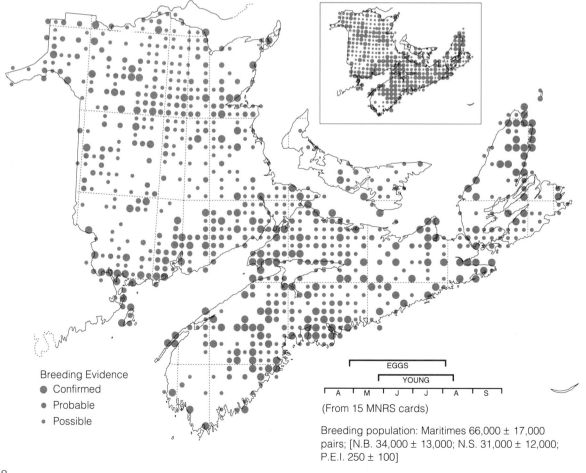

Breeding Evidence
- Confirmed
- Probable
- Possible

EGGS

YOUNG

A M J J A S

(From 15 MNRS cards)

Breeding population: Maritimes 66,000 ± 17,000 pairs; [N.B. 34,000 ± 13,000; N.S. 31,000 ± 12,000; P.E.I. 250 ± 100]

American Goldfinch
Chardonneret jaune

Carduelis tristis

Goldfinches in the Maritimes are near their northern limit, although they breed across southern Canada and almost throughout the United States. They are considered ubiquitous here, because they are easily recognized both by appearance and call-notes, but when breeding they are found mainly in open habitats. Flying birds may be heard over forest and field alike, and some records of "possible breeding" may have involved birds ranging widely in search of feeding or nesting areas. Goldfinches like other cardueline finches are not territorial at any season, and they nest late in the season, so records other than confirmed breeding before mid-July may not have represented birds that had settled to breed. The map for this species is similar to those for Savannah Sparrow and other grassland species, if one discounts peripheral records of "possible breeding." The main foods of Goldfinches are seeds of the family Compositae (daisy-like, including thistles), which grow mainly in open and edge situations. Because of their late breeding schedule, with fledging in August after song of other passerines has ceased, records of confirmed breeding were relatively less frequent (in 26 per cent, 274 of 1047 squares with Goldfinches) than, for example, in Purple Finch (36 per cent) and Savannah Sparrow (56 per cent). Most confirmations involved fledged young (FL, in 108 squares) or nest-building (NB, in 70 squares), the latter a sign that few observers were active late in the summer to confirm this species more conclusively.

In the primaeval forests that formerly covered the Maritimes, Goldfinches were much scarcer than at present. The opening-up of the land by European settlement benefited the species, although reversion of farmland to forest over the past half-century has reversed that trend to a minor extent. No decline is evident in the BBS data since 1966. Neither land-use changes, nor climatic warming, nor environmental pollution, seem likely to restrict the range or numbers of Goldfinches in the Maritimes in the coming decades.

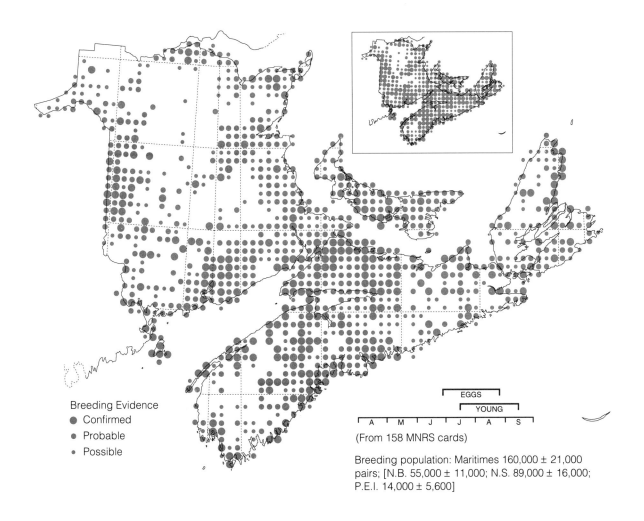

Breeding Evidence
- Confirmed
- Probable
- Possible

EGGS

YOUNG

| A | M | J | J | J | A | S |

(From 158 MNRS cards)

Breeding population: Maritimes 160,000 ± 21,000 pairs; [N.B. 55,000 ± 11,000; N.S. 89,000 ± 16,000; P.E.I. 14,000 ± 5,600]

Evening Grosbeak
Gros-bec errant

Coccothraustes vespertinus

The Evening Grosbeak breeds in a rather narrow strip across Canada in the southern boreal forest, and south in the western mountains to California. Its breeding has been little studied, and the hundreds or even thousands in an area in summer give little idea of its breeding density, though breeding probably occurs in most areas where it summers. Evening Grosbeaks were first seen in the Maritimes in 1913, but their occurrence in summer and breeding were not observed until the 1940s in New Brunswick, and later in the other provinces. They were found throughout our Atlas area, perhaps least frequently in southwestern Nova Scotia, western Prince Edward Island and southern Cape Breton Island. When these birds first wintered in the east, they often fed on the seeds of Manitoba maple, a western tree widely planted in eastern towns, but this association is less obvious in recent decades. Evening Grosbeaks are familiar visitors to winter feeding stations, but their breeding haunts are in conifer forests, the nest usually being high in a spruce tree. They attain high densities during outbreaks of spruce budworm, when the grosbeaks join many other birds in feeding on budworm larvae, and their numbers in an area fluctuate wildly in response to changes in budworm density. Breeding by this species in the Maritimes was confirmed in 217 squares (26 per cent of 850 with grosbeaks), but only five nests were observed in the Atlas period.

This familiar bird is a newcomer to the Maritimes, where it has become common only in the last 50 years. Settlement, with accompanying clearing of forests and later planting of exotic tree species, probably made the region more suitable for wintering by nomadic Evening Grosbeaks, but its breeding habitat, as now understood, was available here before Europeans first came. Their present numbers are probably as many as at any earlier period. Management of forests for conifer pulpwood production will ensure habitat for the species in the future, although global warming may gradually come to favour broad-leafed over conifer forests in this region.

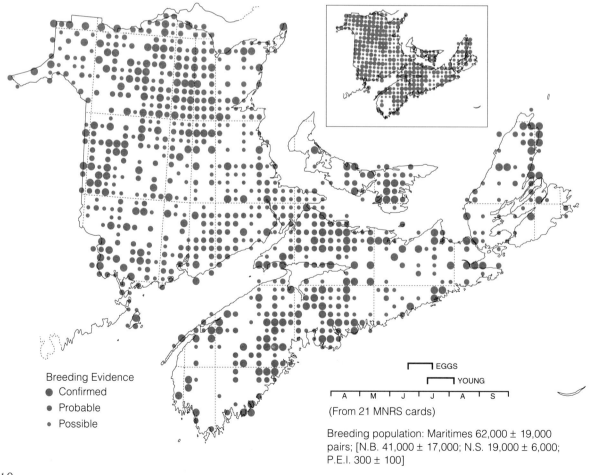

Breeding Evidence
- Confirmed
- Probable
- Possible

EGGS
YOUNG

A M J J A S

(From 21 MNRS cards)

Breeding population: Maritimes 62,000 ± 19,000 pairs; [N.B. 41,000 ± 17,000; N.S. 19,000 ± 6,000; P.E.I. 300 ± 100]

House Sparrow
Moineau domestique

Passer domesticus

Originally native to parts of Europe, western Asia, and north Africa, the House Sparrow has frequented human settlements for many centuries. Following movements of pioneering settlers, it spread north to the tree-line and east across Siberia, and it has been transported deliberately by people to every continent except Antarctica. It is found throughout North America, wherever human settlements occur. In the Atlas study, House Sparrows were found in all settled areas of the Maritimes, including both farmland and urban centres. The species must never have been present in large forested areas in New Brunswick and Nova Scotia, unless around large lumber camps, and it is still absent in all areas without permanent human occupation in the forest regions and along the outer coasts. Its close association with human dwellings for breeding contributed to its easy detection and confirmation; nests, mainly in nest-boxes and in buildings, were detected in 31 per cent of 753 squares with this species. The Atlas data should provide a useful picture of its status here.

House Sparrows came late to the Maritimes. Introductions began about 1850 in the United States and soon afterwards in Canada. Sparrows were brought to Nova Scotia ca.1857 from the U.S.A., but did not prosper, as they were not recorded in New Brunswick, Prince Edward Island, and on Cape Breton Island until the 1880s, apparently by direct spread from the west. In the era of subsistence farming and horse-drawn transport, they did well in agricultural and urban habitats. When motor transport replaced horses, and marginal farms began to revert to forest, the numbers and distribution of rural sparrows diminished. House Sparrows remain one of our most abundant urban birds. Their highly clumped distribution makes extrapolation to population estimates more uncertain than for most species. Even this adaptable bird is not immune to change. Recent tidying-up in towns and cities, termed "urban revitalization," including closing openings in old buildings where sparrows nested, make them less attractive even to these "feathered rats" that were pre-adapted to urban habitats. There is no fear that they will be eliminated completely in future, for this or any other reason, but House Sparrows are still decreasing in the Maritimes, as shown by the BBS and other surveys.

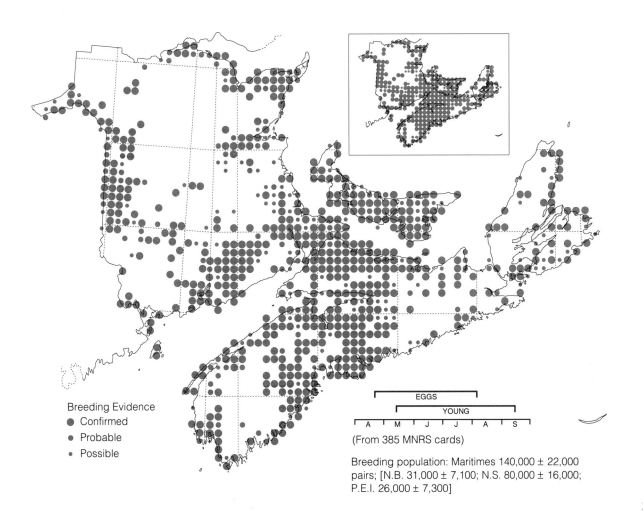

Breeding Evidence
- Confirmed
- Probable
- Possible

EGGS

YOUNG

A   M   J   J   A   S

(From 385 MNRS cards)

Breeding population: Maritimes 140,000 ± 22,000 pairs; [N.B. 31,000 ± 7,100; N.S. 80,000 ± 16,000; P.E.I. 26,000 ± 7,300]

Peripheral and casual species

Least Bittern   Petit Butor
Ixobrychus exilis

Least Bitterns live inconspicuously in dense marshes around freshwater lakes and rivers, a relatively scarce habitat in many areas. From South America through the U.S.A., except in mountain and desert regions, to southern Manitoba, Ontario, and Quebec, they occur locally, seldom seen and under-recorded relative to their true numbers. Records were obtained in 5 squares in the Maritimes Atlas: probable breeding near St. Anne-de-Madawaska (P in 1990) and several records at Red Head Marsh near Saint John (T in 1989), with single sightings at Piries Lake (1989) and Musquash (1989), all in western New Brunswick, and at Amherst Point Bird Sanctuary, N.S. (1990). The only previous records of confirmed breeding were of a nest at Little River, N.B. in 1895, flightless young at nearby Red Head Marsh in 1963, and an unsuccessful nest at Amherst Point in 1982. Birds were seen annually from 1975–84 near Germantown, N.B., but breeding was not proved, and none were detected there during the Atlas period. Given the secretive nature of the birds, the accu-

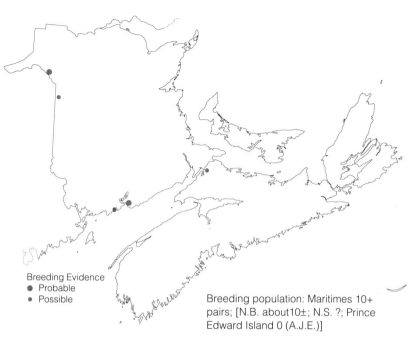

Breeding Evidence
● Probable
● Possible

Breeding population: Maritimes 10+ pairs; [N.B. about10±; N.S. ?; Prince Edward Island 0 (A.J.E.)]

mulated records indicate that breeding occurs at least sporadically in suitable habitat, through Maine to southern New Brunswick, where there may be 10 or more pairs in some years. Notwithstanding recurring records in a few areas, breeding may not occur every year anywhere in our area, and the species' status here is precarious. However, its habitat, especially in artificial impoundments, is increasing.

Snowy Egret   Aigrette neigeuse
Egretta thula

Like other egrets, the Snowy is mainly a bird of the tropics, breeding around marshes from Chile and Argentina to the southern U.S.A. A northward extension up the Atlantic coast reached Maine by 1961, and six colonies, with 200 nests, were found in that state in 1978–84. In the Maritimes it was known earlier as a vagrant, much more regularly since 1960, especially in spring, when it was seen most often in salt marshes. (In passing, we note that during the Atlas period several Little Egrets were identified in spring in coastal habitats in the Maritimes; some of the earlier reports of Snowy Egrets may have involved this very similar species from the Old World.) Starting in 1984, several Snowy Egrets summered in the mixed heron colony on Bon Portage Island, N.S. Breeding was stated to have occurred in 1988, but no details have emerged; however, the long-continued presence of summering egrets in the colony met our minimum criteria for "probable breeding."

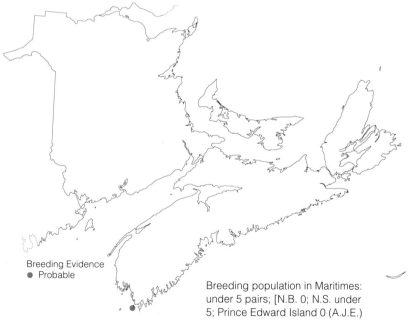

Breeding Evidence
● Probable

Breeding population in Maritimes: under 5 pairs; [N.B. 0; N.S. under 5; Prince Edward Island 0 (A.J.E.)

Although an island bordering one of the coldest coasts in similar latitude seems a strange breeding place for this exotic-looking southerner, Snowy Egrets may now breed in Nova Scotia—if so, the first breeding in Canada.

Glossy Ibis
Ibis falcinelle
Plegadis falcinellus

The Glossy Ibis has a highly fragmented or possibly relict distribution in tropical regions from Australia through Asia and Africa to the Americas, where it now overlaps with the closely related White-faced Ibis in Louisiana. Its breeding range has been extending northward along the Atlantic coast of North America in recent decades, like other tall wading birds. By 1972 it had reached southern Maine, where over 100 pairs nested in three areas in 1978–83. Glossy Ibises had wandered to the Maritimes, usually in spring, in various years since 1865 (in Nova Scotia), although they were not noted in New Brunswick until 1952. The first and only known breeding attempt in Canada was during the Atlas period, in a colony with Double-crested Cormorants on Manawagonish Island, N.B., 1986, an unsuccessful and evidently isolated event. There is no other evidence of Glossy Ibises breeding in the Maritimes, before or since, and the species has yet to establish itself as a breeder here. If global warming comes to pass, it may do so in future.
Ref. 95

Breeding Evidence
● Confirmed

Redhead
Morillon à tête rouge
Aythya americana

The Redhead is a western duck, breeding in fertile marshes on the prairies and in the intermountain region, from British Columbia to California, and from south Mackenzie and Manitoba to New Mexico and Wisconsin. Scattered breeding farther east mostly resulted from introductions, after 1945, in several states around the Great Lakes, although there were a few earlier records. Apart from a few old and unverifiable reports, and one brood on the St. John River in 1944, Redheads have bred in the Maritimes only on recent DUC impoundments in Nova Scotia, first at Wallace Bay in 1978, and at Amherst Point in 1981. The Atlas received reports of pairs (P) in the Amherst area and in central Prince Edward Island, but breeding was not confirmed anywhere in the Maritimes during the Atlas period. As with Ruddy Ducks, Common Moorhens, and Black Terns, Redheads perhaps find conditions here suitable for breeding only during the "first flush" of insect life after an area is newly impounded, natural marshes being insufficiently fertile. They evidently are less frequent than in 1980, and the species has not yet become established here.

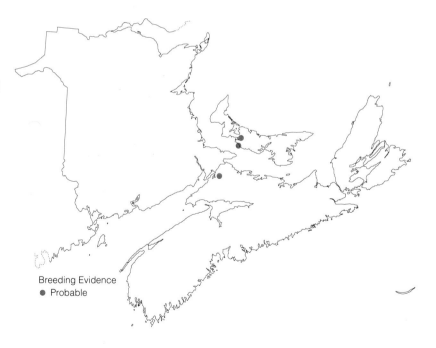

Breeding Evidence
● Probable

Harlequin Duck
Canard arlequin
Histrionicus histrionicus

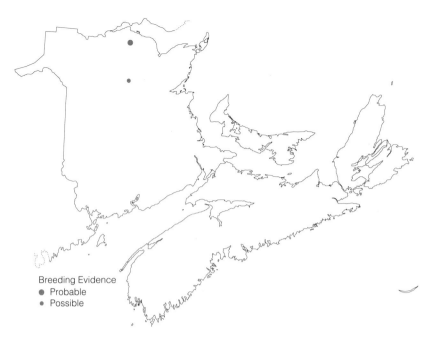

Breeding Evidence
● Probable
● Possible

The Harlequin Duck is a fairly common bird in eastern Siberia and northwest America, far distant from its small breeding populations in Iceland, Greenland, and eastern Canada, where it breeds locally from Baffin Island, N.W.T., south to the Gaspé Peninsula, Que. (max. 100 pairs) and western Newfoundland (a few pairs). Harlequin Ducks have declined in eastern Canada in the past century and are now scarce throughout this area; the eastern population was designated by COSEWIC as endangered in 1990. Harlequin Ducks frequent rocky shores in winter, but usually breed by rapid, rocky rivers inland from the coast. They have never been known to breed in the Maritimes, although small numbers winter here regularly. During the Atlas, records were obtained in suitable habitats on the Nepisiguit River (H, in 1988) and on the Benjamin River (P, in 1989), both in northern New Brunswick. The species had been seen in two previous summers at the former location without other evidence of breeding. Most probably these few summer records of adult-plumaged birds were of strays from the small Gaspé population. There is no other evidence that the species breeds in the Maritimes at present.

Ref. 60

Ruddy Duck
Canard roux
Oxyura jamaicensis

Breeding Evidence
● Confirmed
● Possible

This bizarre little duck is mainly a western species, breeding in fertile marshes from British Columbia to Manitoba and the northern prairie states, and locally southward through the western U.S.A. to the West Indies and northernmost South America. In recent decades it has bred very locally farther east, especially on sewage lagoons, to Lac St-Pierre, Que.; no records were obtained in the Maine atlas. The only confirmed breeding in the Maritimes, except for a vague report of a brood at Black Pond, Prince Edward Island, in 1962, was at Amherst Point, N.S., annually in 1979–84, with one report (for 1986, not submitted until 1990) in the Atlas files. No other Ruddy Ducks were seen at Amherst Point during the Atlas period, when three sightings (H) in Prince Edward Island were the only other reports. Most recent records have been on DUC impoundments, which simulate the prairie slough habitat preferred by these birds. The Ruddy Duck seems not to have become established here as a breeder, perhaps because few marshes attain suitable fertility levels except in the first three to five years after flooding.

Turkey Vulture
Urubu à tête rouge
Cathartes aura

Turkey Vultures are familiar birds in the United States, and beyond through Central and South America, but they breed north only into the southern edge of Canada, where small numbers are scattered from British Columbia to Ontario, most migrating south for the winter. Their breeding range has spread north and east in recent decades, reaching Camden, Maine, by 1982, with many records in southwest Maine. Sightings in the southwest Maritimes have become so frequent that not all are reported, and breeding was expected to be confirmed during our Atlas project. Most reports in 1986–90 clearly represented spring migrants overshooting their New England range (15 in New Brunswick, 19 in Nova Scotia, from 5 March through 21 May); 7 records from 28 May to 18 July seemed not to belong to either spring or fall migration periods. The 3 in the southwest plausibly might represent breeding areas, as Turkey Vultures are regular in extreme western Nova Scotia (whence 15

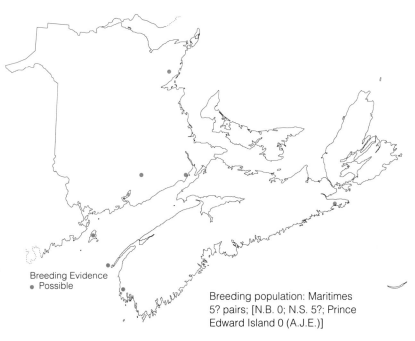

Breeding Evidence
• Possible

Breeding population: Maritimes 5? pairs; [N.B. 0; N.S. 5?; Prince Edward Island 0 (A.J.E.)]

of the spring migration records also came). As nests of this species are rarely found, and most birding effort in southwest Nova Scotia was put forth during the migrations, it is not impossible that a few pairs of Turkey Vultures already breed in Nova Scotia, despite the failure to confirm this during the Atlas period.

Golden Eagle
Aigle royal
Aquila chrysaetos

The most widespread of the world's large eagles, this species is scarce in North America except in the western mountains. It breeds locally across northern Canada, and south to Mexico, as well as in Eurasia. Remote areas, usually with cliffs or mountains, are its usual haunts, as it has been persecuted, often to local extirpation, as a predator of domestic animals. Breeding in the Maritimes has never been proved, although the more frequent records in the 19th century suggest it may have done so. It has bred for many years in a few locations in Maine and in the Gaspé Peninsula, so a few may possibly breed in our region. The Atlas received six reports, two involving pairs, all from remote areas seldom visited by bird students, including some sighted during aerial forest surveys; other second-hand reports lacked details and were not entered into the Atlas files. Conclusive evidence of breeding, if any had occurred, would have required a directed search, which was not practicable

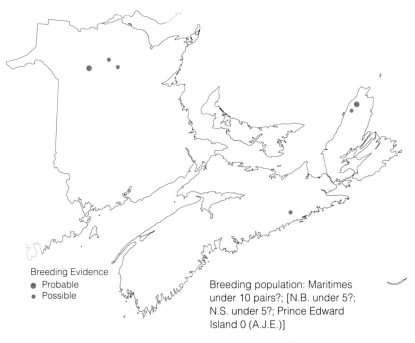

Breeding Evidence
• Probable
• Possible

Breeding population: Maritimes under 10 pairs?; [N.B. under 5?; N.S. under 5?; Prince Edward Island 0 (A.J.E.)]

during the Atlas period. Continued fragmentation of forests by lumbering will encroach on potential nesting areas, even if it may make foraging easier for the birds; disturbance through subsequent use of logging roads makes it unlikely that breeding by Golden Eagles will ever be frequent in the Maritimes.

Yellow Rail
Râle jaune
Coturnicops noveboracensis

The Yellow Rail is among the least known of North American birds, even among rails, which are notoriously secretive. This bird occurs in summer across Canada from Alberta to the Maritimes, and south into the U.S.A. from North Dakota to Massachusetts, with an isolated population in California and Mexico. The known range shows more gaps than records almost everywhere. Yellow Rails call, making a distinctive sound like tapping stones together, mainly late at night when few people are out in the damp grassy meadows and marshes frequented by these birds. They are seldom detected even where known previously to occur. In the Maritimes a nest with eggs taken near Milltown, N.B., in 1881 is the only proven breeding. Yellow Rails were found in the Midgic and Jolicure marshes near Sackville, N.B., in many summers since 1949, which suggests breeding there. During the Atlas period, one calling bird was detected near Sackville (1987), and single birds were flushed

Breeding Evidence
• Possible

Breeding population: Maritimes under 100? pairs; [N.B. under 50; N.S. under 50; Prince Edward Island 0 (A.J.E.)]

(one or both also heard) at Amherst Point in 1986 and 1989. Several were seen and heard at Grand Lake Meadows, N.B., in 1989 and 1990. It is likely that a few Yellow Rails breed in the Maritimes at present, but the data available do not allow adequate representation of either their distribution or numbers.

Laughing Gull
Mouette à tête noire
Larus atricilla

This is the breeding black-headed gull of the coasts from northernmost South America, the West Indies, and Mexico to California, the Gulf Coast states, and New England. The Maine atlas showed breeding in 9 squares, with over 230 pairs in 1977. In the Maritimes, the only record (H) in the Atlas period was from the large tern colony at Machias Seal Island, N.B., less than 100 km from the easternmost record in Maine. Breeding attempts recur at Machias Seal Island every few years, with nests noted in 1948, 1966, and 1981. Laughing Gulls bred regularly at several sites along the Atlantic coasts of Nova Scotia before 1940, but nesting has not been reported there since 1962, perhaps as a result of the increased numbers of large gulls breeding here in recent decades. The species is not known to have bred in Prince Edward Island, and it probably does not breed regularly

Breeding Evidence
• Possible

anywhere in the Maritimes at present. Its return here seems unlikely while large gulls provide ubiquitous competition.

Common Murre
Marmette de Troïl
Uria aalge

The Common Murre breeds on islands and sea-cliffs along coasts in the low arctic and boreal zones of the north Atlantic and north Pacific Oceans. It formerly bred in large numbers in eastern North America, where it was almost extirpated by unrestricted human exploitation of eggs and adult birds during the 1800s. The Maritimes shared in this dismal picture, with breeding ended before 1925 (in Nova Scotia), but numbers here were probably far fewer than in Newfoundland and in eastern Quebec. Since the 1920s, numbers have partly recovered in Newfoundland and Labrador. The discovery in 1981 of adults with eggs at Yellow Murr Ledge, N.B., with repeated sightings of birds ashore at Machias Seal Island, and a sighting of flightless young off Grand Manan Island in 1973, give some hope of re-establishment here. The only confirmed breeding record during the Atlas period was of adults with flightless young off Yellow Murr Ledge in 1988. As elsewhere, murres

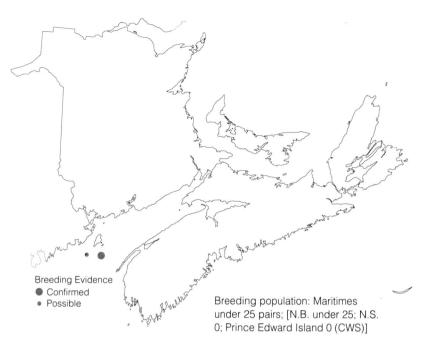

Breeding Evidence
● Confirmed
• Possible

Breeding population: Maritimes under 25 pairs; [N.B. under 25; N.S. 0; Prince Edward Island 0 (CWS)]

depend on suitable nest-sites secure from land predators, close to adequate stocks of small fishes. Increased tourist visitation to offshore islands and increased commercial exploitation of small fishes pose serious threats to hopes of recovery of all seabird populations in our area.

Yellow-billed Cuckoo
Coulicou à bec jaune
Coccyzus americanus

The Yellow-billed Cuckoo breeds in open woods and shrubland across the U.S.A. and in Mexico and the West Indies. It is a regular breeder in southern Ontario, but breeds only sporadically elsewhere in Canada, in southern Manitoba and Quebec. The eggs collected near Saint John (1899), the only presumptive evidence of this species breeding in the Maritimes, are now believed to be of Black-billed Cuckoo. However, records of Yellow-billed Cuckoos were obtained across southern Maine in 1978–83, including probable breeding in 3 squares adjoining the New Brunswick border. The one record received in the Maritimes Atlas, a single sighting near Centreville, N.B. (1988), about 40 km north of the nearest record in Maine, was in an area with similar climate and vegetation. The limit of regular breeding evidently lies very close to the border of our region, and a few pairs may

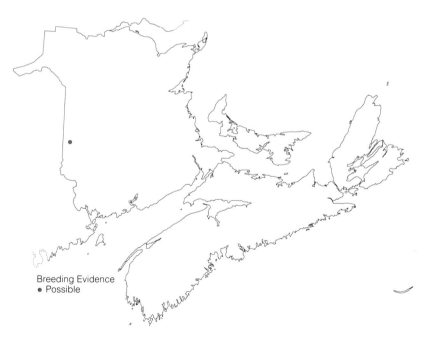

Breeding Evidence
• Possible

breed in New Brunswick in some years. Climatic warming may allow the species to become better established here.

Eastern Screech-Owl
Petit-duc maculé
Otus asio

Screech-Owls are resident wherever they occur regularly. This species breeds throughout the eastern broad-leafed forests of the U.S.A., south just into Mexico and north just into southern Canada. It has not been proved to breed in Maine for many years, although it was detected in 7 squares there during 1978–83. Similarly, there are no proven records of breeding in New Brunswick, where five sight records were obtained in 1965–73. Only one specimen (1892) and one report (1974) are known in Nova Scotia. The Atlas produced four reports, three in New Brunswick: single records (H) west of Welsford and near Lakeville Corner in 1988, and sightings west of Minto in October of both 1988 and 1990. A call, believed to be of this species, was heard near St. Peter's, P.E.I., in May 1990. The New Brunswick records are in the same part of that province as earlier reports, which lends them some additional credibility; even the October reports, and earlier records from

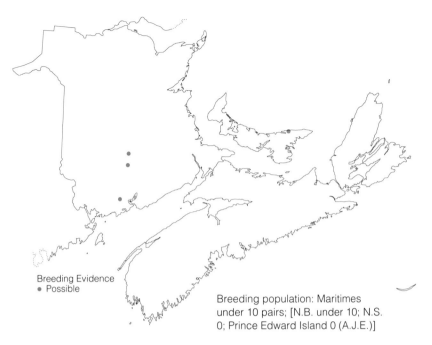

Breeding Evidence
● Possible

Breeding population: Maritimes under 10 pairs; [N.B. under 10; N.S. 0; Prince Edward Island 0 (A.J.E.)]

Christmas bird counts, may strengthen the picture for a supposedly resident species. In view of the poor representation of all owls in the Atlas, it is possible that a few pairs of Screech-Owls may breed in New Brunswick. Proving this remains a challenge for bird students.

Northern Hawk-Owl
Chouette épervière
Surnia ulula

This is a bird of the boreal forests, breeding in North America, from Alaska and British Columbia to Labrador and Newfoundland, as well as in Eurasia. Unlike most owls, Hawk-Owls regularly hunt by day, at all seasons, perching for long periods on tree stubs or branches in open woodland, so they are likely to be detected more readily than other, more numerous owls. There are occasional isolated cases of breeding well south of the main boreal range, e.g., near Ottawa, Ont., and the few records in the Maritimes seem likely also to be exceptional occurrences. The only nest, near Tabusintac, was found 6 June 1925 (but not reported until 1944!), and there were records of flying young in August 1878 at Point Lepreau and June 1963 on Grand Manan Island, all in New Brunswick. The Atlas accepted three more reports—a family group (FL) 14–24 August 1988 and a mating pair (C) 4–11 May 1986, also in New Brunswick, and a lone bird (H) 15–25 April 1990 in Nova Scotia—but no other sight-

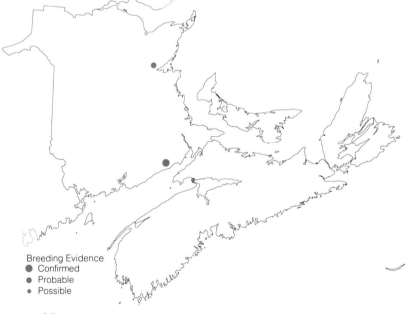

Breeding Evidence
● Confirmed
● Probable
● Possible

ings were reported despite far more time spent by knowledgeable observers in suitable habitats (cut-over forests) in New Brunswick than in the past. Our owls were all under-recorded, but Hawk-Owls are certainly very scarce and probably only sporadic breeders in the Maritimes.

Red-headed Woodpecker
Pic à tête rouge
Melanerpes erythrocephalus

Red-headed Woodpeckers breed throughout the eastern U.S.A., from New Mexico
and Florida north to southern Manitoba
and southern New Hampshire. Although
vagrants appear farther northeast almost
every year, no breeding is known to have
occurred in Maine since 1927, and the
only proven breeding in New Brunswick
was in 1881. This species lost ground
everywhere after the arrival of the Starling,
which frequents similar habitats and
which usurped nest-cavities faster than the
woodpecker could make new ones; however, Red-headed Woodpeckers never
bred regularly here, even before Starlings
arrived in the 1920s. The Maritimes Atlas
produced only four records: two in southern New Brunswick, with birds southwest of McAdam (T)
17–30 May 1987, and near Apohaqui (H) 27 June 1986;
and two in Nova Scotia, one in Amherst (H) in early June
1987, and one near Pubnico (H) in June 1988. The frequent appearances of these striking birds in autumn, with

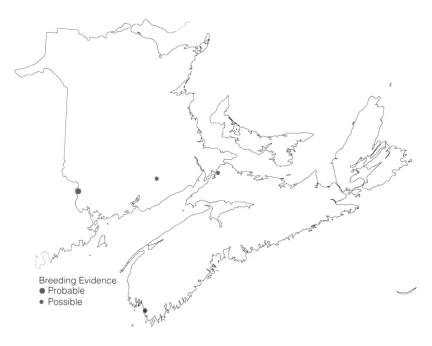

some successfully over-wintering at feeders and being
seen subsequently in early summer, tantalize us with the
idea that they may breed here again. However, there is
no solid evidence that Red-headed Woodpeckers now
breed in the Maritimes.

Tufted Titmouse
Mésange bicolore
Parus bicolor

The titmouse is a permanent resident
throughout its breeding range in the
hardwood forest region of the eastern
U.S.A., north to extreme southern
Ontario. During the Maine atlas study it
was first shown in 1978 to breed in
Maine, where scattered records, including confirmed breeding, were mapped
northeast to the Bangor area, far beyond
the previously known range limit (cf.
A.O.U. 1983). The Maritimes Atlas produced only one record, two birds at
Florenceville, N.B., on 22 June 1986,
"the first [in the province] for several
years." No subsequent records were
obtained, there or in nearby squares, so
it seems unlikely that breeding was
attempted, even though 2 birds—which need not equal
a pair—were seen. This location is 200 km northeast of
Bangor. Given the slow northward spread in Ontario,
the recent colonization of Maine, and the very few previous records in New Brunswick, we would not expect

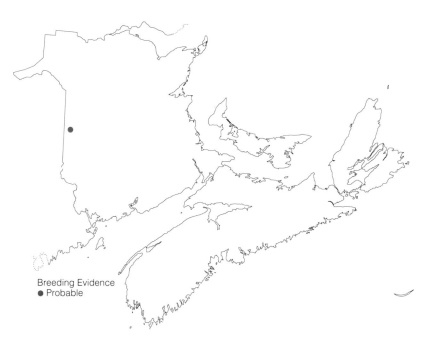

titmice to breed here yet, although with global warming
they may do so in the future.

House Wren
Troglodyte familier
Troglodytes aedon

The House Wren breeds all across the U.S.A from California and Georgia north into Canada, reaching 58° N in northeast British Columbia and Alberta but only to about 47° N, and mostly near human settlements, in Ontario and Quebec. It was regular in the southwest third of Maine, with scattered records of confirmed breeding east and north to approximately 70 km from the boundary with New Brunswick. Breeding in the Maritimes had been proved only in Fredericton, N.B., with nests in 1938 and 1946, plus an undetailed statement of nesting from Woodstock, N.B., in 1879. The MNRS received no nest records of this species since its start in 1960. In the Atlas period, only eight reports were received: unpaired singing males defended nest-boxes and built nests (no eggs laid) near Antigonish (1988) and at Lower West Pubnico (1989) in Nova Scotia, and near Nackawic (1990) in New Brunswick; there were also repeated sightings (T) in Saint John West (1990) and St. Stephen (1988) in New Brunswick. Another such bird sang for two weeks at Diligent River, N.S., in August 1989, improbably late for

breeding (not mapped), and thus, like various single sightings, was probably a wandering bird. Breeding in New Brunswick evidently occurred in several years in the distant past, but House Wrens apparently did not breed in the Maritimes during the Atlas period; the males that built nests would have attempted to do so had mates appeared in those areas. Conceivably, global warming may permit more regular breeding in the years to come.

Sedge Wren
Troglodyte à bec court
Cistothorus platensis

This little wren has a fragmented distribution on both continental and local scales. It has three separate breeding populations, in South America, in Central America, and from Alberta and Quebec south to Arkansas and Virginia. The main range of the northern stock extends from Michigan to Saskatchewan, where it frequents damp meadows and sedge marshes with low bushes, sometimes at high densities, e.g., 52 mapped territories in 18 ha in Manitoba in 1972. It was always scarce in the east. Groups of singing males occurred in New Brunswick near Sackville and Jolicure almost annually from 1949 into the early 1970s, and near Kingston in 1965. Breeding was never confirmed, and none of these sites seems to have been used since 1975. The only reports to the Atlas were of single singing birds (H) in New Brunswick, near Albert in 1988, and near Penobsquis and near Sackville (several kilometres from the earlier location), in 1989. Given the increases in numbers of

observers and in field activity here, the meagre recent evidence of Sedge Wrens in the Maritimes suggests that they no longer breed here, as was concluded also from the atlas project in Maine. Ongoing drainage of damp meadows is one possible influence, but the species never was well-established here.

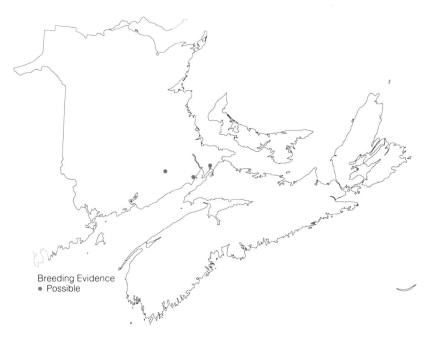

Breeding Evidence
● Probable
● Possible

Breeding Evidence
● Possible

Blue-gray Gnatcatcher
Gobe-moucherons gris-bleu
Polioptila caerulea

Gnatcatchers are southern birds: this species breeds mainly in Central America and the southern U.S.A., with a northeastward extension to Wisconsin, southern Ontario, and since 1979 to southern Maine. It is a bird of broad-leafed forest and thickets. The Maritimes Atlas received an extraordinary record, reported independently by two observers (one of whom knew the species from winters in Florida), of an adult Blue-gray Gnatcatcher feeding newly fledged young in a garden in Sackville, N.B., in 1989. Sightings of this species in spring have been reported almost annually along the Bay of Fundy coast of New Brunswick in recent years, which lends some slight credence to this record. As the edge of the main range north of Bath, Maine, is at least 500 km southwest, the Sackville record was clearly an exceptional case. Climatic warming might allow such occurrences to become more frequent in future.

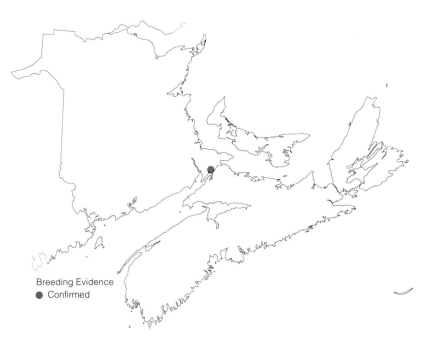

Breeding Evidence
● Confirmed

Bohemian Waxwing
Jaseur boréal
Bombycilla garrulus

The Bohemian Waxwing breeds in the northern boreal and subarctic regions of Eurasia and in western North America from Alaska and Washington (mountains only) east to northern Ontario, and probably farther east in central Quebec. In winter it appears, sometimes in large numbers, far outside its breeding range, hence the name "Bohemian," indicating a vagrant. The literature from both Europe and America gives no evidence of these waxwings breeding in areas reached during their winter irruptions. In the Maritimes, the Atlas received two documented summer reports, one (H,1989) of three birds seen in August, without other evidence of breeding. The other (1987) was a lucid description of adults visiting a nest with young in a young chestnut tree, apparently in or beside a house garden. The observer, then preoccupied with a family picnic, returned later for more details and photographs, to find the nest gone and its supporting branch broken. It seems scarcely conceivable that Bohemian Waxwings nested in Nova Scotia in 1987, as the habitat described is highly atypical for nesting by these birds, and the location is over 1000 km outside their known range. The species should be considered as no more than hypothetical in our breeding avifauna, on the evidence received by the Atlas.

Breeding Evidence
● Confirmed
• Possible

221

Loggerhead Shrike
Pie-grièche migratrice
Lanius ludovicianus

This shrike has become rare throughout
its Canadian range in recent decades and
was designated as "threatened" by
COSEWIC in 1986. It breeds from Mexico
through the U.S.A. to Alberta, Quebec,
and formerly the Maritimes, mainly in
open country with hedges or scattered
trees, habitats that were always scarce
here. In Maine, sightings of Loggerhead
Shrikes were noted in only 4 squares in
1978–83, and there are no recent breeding records there. The picture in the
Maritimes is similar, with only two summer sightings of shrikes, neither positive-
ly identified to species*, in the Atlas
period, near Centreville, N.B. (11 June
1986), and Montague, P.E.I. (15 June
1988). No breeding has been reported since 1972 (FL,
west of Moncton, N.B.); eggs were collected in 1888 and
1900, with occasional reports of family groups through
1946. Breeding was confirmed in Nova Scotia for the
first time in 1969, but the species is not even sighted in
that province most years. It is not known to have bred

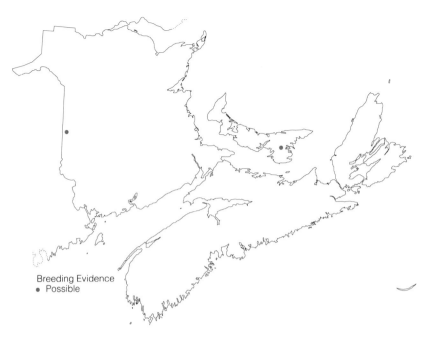

Breeding Evidence
• Possible

in Prince Edward Island and evidently no longer breeds
anywhere in our region.

*The possibility that these were Northern Shrikes summering 1000
km south of their known breeding range seems too remote to be considered seriously.

Rufous-sided Towhee
Tohi à flancs roux
Pipilo erythrophthalmus

This large sparrow frequents brushy
habitats and wood-edges from Central
America to southern Canada (British
Columbia to Quebec). The Atlas received
five records, with probable breeding only
near Bocabec, N.B. (T, 1988); the others
were single sightings (H) scattered in all
three provinces. Only the first of these is
within 100 km of the limit of regular
breeding in Maine. Towhees have never
been found breeding in the Maritimes,
despite annual occurrences in the last 30
years. It seems likely that the Atlas
records all involved strays that lingered
into summer, as happened several times
in the past (e.g., near Dartmouth, N.S., a
singing male was on a census plot 21
May-21 July in 1957). If global warming continues, this
species may eventually spread into the Maritimes as a
breeding bird.

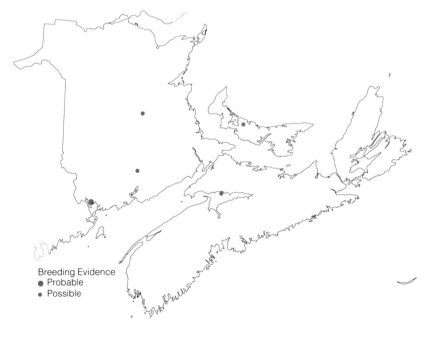

Breeding Evidence
• Probable
• Possible

American Tree Sparrow
Bruant hudsonien
Spizella arborea

The Tree Sparrow breeds from Alaska to southern Labrador, in scrubby woodland of the subarctic and tundra-edge zones, with southward penetration in the mountains to northern British Columbia. It has never been confirmed as breeding on the island of Newfoundland, although known from Brador, Que., and Battle Harbour, Labrador, just across the Strait of Belle Isle to the north. The Maritimes Atlas received one report, from the wind-blasted barrens of Cape Breton Highlands, of adult Tree Sparrows watched repeatedly plucking small spruce buds, perhaps containing insect larvae, and carrying them down among the dense scrubby cover, returning without them. Unfortunately, no attempt was made to confirm that a nest with young was present, as the evidence indeed suggests. In the absence of more positive breeding evidence, both at the reported site and from intervening areas in Newfoundland, we conclude that breeding by Tree Sparrows in Cape Breton Island, at least 600 km southwest from the nearest known

Breeding Evidence
● Confirmed

breeding, was exceptional, if it occurred here. Although the evidence satisfied one criterion for "confirmed breeding," this one really presupposes firmer evidence available in some nearby squares; in the isolated context of this record, "probable breeding" would be a more realistic assessment of this species' status in the Maritimes during the Atlas period.

Clay-coloured Sparrow
Bruant des plaines
Spizella pallida

The Clay-coloured Sparrow is characteristic of shrubby areas in the prairies of Canada and northern U.S.A., breeding locally and less commonly in similar habitats west into British Columbia, north into Mackenzie, and east through southern Ontario to southwest Quebec. No breeding evidence has been obtained farther east, although occasional singing birds have been reported, e.g., two in Maine during their 1978–83 Atlas period. Similarly, our Atlas received three records of singing birds in New Brunswick, all of which remained for two weeks or more (T), near Saint John (1987), St.-Simon (1989), and Shediac River (1990). The latter bird was reported near a nest attended by a Chipping Sparrow, but this seems likely to have involved response to similar begging calls rather than an interspecific association. The song, three or four "bzzzz" notes in series, though quiet, is so distinctive that few of

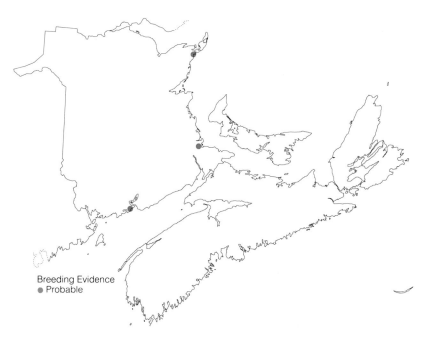

Breeding Evidence
● Probable

these birds would pass unnoticed within hearing of an observer. Clay-coloured Sparrows are not at present breeding birds in the Maritimes and are not likely to become established here in the near future.

Field Sparrow
Bruant des champs
Spizella pusilla

This species breeds in the eastern U.S.A. from Texas and Georgia northward, and in southern Ontario, southwest Quebec, and southern Maine. Despite its name, its habitat is in shrublands with scattered taller trees and only few and small open areas. Field Sparrows nested at Fredericton in 1972, and other singing birds have visited areas in New Brunswick and Nova Scotia for extended periods in the nesting season, without other evidence of breeding, in recent decades. In view of the earlier records, there were surprisingly few reports of Field Sparrows during the Atlas period, none of confirmed breeding. The continued presence of singing birds (T) was detected in 1988 at St. Stephen and Brockway and in 1986 near Pokiok, with a single record (H) at Pennfield (1987), all in southwest New Brunswick. It is possible that Field Sparrows bred in New Brunswick during the Atlas period, despite our failure to confirm breeding, as

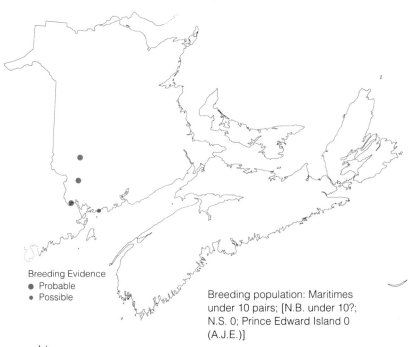

Breeding Evidence
● Probable
● Possible

Breeding population: Maritimes under 10 pairs; [N.B. under 10?; N.S. 0; Prince Edward Island 0 (A.J.E.)]

this region is not far beyond the continuous and regular range of the species in southern Maine. Global warming may allow these birds to breed here more regularly in the future.

Common Redpoll
Sizerin flammé
Carduelis flammea

Redpolls breed around the arctic of Eurasia and America, in tundra and sub-arctic zones. Their American breeding range extends from Alaska to Baffin Island and Newfoundland, south to the closed boreal forest, but with a very few isolated cases of breeding in prairie wintering areas. In the Maritimes, redpolls have been strictly winter visitors, excepting a few that summered on Sable Island in 1968. Two sightings on Cape Breton Highlands in July 1990 (H) were in open spruce-shrub habitat comparable to breeding areas on the island of Newfoundland, 400 km northeast. As the species has been known to breed sporadically in wintering areas elsewhere, it is remotely possible that a few Common Redpolls might breed on the Highlands in some years, despite the lack of any previous summer records there. Global warming,

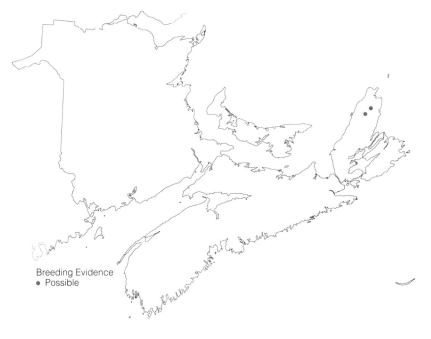

Breeding Evidence
● Possible

if it develops, is likely to discourage such northern species from becoming established farther south in future.

Additional species (not mapped)

(a) Species known to have bred in the past for which the Atlas received no records considered to represent even possible breeding (Province and date of last known breeding, and nearest recent regular breeding area elsewhere with distance away, in parentheses)

Horned Grebe   Grèbe cornu
Podiceps auritus
(New Brunswick, 1873. Not regularly east of prairies, 2300 km west)

Red-necked Grebe   Grèbe jougris
Podiceps grisegena
(New Brunswick, 1869. Luther Marsh, Ont., 1100 km west)

Northern Gannet   Fou de Bassan
Sula bassanus
(Nova Scotia, 1880; attempted breeding at Kent Island, N.B., in 1974, and sites occupied in summer, without further evidence, in Grand Manan area, N.B., in 1979 and 1982. Bonaventure Island, Que., 400 km northeast)

Bufflehead   Petit Garrot
Bucephala albeola
(New Brunswick, before 1900. North of Sudbury, Ont., 1300 km west)

Willow Ptarmigan   Lagopède des saules
Lagopus lagopus
(Nova Scotia, 1970s [introduced 1968–69]. Newfoundland, 150 km northeast)

Northern Bobwhite   Colin de Virginie
Colinus virginianus
(Nova Scotia, 1954–55 [introduced 1952]; escaped birds occasionally since. Massachusetts, 500 km southwest)

Passenger Pigeon   Tourte
Ectopistes migratorius
(New Brunswick, 1879)[Extinct 1914]

(b) Species with no previous breeding reported, for which records received by the Atlas were considered not to represent breeding (Location and evidence, plus distance to nearest known breeding area and latest known date there, in parentheses)

Manx Shearwater   Puffin des Anglais
Puffinus puffinus
(St. Paul Island, N.S., burrow of suitable size, no birds seen, heard, or smelt in area. Southeast Newfoundland, 1980s, 350 km northeast)

Eurasian Wigeon   Canard siffleur d'Europe
Anas penelope
(Grand Falls, N.B., and Jemseg, N.B., both adult males accompanied by unspecified female wigeons. Iceland, 1980s, 3200 km northeast).

Bonaparte's Gull   Mouette de Bonaparte
Larus philadelphia
(New River Beach, N.B., fledged young with adult on salt water. Northwest Maine 1982, 300 km northwest; perhaps not regularly nearer than northwest Quebec, 800 km northwest)

Common Black-headed Gull   Mouette rieuse
Larus ridibundus
(Musquodoboit Harbour, N.S., broken egg [identity uncertain] found in abandoned tern colony; this species considered by observer more likely than Laughing Gull. West Newfoundland, 1985, 500 km northeast)

Louisiana Waterthrush   Paruline hochequeue
Seiurus motacilla
(Welsford, N.B., single sighting of singing adult. Southwest Maine 1980, 350 km southwest)

Dickcissel   Dickcissel
Spiza americana
(St-Cyrille, N.B., single sighting of singing male. New York, 1980s, 900 + km southwest; occasionally Massachusetts 500 km).

(c) Species recently introduced, questionably wild, and as yet uncertainly established.

Chukar   Perdrix choukar
Alectoris chukar
(Introduced 1989, Lower West Pubnico, N.S., nested successfully in a garden in 1990; no introductions east of the Rockies are known to have persisted.)

Sharp-tailed   Gélinotte à queue fine
Grouse
Tympanuchus phasianellus
(Introduced 1988, northeast Prince Edward Island, noted in 1989 and/or 1990 in the same or adjacent squares, no breeding confirmed as yet; the natural range limit is 1300 km to the west, near the Quebec-Ontario border.)

Wild Turkey   Dindon sauvage
Meleagris gallopavo
(Introduced 1986–87, Grand Manan Island, N.B., broods seen 1988, and birds present through 1990; flightless young on Digby Neck, N.S., in 1988 were from an unauthorized release or escaped from captive stock; the natural range ended in southern Maine, with established stocks northeast to Penobscot Bay.)

VII
Distribution patterns and faunal elements

One purpose of mapping species distributions is to see which species occur in each area. Of particular interest is seeing which species have similar patterns of presence and absence, as these often are responding similarly to particular aspects of their environment. In earlier chapters we discussed some of the factors that may influence bird distribution within the Maritimes and our ability to assess it. Here we examine the distribution patterns set out in the species maps and correlate them with environmental factors.

General effects of sampling and coverage on patterns of bird occurrence and detection

The mapped distribution shows where a species was detected; this reflects where the species occurs, but is modified by the effectiveness of coverage, which in turn is affected by the detectability of the species. First we must discuss the influence of coverage.

Unlike the atlas projects in southern Ontario and New York State, which had complete coverage of their areas, we achieved a useful level of coverage in only about two-thirds of the squares in our region. We achieved complete coverage in essentially all priority squares, but in only about one-quarter of the other squares. Many of the latter received short surveys (one to five hours) only, and still others had only cursory visits (often focused on one or a few species). Our mapped data, collected on a grid of 10 x 10 km squares, show some gaps where a species was sought but not found and many other gaps where a square was sampled inadequately. If the data are mapped (as in the small inset of each species map) on a grid of 20 x 20 km squares, each of which includes a completed priority square (10 x 10 km), the distribution pattern appears more continuous than on the main (10 x 10 km grid) maps. The maps using a 20 x 20 km grid, by standardizing the coverage, depict more clearly (and more representatively) the real distributions for widespread species, but they may be misleading for some other birds.

In the Maritimes, many habitats recur in most of the 10 x 10 km squares used as sampling units, so the birds associated with those habitats appeared to occur almost everywhere, *if* they were relatively easy to detect where present. Many scarcer or less easily detected species were also widely distributed in suitable habitats, and they would have been found more widely if we had achieved more complete coverage within the 10 x 10 km grid. Birds of scarce habitats would appear less broadly distributed if a smaller square size (e.g., 5 x 5 km) had been used, because the scarce habitat would be less likely to occur in every square.

Table 3 summarizes the relative proportions of scarce species among the songbirds plus woodpeckers vs. the other non-passerines. A large proportion of our land birds breed across the Maritimes: 73 species out of a total of 116 songbirds plus woodpeckers occurred in at least 400 squares each. These species used a variety of habitats, or they used a few common habitats, usually forest or edge, that occurred in most squares; they occurred wherever suitable habitat was encountered, and thus all across our region.

The picture was quite different for the other, non-passerine, birds. Of 98 species with records, only 22 were found in more than 400 squares, and 32 were in fewer than 20 squares. These birds are mostly larger, scarcer, and more specialized in habitat use than the passerines. Either availability of habitat was limiting for many of these birds, or the observers failed to sample the suitable habitats intensively enough for detection of scarce or secretive birds. As coverage affects detection of scarce species more than of common ones, we must consider habitat specialization among birds.

Table 3.

Frequency of detection of non-passerine *vs.* passerine & woodpecker species by numbers of squares. Of 1682 squares, 1539 (91 per cent) had one or more species detected, and 1208 (72 per cent) had at least 30 species detected during the Atlas project.

Frequency class	Number of species in frequency class	
(Number of squares)	Non-passerines	Passerines & Woodpeckers
≥1300	0	1
1200–1299	0	3
1100–1199	0	13
1000–1099	0	10
900–999	2	9
800–899	2	9
700–799	3	13
600–699	3	7
500–599	5	4
400–499	7	4
300–399	7	3
200–299	8	7
100–199	12	8
50– 99	8	4
30– 49	3	5
10– 29	15	4
1-9	28	16

Broad-scale zonation

One simple grouping of bird species depends on their overall range limits. If we leave aside habitat requirements in our first segregation, the birds found breeding in the Maritimes may be grouped, on the basis of the distributions mapped in the Atlas, as:

- southern species that reach their northern range limits within the Maritimes (about 48 species);

- northern species that reach their southern range limits here (about 21 species); and

- species that breed here but show no range limits within the Maritimes other than those imposed by the seacoast and by habitat, the typical birds of our transitional region (about 106 species).

Range limits reflect factors that restrict where birds can breed successfully. Summer temperature is one factor that changes most obviously from the southern interior of North America to the northeast coasts of the Maritimes. With temperature, there is a change in forest cover, from temperate mixed forests to boreal conifer forests to subarctic krummholz. The birds change too, and generally decrease in variety in the cooler areas. Only a part of this gradient is found in the Maritimes, but its effect on bird variety is striking. A parallel gradient in coastal water temperatures, less obvious to land-based humans, also exists. The temperate to boreal to subarctic zonation is more easily related to temperature than to other factors.

The zonal classification leaves us with a number of unassigned species. On the one hand are those that were found in too few squares to allow allocation with any confidence — about 23 species, in fewer than 8 squares each. On the other hand are those with distributions that shared no common patterns among themselves or with the other groupings, although a few of these reached north or south limits in the Maritimes— about 16 species, which may be termed the "misfits."

This level of grouping is especially useful for comparisons on a continental scale. To examine further the bird distribution patterns within and extending beyond the Maritimes, we must superimpose on it a segregation based upon habitat.

Habitat groupings

(a) Forest and edge habitats

People who come to the Maritimes from other parts of the continent are often surprised to find that this long-settled region is still mostly forested and that coniferous trees make up a larger proportion of the forests than in similar latitudes not far to the west. They have to be reminded that nearly all of eastern North America was forested four centuries ago; forests are the major natural habitats of our region and also of those that adjoin it. However, the predominance of conifers here compared to regions to the west and southwest is a feature of broad-scale zonation, reflecting the transition from southern to northern forests.

Many birds associated with forests use openings in the forest, or edges where forest adjoins other habitats, rather than or as well as the forest interior. Because of the frequent overlaps in use of forest and edge habitats, our first and largest grouping includes both, comprising all bird species chiefly associated with woody vegetation (i.e., trees and shrubs).

The numbers of squares in which members of a group were found, in comparison to the 1208 squares in which 30 or more species were noted, give a hint as to how widely these species are distributed, assuming that all are detected roughly in proportion to their numbers (see comments on detectability in Chapter V); species noted in over 400 squares are fairly common to abundant, and widespread, whereas those found in fewer than 100 squares are usually scarce or local. Among common species, frequent confirmation of breeding suggests that a species is easier to detect than one that is less often confirmed [compare (2)a and (2)b].

(1) *Temperate forest species*, reaching northern limit of range within the Maritimes

All of these birds are found in predominantly broad-leafed woodlands, or their edges; conifers, especially hemlock and white or red pines, may be present too, but usually are minor or local elements of the forest cover in these areas.

a. Mostly in southwest New Brunswick, other records mostly in southwest Nova Scotia (in 17–62 squares, except * in 5–9 squares):

*Cooper's Hawk, Red-shouldered Hawk, Whip-poor-will, Willow Flycatcher, *House Wren, Brown Thrasher, Pine Warbler, Northern Cardinal, Indigo Bunting, House Finch.

Among these, the Pine Warbler is restricted to stands of white or red pines, and the House Wren is a cavity-nester that does not excavate its own holes in trees (or nest-boxes); thus, they are restricted in distribution by habitat or nest-site requirements, in addition to their zonal habitat limitations. The House Finch is still expanding its range and eventually will show a wider distribution here, probably clumped in urbanized areas.

b. More generally in southwest New Brunswick, scattered records to north and east (in 82–243 squares):

Black-billed Cuckoo, Eastern Phoebe, Great Crested Flycatcher, Purple Martin, Eastern Bluebird, Wood Thrush, Warbling Vireo, Scarlet Tanager, Northern Oriole.

In this group, Great Crested Flycatcher, Purple Martin, and Eastern Bluebird are all secondary cavity nesters using pre-existing tree-holes or nest-boxes, which further restricts where they occur within their overall ranges.

c. In all New Brunswick regions but scarcer or absent in eastern Nova Scotia (in 400–819 squares):
Broad-winged Hawk, Veery, Chestnut-sided Warbler, Black-throated Blue Warbler.

(2) *Transitional species*, not reaching northern or southern limits within the Maritimes.

These are the common birds of the Maritimes forests and edges, using a wide variety of conifer and mixed forests, and edge situations associated with them, throughout our region.

a. Nearly everywhere, well-detected (in more than 780 squares, more than 30 per cent confirmed):
Ruffed Grouse, Downy Woodpecker, Hairy Woodpecker, Northern Flicker, Tree Swallow, Blue Jay, Common Raven, Black-capped Chickadee, Red-breasted Nuthatch, Golden-crowned Kinglet, Ruby-crowned Kinglet, Swainson's Thrush, Hermit Thrush, American Robin, Cedar Waxwing, Red-eyed Vireo, Northern Parula Warbler, Magnolia Warbler, Yellow-rumped Warbler, Black-throated Green Warbler, Blackburnian Warbler, Black-and-White Warbler, American Redstart, Common Yellowthroat, Song Sparrow, White-throated Sparrow, Dark-eyed Junco, Common Grackle, Purple Finch.

b. Probably nearly everywhere, but less well-detected (in 168–1035 squares, only five species with more than 32 per cent confirmed):
Sharp-shinned Hawk, Northern Goshawk, Red-tailed Hawk, Merlin, American Woodcock, Great Horned Owl, Barred Owl, Saw-whet Owl, Common Nighthawk, Ruby-throated Hummingbird, Black-backed Woodpecker, Olive-sided Flycatcher, Eastern Wood-Pewee, Yellow-bellied Flycatcher, Gray Jay, Boreal Chickadee, Brown Creeper, Winter Wren, Solitary Vireo, Tennessee Warbler, Nashville Warbler, Bay-breasted Warbler, Ovenbird, Canada Warbler, Chipping Sparrow, White-winged Crossbill (widespread only in 1988), Pine Siskin, Evening Grosbeak.

These species were, except for Solitary Vireo, Tennessee Warbler, Ovenbird and Chipping Sparrow, less often detected than any in the preceding group and were also less often confirmed as breeding. Specific habitat or nest-site restrictions apply in many cases, but are so varied as to reduce the value of generalizations.

c. Found in all regions, but less general on the Atlantic slope of Nova Scotia where broad-leafed trees are scarcer (in 714–901 squares):
American Kestrel, Yellow-bellied Sapsucker, Least Flycatcher, Eastern Kingbird, Rose-breasted Grosbeak.

The Kestrel is a secondary cavity-nester, and thus somewhat more restricted in distribution than the others in this group.

d. Found in all regions, but nearly absent in Prince Edward Island, where forests were almost all removed in the past and not yet re-established (in 236–649 squares):

Spruce Grouse, Chimney Swift, Pileated Woodpecker, Northern Waterthrush.

The Chimney Swift depends on large, hollow trees for nesting, although secondarily using chimneys and empty buildings. The Pileated Woodpecker, a large bird, needs large trees or stubs in which to excavate its nest-cavities. Both of these species depend on relatively large areas of mature woodland, and thus are more restricted than other species of the transition zone.

(3) *Northern forest birds*, reaching southern limit of range within the Maritimes.

These birds are characteristically associated with cool conifer habitats (forests or bogs), or with successional poplar or birch habitats associated with the cool conifer forest.

a. Found in most regions, but scarce to almost absent in southwest Nova Scotia (in 409–669 squares):
Mourning Warbler, Wilson's Warbler, Lincoln's Sparrow.

Wilson's Warbler was less easily detected than the other species in this group and was anomalously absent from Prince Edward Island, but it seems to fit here better than in the other groupings.

b. Only in cool regions of northwest New Brunswick, along cool coasts, and north Cape Breton Island (in 10–232 squares):
Greater Yellowlegs, Boreal Owl, Three-toed Woodpecker, Gray-cheeked Thrush, Philadelphia Vireo, Blackpoll Warbler, Fox Sparrow.

The Boreal Owl is a secondary cavity-nester, and thus restricted within its range; it was not recognized as a regular breeding species in northern bogs until 1988 and was poorly detected. The Philadelphia Vireo also was poorly detected, owing to confusion of its song with the widespread Red-eyed Vireo; the scattered records farther south complicate the classification of its distribution pattern.

(b) Open lands
The birds associated neither with trees or shrubs, nor with wetlands or shores, may be grouped loosely as birds of open lands, including agricultural fields, coastal barrens, dry open bogs, or recent burns and clear-cut areas. In the Maritimes, these habitats have largely arisen from human actions, but small open areas of many types were present before European settlement, and the birds that use them were there too. The scarcity of farmlands here, noted above, is another reflection of the transitional nature of our region; agriculture becomes increasingly marginal as one moves from warm southern areas to cooler northern regions with less fertile soils. The species in the open lands are:
(1) *Southern species*, reaching northern limit in the Maritimes.

These birds were found in few regions, and only in

farmlands and other open areas (in 9–94 squares):

Upland Sandpiper, Northern Rough-winged Swallow, Vesper Sparrow, Eastern Meadowlark.

These species share a need for open areas for foraging and nesting, but the sandpiper and meadowlark use mainly lush fields with long grass, whereas the sparrow is characteristic of areas of low vegetation, including over-grazed pastures (now scarce) and blueberry fields. The swallow nests in rock crannies and forages over water as well as in open areas.

(2) *Transitional species*, not reaching northern or southern limits.

This assemblage includes several groups that share the habit of foraging in open areas, but use a wide variety of nesting situations.

a. Found in few regions, and only in open areas with short vegetation (cf. Vesper Sparrow above):

Horned Lark (in 67 squares).

b. Found in all regions, but only in areas with farmland or other open areas, nesting in various situations (in 498–1115 squares):

Northern Harrier, Killdeer, Mourning Dove, American Crow, Bank Swallow, Gray Catbird, European Starling (introduced), Yellow Warbler, Savannah Sparrow, Bobolink, Brown-headed Cowbird, American Goldfinch.

In this group, the Harrier, Killdeer, Savannah Sparrow, and Bobolink nest only on the ground, all the others using elevated sites in trees or shrubs. The Bank Swallow excavates burrows in cliffs and cut-banks and forages over water as well as in open areas. The Starling also nests in buildings (see below), but it uses tree cavities away from human habitation even more, so it is placed in this group rather than the next.

c. Found in all regions, but only in areas with farmland and other open areas, nesting on buildings and other structures, mostly in urban areas or farms (in 562–1159 squares):

Rock Dove (introduced), Cliff Swallow, Barn Swallow, House Sparrow (introduced).

Rock Doves and Cliff Swallows increasingly use highway bridges and overpasses, sometimes remote from human settlement, for nesting, whereas House Sparrows remain associated with towns or farms at all seasons. Barn Swallows nest almost entirely on artificial structures, even far into the forest, but they have no affinity for trees, and they forage mostly in open areas (and over water), so they are most appropriately placed here.

(3) *Northern species*, reaching southern limit. In uncultivated open land:

Short-eared Owl (29 squares) is the only species that fits here.

(c) Freshwater wetlands

The birds of inland waters make up a third group based on habitat, here including not only water birds in the marshes, but also some shorebirds and perching birds using vegetation along the shores or emerging from the waters. Again in wetlands, there is a transition in habitats, from more fertile marshes in warmer areas to the south to sterile bogs and rocky lakes in the cooler and more northern lands; the Maritimes wetlands include examples all along this gradient. A few of these species also breed in coastal habitats (see next section), but their breeding populations in the Maritimes are mainly inland. The species here are:

(1) *Southern species*, reaching northern limit in the Maritimes.

All these birds are more general farther to the west and southwest and are restricted in their distribution here.

a. Found in few regions, and only in fertile fresh marshes in open areas, mainly lower St. John River, northeast coast New Brunswick, New Brunswick–Nova Scotia border, eastern Prince Edward Island (in 21–332 squares, except those marked * in 6–14 squares):

Pied-billed Grebe, *Least Bittern, *Green-backed Heron, Blue-winged Teal, Northern Shoveler, Gadwall, American Wigeon, Virginia Rail, Sora, *Common Moorhen, American Coot, *Wilson's Phalarope, *Black Tern, *Marsh Wren.

b. Found now in most regions, scarce or absent except in southwest New Brunswick before widespread erection of nest-boxes and introductions of Wood Ducks (in 158–290 squares):

Wood Duck (partly introduced), Hooded Merganser (not in Prince Edward Island).

These birds depend for nesting on large tree cavities, which originally were scarce except on the St. John River flood-plain; with human assistance, they have spread beyond their former range, but not beyond our region.

(2) *Transitional species*, not reaching northern or southern limits.

These species are found in most or all regions in the Maritimes and are mainly associated with inland waters, but are not restricted to fertile marshes (those marked * breed also in coastal habitats to varying extent, as well as inland) (in 257–1008 squares):

Common Loon (not in Prince Edward Island), American Bittern, *Green-winged Teal, *American Black Duck, *Mallard (mostly introduced), Ring-necked Duck, Common Merganser (not in Prince Edward Island), *Spotted Sandpiper, Common Snipe, *Belted Kingfisher, Alder Flycatcher, Swamp Sparrow, Red-winged Blackbird.

This group is heterogeneous, including birds of open lakes, rapid rivers, bogs, marshes, and water-edge shrubbery. Common Mergansers nest mostly in large tree cavities in this region. Alder Flycatchers also breed far from open water, but only where the shrubs characteristic of water-edges grow, mainly in relatively damp sites.

(3) *Northern species*, reaching southern limit.

These birds breed around freshwater wetlands, Northern Pintail (92 squares) and Common Goldeneye (167 squares) being the only species that seem to fit here. The latter requires large tree cavities for nesting and thus, before nest-boxes, was most common in flood-plain forests. The Pintail uses mainly wetlands in open areas.

(d) Seacoast habitats

Birds associated with the sea and its coasts make up our final habitat grouping of breeding species. It is customary to distinguish pelagic species or "primary seabirds" (ones that feed at sea, often far from land, and come ashore only to breed) from coastal birds or "secondary seabirds" (ones that forage in shallow waters near shore, on the beaches, and in the salt-marshes of estuaries and lagoons). Some of the latter group also breed around inland waters, and their inclusion here reflects the fact that in the Maritimes most of the species' population are coastal rather than inland breeders; the same species may breed by lakes in the interior of the continent. The birds in these groupings are:

(1) *Southern species*, reaching northern limit in or slightly beyond the Maritimes.

These birds breed mainly on coasts and islands, and thus are not found in all regions (in 57–414 squares, except those marked * in 10 squares):

Great Blue Heron (some inland, but mostly coastal), Piping Plover, Willet, *Roseate Tern, Sharp-tailed Sparrow (a few inland, by rivers).

The Willet and Sharp-tailed Sparrow are associated with salt marsh habitats for foraging and brood-rearing, although Willets nest mainly in nearby uplands; these species, plus Piping Plover, also have separate populations breeding inland in the Prairie Provinces. Roseate Terns are too scarce to show any pattern, but they reach their northern limit in our area.

(2) *Transitional species*, not reaching northern or southern limits.

Birds in this group were found in most regions, mainly along coasts and islands (a few, marked *, also nest inland) (in 16–650 squares):

*Double-crested Cormorant, Great Cormorant, *Red-breasted Merganser, *Osprey, Semipalmated Plover, *Ring-billed Gull, *Herring Gull, *Great Black-backed Gull, *Common Tern, Black Guillemot.

This group includes seabirds that nest on cliffs, on sand or gravel beaches, and in trees, with a shorebird and a duck nesting on beaches, and one tree-nesting raptor, all of which obtain their food mostly from shallow waters or shores in breeding season. The gulls secondarily use human wastes of many kinds, especially at other seasons. All the species marked * breed also in parts of the continent far from the sea, so their requirement is for open water rather than for salt water *per se*; all are mainly coastal-breeding birds in the Maritimes, although many Ospreys nest inland and commute to coastal waters to forage.

Great Cormorant, Red-breasted Merganser, Semipalmated Plover, and Black Guillemot breed only a little farther south than this region and also far to the north, but their ranges here are not restricted to cooler coastlines, so they are placed here rather than with the next group. The world distribution of Leach's Storm-Petrel (see below) suggests that it belongs here, but its mapped range puts it in the northern group associated with cool coasts.

(3) *Northern species*, reaching southern limit.

This group bred only along cooler coasts and islands, near Grand Manan Island or on the Atlantic shores of Nova Scotia (in 27–105 squares, except species marked * in 2–6 squares):

Leach's Storm-Petrel, Common Eider, *Least Sandpiper, *Black-legged Kittiwake, Arctic Tern, *Razorbill, *Common Murre, *Atlantic Puffin.

Inclusion of Least Sandpiper with this group, otherwise comprising only pelagic seabirds, arises from its never having been found breeding on the Gulf of St. Lawrence or Northumberland Strait coasts of the Maritimes (compare with Semipalmated Plover). All of these species, except Arctic Tern (on Sable Island), had been reduced to relict stocks in our area by 1900, and most have not yet recovered their former numbers nor re-occupied all of their potential range.

This classification is arbitrary, and another person would assign a few species to different categories. Nevertheless, these groupings provide useful perspective and are based on the Atlas maps rather than being purely intuitive. Table 4 summarizes the numbers of species by habitat group in the three major zones.

Table 4.
Numbers of bird species by habitat and zonal groupings.

Zone	Forest & Edge	Open lands	Freshwater wetlands	Coastal habitats
Southern (48)	23	4	16	5
Transition (106)	66	17	13	10
Northern (21)	10	1	2	8

The species not listed in the groupings described above include the "misfits," which are listed below with a summary of the distribution anomaly, the number of squares with the species, and in [ ] its most plausible placement in the classifications above:

Black-crowned Night-Heron: discontinuous here, near north limit of range (48) [(d)(1)]
Canada Goose: extirpated, and re-introduced locally (104)[(c)(2)]
Greater Scaup: recent, isolated, atypical habitat (5)[(c)(3)]
Bald Eagle: discontinuous here (328) [(d)(2)]
Peregrine Falcon: extirpated, and re-introduced locally (11) [??]
Gray Partridge: introduced, declining (51) [(b)(1)]
Ring-necked Pheasant: introduced, ongoing releases (230) [(b)(1)]
Solitary Sandpiper: recently detected, discontinuous here, at south limit of range (10) [(c)(3)]
Long-eared Owl: probably widespread, so poorly detected that no pattern is obvious (35) [(a)(2)?]
White-breasted Nuthatch: patchy, in broad-leafed woodlands, near north limit of range (202) [(a)(1)?]
Northern Mockingbird: discontinuous here, related to winter feeding, near north limit of range (139) [(a)(1)?]
Cape May Warbler: patchy, owing to very uneven detection, near south limit of range (422) [(a)(2)]
Palm Warbler: discontinuous here, avoiding steep terrain, near south limit of range (434) [(a)(3), or (c)(3)]
Rusty Blackbird: uneven distribution, near south limit of range (446) [(a)(3), or (c)(3)]
Pine Grosbeak: uneven distribution, near south limit of range (339) [(a)(3)]
Red Crossbill: uneven and sparse distribution, perhaps owing to former reduction of food trees (163) [(a)(1)].

Each of these species has a unique distribution pattern that would extend the variability of other groupings with which they might be placed.

Finally, several species were detected too seldom to be treated as belonging to the regular avifauna of the Maritimes. Most of these are near or beyond their southern or northern range limits, and none is thought to breed regularly, if at all, in the Maritimes (all detected in 1–7 squares only):

Snowy Egret, *Glossy Ibis, *Redhead, Harlequin Duck, *Ruddy Duck, Turkey Vulture, Golden Eagle, *Yellow Rail, *Laughing Gull, Yellow-billed Cuckoo, Eastern Screech-Owl, *Northern Hawk-Owl, *Red-headed Woodpecker, Tufted Titmouse, Sedge Wren, *Blue-gray Gnatcatcher, *Bohemian Waxwing, *Loggerhead Shrike, Rufous-sided Towhee, *American Tree Sparrow, Clay-coloured Sparrow, *Field Sparrow, Common Redpoll.

The species marked * were reported to have bred at least once in the Maritimes, during or before the Atlas period, but such breeding was sporadic in time and was not continuous with the main ranges of these species. A few other species, also reported to the Atlas, seemed not to warrant consideration in any category; they probably were correctly identified, but they are best treated as wandering individuals only (see list at end of the species accounts). A few species in the list immediately above might be treated similarly.

Faunal groupings

Zoogeographers recognize America north of the Mexican border as a region or "realm" whose fauna overlaps more extensively with that of northern Eurasia than with tropical America. Within North America, ecological zonation schemes mostly place the Maritimes somewhere on a gradient from temperate broad-leafed forest to boreal conifer forest to subarctic dwarf forest, but the placement varies with the criteria used to define the transition. Grouping of bird species showing similar distribution patterns, as above, showed that far more of our breeding birds also breed both farther north and farther south than reach their northern or southern limits here. There is not an abrupt transition from southern to northern avifaunas; many species, referred to above as the "transitional" group, are shared over a wide area that includes our region. Only the smaller groups of species that reach their range limits within the Maritimes may be termed unambiguously as belonging to southern or northern bird groupings.

As more species were placed in the southern grouping than with the northern one, we might expect that greater diversity, i.e., more species per square, would be found in the southwest parts of the Maritimes than elsewhere in our region. For squares with relatively complete coverage and similar effort, this may be recognized in the Atlas data, but so many factors complicate the picture that few generalizations are warranted. Much the highest numbers of species, averaging over 95 per square, were found in southwest New Brunswick, owing to the many birds that penetrate our region only there. Most areas averaged between 84 and 91 species per square. The lowest diversity, averaging under 83 species per square, was found in the northwest and north-central highlands of New Brunswick and on the Atlantic slope of Nova Scotia. Most regions had inland squares with no farmland or marshes that averaged under 70 species per square, even when completed; the mean diversity often varied greatly between adjacent squares. The influence of cooler temperatures, mostly expressed through conifer forest cover, is exerted by proximity to the cold ocean and by greater elevations as well as by higher latitudes, so many southern birds reach their range limits to the east, as well as northward, within the Maritimes.

A common impression among people visiting our region from southern Ontario and the eastern United States is that the Maritimes has a greater variety of forest birds, and especially of breeding warblers, than other equivalent areas. This is a misconception, perhaps arising from a belief that the Maritimes, being a "settled" region, is more similar to other settled and predominantly agricultural parts of southern Canada and New England than to the forested "hinterlands" of central Ontario and central Quebec. Actually, on the basis of its birds, the Maritimes is much more nearly equivalent to the northern hinterlands. It has different breeding birds, but not more species, than areas farther west at the same latitude.

In the Maritimes Atlas, we found fewer, rather than more, species per square than in most of southern Ontario. In part, this resulted from less intensive coverage here as, other things being equal, the more time spent in a square the more species were found. Another factor reducing our species per square counts compared to southern Ontario was the preponderance of infertile soils and unproductive conifer forests here (Black and Maxwell, 1972). Although many species of warblers are characteristic of eastern boreal and subboreal conifer forests (Mengel 1964), few are completely restricted to these habitats, and many breed across Ontario and locally farther south. In contrast, a wide array of birds breeding north into southern Ontario, associated with the eastern broad-leafed forests and shrubby edges and openings in and around them, find the cooler and less fertile forests of our region unacceptable. Our forested squares, with no farmland or coast included, had many of the lowest species totals of all well-worked areas in the Maritimes.

Geographic dispersion probably does not enter the picture. In moving from a continent to a peninsula to an island, the ease of access decreases, and the number of species present may also decline. For birds, access is much easier than for mammals, amphibians, or freshwater fishes, not to mention plants; all bird species breeding in New Brunswick have also reached Nova Scotia and Prince Edward Island, though not all have established breeding populations there. The long lists of weather-assisted vagrants recorded in the Maritimes in spring show that access here by birds is easy, and has not placed limits to the number of species that settled here.

The areas of Prince Edward Island and, still more, of Nova Scotia are large enough that no one species there has occupied several ecological niches generally used by different species in New Brunswick, as may occur on small remote islands. The history of human activities in Prince Edward Island explains best the smaller variety of bird species there: nearly all of that province's forests were cleared for agriculture in the recent past, thus eliminating the entire populations of many forest birds over wide areas. Some species eventually will re-establish themselves in Prince Edward Island, as Pileated Woodpecker and Broad-winged Hawk may be starting to do, if the regenerating forests are allowed to grow to near-maturity over sufficiently large areas. The other differences in species richness between New Brunswick and Nova Scotia/Prince Edward Island probably arose from temperature and local habitat distrib-

ution, as outlined above for the various groupings.

Most of our squares include conifer and mixed forests of a wide range of ages, with scrub woods and shrublands. Broad-leafed trees tend to be scattered in mixed forest as often as, or more often than, forming pure stands, but these too are widespread. All birds of hardwoods except the more strictly southern species are found in most parts of the Maritimes. These forest and edge habitats provide a "core" group of species, found almost everywhere. The inclusion or not of farmlands and settlement, even as "ribbon development," is an important distinction between squares with higher vs. lower species counts, as a number of birds associated with these open habitats are scarce or lacking in many forested squares here. Despite the variety of birds associated with some fertile wetlands, the generally distributed minor damp areas along brooks or rivers added usually only two or three species to a square. Thus, in contrast to southern Ontario, few squares in the Maritimes were predicted to have over 100 species, most having between 80 and 95 species. The numbers found seldom fell outside these limits, and only 21 squares (Table 1) had 115 or more species recorded. These squares averaged 114 hours of field study, far above the overall mean of 28 hours, and thus were better studied but not necessarily richer in species than the others.

As noted above, the forest bird fauna of the Maritimes is very similar to those of other boreal regions of Canada east of the Rockies. Comparison of the transitional groupings (above) with the species lists for various forest habitats summarized in *Birds in boreal Canada* (Erskine 1977) shows almost complete agreement for the groups that were adequately sampled in both studies. The distinctively "northern" forest species in the Maritimes Atlas mostly were identified as characteristic of the boreal-subarctic ecotone rather than of the main boreal forest. The "southern" forest species groupings in our classification include some that extend far south through the temperate broad-leafed forest region, along with others that are quite representative of the transition from temperate to boreal forests.

The widespread occurrence in the Maritimes of trees characteristic of the temperate-boreal transition, especially eastern hemlock, white pine, sugar maple, beech, and yellow birch, though interspersed with larger amounts of spruces and balsam fir, suggests placement of this region in a transitional position. Most of the temperate forest tree species extend north beyond the Maritimes, in favourable situations, to the Gaspé peninsula, some even occurring on the island of Newfoundland. Nevertheless, the forest birds here are a boreal assemblage, only in the locally hardwood-dominated regions of southwest New Brunswick approaching the avifaunal composition of the temperate-boreal transitional zone in Ontario and Quebec.

The forests of the boreal-subarctic ecotone differ from those of the main boreal forest more through losing species less tolerant of cold than by replacement with distinctively northern plants, but they often feature more stunted growth forms. In the Maritimes, only the dwarfed and patchy tree cover of the top of the Cape Breton plateau resembles the subarctic forests of Newfoundland and Labrador. The occurrence of northern birds in the Maritimes only occasionally corresponds to a typically "subarctic aspect" of the vegetation; "islands" of such vegetation also occur in suitably cool "frost pockets," often in bogs, over much of the Maritimes, sometimes harbouring some northern birds as well. Thus, zonation based on the birds places the Maritimes much farther "north" along the temperate-boreal-subarctic gradient than does the tree cover.

Our forest birds are a boreal assemblage, but they make up only part of our avifauna. Few species of open lands or of freshwater wetlands could be identified as reaching their southern limits here. Farmland is nearly absent farther north, and the few birds that use other open areas there are mostly widespread species. Similarly, fertile wetlands, and also artificial impoundments, are even less frequent farther north than in our area; the greatest diversity of northern water birds is encountered only as one approaches the tundra-edge, far beyond the Maritimes.

The coastal birds provide a slightly more promising group with which to illustrate temperature-related ecological zonation; the main obstacle is that several species here (and elsewhere) occur only as remnant populations in a few relict colonies, owing to excessive exploitation and disturbance in the past. On the basis of temperature-salinity relationships, combined with distributions of marine invertebrates, seabird workers recognize a boreal zone that includes our region (Brown, 1986). The adjacent oceanographic zones, termed "subtropical" and "low arctic," extend south to Florida and north to Baffin Island, respectively, and thus correspond roughly to the latitudinal spans of the terrestrial zones "temperate forest" and "subarctic dwarf forest" that adjoin the boreal forest.

The ranges of coastal birds inevitably are discontinuous, owing to the needs for suitably safe nesting sites, on appropriate substrates, within commuting range of sufficient and accessible food supplies. Thus, zonal boundaries are not very obvious, and some species have mapped ranges placing them in a different group than where they would be expected on the basis of wider knowledge. These inconsistencies arise partly because some species' ranges represent only remnants of former wider distributions, and partly because the oceanographic boreal zone extends rather little beyond our region, either to north or south; thus, many coastal birds close to their range limits may be found in situations that seem somewhat aberrant ecologically, e.g., Black Guillemots tolerating warmer water than usual in Prince Edward Island to take advantage of adequate food near suitable nest-sites. Such tolerances are usually flexible, to some degree. Our coastal and forest birds agree in indicating that our region has a boreal avifauna, rather than one that is transitional between temperate and boreal as suggested by forest cover.

VIII
Estimates of populations of Maritimes birds

Background and development

Distribution and abundance are related concepts, as noted in Chapter II. Organizers of the first modern bird atlas project, covering the British Isles (Sharrock, 1976), recognized the desirability of population estimates, but decided against adding the collection of such data to what was already a daunting task. As no indices to bird density were collected during the field effort there, the compiler later derived population estimates from other sources and included them in the final publication. Soon after the first three atlasses were published in 1976, it was generally accepted that all later atlas projects should attempt to include numerical indices to bird abundance in the data-collection process. The first Canadian atlas project, in Ontario, assembled "abundance indices" along with the distributional data collected by their co-operators and presented summaries of these density data in the accounts for each species.

Following the lead of the Ontario project, the Maritimes Atlas also requested abundance indices from its observers, in addition to the bird distribution data basic to the project. The question of whether anything further could be done with these indices was discussed by the Steering Committee, starting early in 1989. We recognized that great precision could not be expected in extrapolations from these indices. But even "order of magnitude" estimates would be an advance over the present lack of published population figures for nearly all of our birds except waterfowl and seabirds.

Our first attempt, in June 1989, at obtaining population estimates from the abundance indices involved manual extrapolations covering three species in one well-surveyed region. Computer extrapolation was seen as essential for the common and widespread species, but an acceptable statistical treatment, with programming to implement it, did not emerge until January 1991 (see Appendix D). The first trial runs were encouraging, as population estimates of the few species for which previous estimates existed were in reasonable agreement with the earlier figures. With the printers' deadline for receipt of the book manuscript then only months away, there was little time for refining the statistical treatment and computer program, and only a few adjustments were made to the original version. Roughly half of the breeding species proved amenable to computer extrapolation, the others being colonial water-birds or scarce species, or so unevenly dis-

tributed that the extrapolation program produced unrealistically high figures.

Few atlassers had much understanding of, or confidence in, the abundance indices at the start of the Atlas project. Many people dismissed them as useless or incomprehensible, or both, and observers who made only one or two visits to a square often hesitated to extrapolate this limited experience across the whole square, even if its coverage was considered "complete." Overall, only about 40 per cent of the records showed abundance indices. Some were wildly unrealistic, often because the instructions were not understood, and these were not used at all. Some other abundance indices were later deleted or altered to more realistic levels; the latter process was avoided, as being too subjective, until we actually began to use the indices. The unrealistically high indices that remained in the database when the calculation of populations began caused much perplexity. To nearly everyone's surprise, the vast majority of the abundance indices appeared plausible, but we will never know what proportion of those were seriously in error.

The abundance indices were subjective, but they represented the judgments of people who had visited those squares. The species for which computer extrapolation of abundance indices was done usually had such indices for 40–55 per cent of the squares in which the species was found.

Problems with the population extrapolations

Differences in the detectability of species and in the intensity of sampling effort carried over into the estimation of numbers. Birds that frequent habitats altered by people, including the roadside edges, were detected more readily and thus *appeared* to be more abundant than those of less disturbed and more remote areas. The estimated population of Alder Flycatchers, a roadside species, relative to that of Yellow-bellied Flycatchers, a bird that uses the much more extensive conifer forest habitats, often distant from roads, illustrates this effect. Under-represented species might be reported as scarce birds (abundance index 2 or 3); they included Leach's Storm-Petrel, our only colonial seabird known to have over 10,000 breeding pairs (index 6) in one square, and Bay-breasted Warbler, often the most common warbler of spruce/fir stands in New Brunswick, where it sometimes achieves densities on 10–20 ha plots equivalent to index 6, but suitable habitat hardly ever is continuous over all of a 100 km² square.

Many squares that showed abundance indices for some species were sampled incompletely, and several to many species that presumably occurred there were missed. The extrapolation process treated absences in such uncompleted squares as if these had not been sampled at all; i.e., it attributed the species to the same proportion of them, at the same average density, as was found among completed squares. This would tend to over-estimate the population, as real absences should have been relatively more frequent in these squares (in none of which a species was found by partial coverage) than in a sample of unworked squares.

The greatest uncertainty in the population estimates arose from the original observer's abundance indices, which might be one class or more too large or too small for any species. From my experience with breeding bird censuses of measured plots, I judged that solitary-nesting species in the Maritimes rarely occur at densities of $10,000+$ pairs/100 km^2 (abundance index 6), and that densities one-tenth of that level (index 5) appear rather seldom, even among common small birds; index 5 should not be the most frequent value reported in a region. Most resident species, such as grouse, woodpeckers, jays, crows, and Boreal Chickadees, which are limited by the carrying capacity of their habitats in winter, seldom attain even abundance index 4 ($100+$ pairs/100 km^2). Ducks seldom occur at abundance index 3 throughout a square, although they reach equivalent densities on small fertile wetlands. These broad generalizations gave me baselines against which to recognize implausible values while editing the abundance indices to create new working files. The original data files were kept unchanged.

The relatively few implausible records that had to be adjusted included all 6s changed to 5s, many 5s to 4s, some 4s to 3s, and a few 3s to 2s. We assumed that other errors would cancel out, with over-estimates from one cause balancing under-estimates from another. The logarithmic scale used meant that any over-estimates retained influenced unduly the extrapolated estimates, so all changes effected in the editing were reductions. Editing the abundance indices was time-consuming and necessarily arbitrary, but having all the adjustment done by one person increased the likelihood that the original variation resulting from different observer perceptions had been replaced by an individual bias in the estimates that was reasonably consistent within a given species as well as between species.

How the population estimates were obtained

The population estimates for each species were obtained as follows:

(a) computer extrapolation*, from the abundance indices provided by observers (see Appendix D for details), was used for most species found in 200 or more squares; confidence intervals are shown for these estimates, which may be recognized thereby;

(b) population estimates derived from independent surveys, by CWS, other government agencies, or individuals, were used for most waterfowl and colonial water birds. These were judged to be more representative than the computer extrapolations for these species. These estimates are identified by (CWS) or (other name);

(c) for species found too seldom or too unevenly to permit meaningful computer extrapolation, and for which no independent estimates existed, I estimated populations by manual extrapolation from the abundance indices for most species found in 30–200 squares, or by intuition and personal experience for scarcer species; these estimates are identified by (A.J.E.).

With species treated under (b) and (c), the Maritimes population estimate was simply the sum of the provincial figures, and no confidence intervals were attempted for these estimates.

As noted above, the earliest computer estimates, based on unadjusted abundance indices and without the later refinements of the computation program, were encouragingly similar to previous estimates (also given as breeding pairs) for the same species, thus:

Species	Atlas estimate	Earlier estimate	Source
Black Duck	32,500	31,700	1
		24,300	2
Starling	511,000	647,000	3
White-throated Sparrow	2,670,000	2,700,000	3
Red-winged Blackbird	452,000	280,000	3

Source: 1. Erskine 1987   2. Erskine et al. 1990   3. Erskine 1980.

Of the earlier estimates, Starling and Red-winged Blackbird were less well-suited to the extrapolation method then used than was White-throated Sparrow. The remarkable agreement between the different estimates for White-throated Sparrow and Black Duck was undoubtedly fortuitous, and the level of agreement with Starling and Redwing was fully as good as I had hoped. The population estimates finally included in the species accounts for these species were lower than those originally obtained, owing to the editing downward of high values that inflated the first estimates excessively.

---

*The Prince Edward Island estimates often showed much wider confidence intervals than those for the larger provinces, reflecting smaller samples as well as great variation in the abundance indices. When any provincial estimate was very much out of line, the sum of the three provincial totals for a species might differ markedly from the total Maritimes estimate, which was calculated independently by the computer.

These revised estimates are still of the same order as the earlier figures, in as good agreement as can be expected from the type of indices assembled in the Atlas project.

The enthusiasm engendered by the initial successes soon evaporated when the extent of necessary adjustments of the species abundance files became apparent. Owing to the file structure used (see Appendix C), the full power of the computer in sorting and manipulating data could not be used here without additional programming, and each species file had to be adjusted separately. The adjustments usually involved the same squares and observers across several of the more abundant species. Observers who surveyed many squares in several years usually improved with practice, and far fewer implausible abundance indices were generated in 1990 than in the preceding years.

At the end of three weeks spent running and re-running population estimates, I began to worry that I might have been adjusting the reported abundance indices too much, although I was reasonably confident about the relative abundances of most species. As a check on my estimates, I turned to the British atlas (Sharrock, 1976), the only one available that included population estimates. One would expect that average densities in the British Isles might be higher for urban and open-country species, and lower for forest birds, than in the Maritimes, owing to the relative availability of these habitats. Only about one-third of the species, mostly non-passerines, occur in both areas, with another third having closely related "replacement species" in the other area; many of these were unsuited to computer extrapolations. To increase the sample for comparison, I selected ecologically equivalent species for groups to which no near relative was available. The British atlas comprised 3862 squares, vs. 1682 in the Maritimes, so I used 40 per cent of each of Sharrock's estimates to represent the Maritimes population, if the same average densities prevailed here as in the British Isles (Table 5). There were a few obvious discrepancies, mostly where a species such as the Winter Wren occupies a much wider ecological niche in Britain than in America. Nevertheless, the overall agreement was striking and strongly confirmed that the Maritimes estimates give a reasonable "order of magnitude" picture of relative abundance of our bird species.

One further comparison was made, between abundance and frequency of detection (Table 6 vs. Table 3). I had remarked earlier, in Chapter IV, that the species detected in the greatest numbers of squares were not always very numerous birds, as the ease of detection and recognition also affected the frequency of detection. The graph of estimated populations against numbers of squares with a species (Figure 12) confirmed this difference. It also illustrated which species were over- or under-represented relative to the others. The over-represented birds, that is, those whose estimated populations were smaller than would be expected from the numbers of squares in which they were found, were all very familiar species, found widely in rural and suburban residential areas. The under-represented species included many of those forecast to be poorly detected overall, plus the colonial seabirds. For most species, however, the relationship between population and frequency of detection was fairly predictable.

In conclusion, the exercise of deriving population estimates for the birds in the Maritimes, using the abundance indices provided by the volunteer observers, achieved its objective of giving reasonably representative figures. We emphasize yet again that perspective rather than precision is the purpose of such estimates. Other atlas projects can be reassured by this example that the collecting of abundance indices by volunteers will not be a futile undertaking. The better perspective made possible by such population estimates should enhance the use in bird conservation of our Atlas and those that follow ours.

Table 5.
Comparison of population estimates of Maritimes birds with those for the same (or closely related or ecologically equivalent) species derived from the British/Irish atlas*.

Maritimes Species	Population (pr)	British Isles Species	Population (pr) x 0.4
Pied-billed Grebe	800	Dabchick	3,500–7,000
Great Blue Heron	5,000	Grey Heron	2,500–5,000
American Black Duck	30,000	Mallard	28,000–50,000
Green-winged Teal	5,500	(same species)	1,400–2,700
Ring-necked Duck	12,000	Tufted Duck	2,500
Red-breasted Merganser	1,000	(same species)	800–1,000
Common Merganser	1,200	(same species)	400–800
Red-tailed Hawk	3,800	Buzzard	3,000–4,000
Sharp-shinned Hawk	3,500	Sparrowhawk	6,000–8,000
Merlin	850	(same species)	250–350
American Kestrel	9,000	Kestrel	40,000
Sora	1,700	Water Rail	800–1,500
Spotted Sandpiper	15,000	Common Sandpiper	20,000
Common Snipe	35,000	(same species)	30,000–40,000
American Woodcock	35,000	Woodcock	4,000–10,000
Black Guillemot	1,150	(same species)	3,000
Rock Dove	30,000	(same species)	40,000
Barred Owl	3,600	Tawny Owl	20,000–40,000
Long-eared Owl	1,200	(same species)	1,500–4,000
Short-eared Owl	100	(same species)	400
Chimney Swift	20,000	Swift	40,000
Northern Flicker	95,000	Green Woodpecker	6,000–12,000
Hairy Woodpecker	40,000	Greater Spotted Woodpecker	12,000–16,000
Downy Woodpecker	50,000	Lesser Spotted Woodpecker	2,000–4,000
Barn Swallow	250,000	(same species)	200,000–400,000
Bank Swallow	90,000	(same species)	>100,000
American Crow	125,000	Carrion/Hooded Crows	400,000
Common Raven	25,000	(same species)	2,000
Black-capped Chickadee	375,000	Coal Tit	400,000
White-breasted Nuthatch	6,500	Nuthatch	8,000
Brown Creeper	20,000	Treecreeper	60,000–120,000
Winter Wren	130,000	(same species)	4,000,000
American Robin	875,000	Song Thrush	1,400,000
Yellow Warbler	270,000	Chiffchaff	120,000
Common Yellowthroat	650,000	Whitethroat	275,000
Golden-crowned Kinglet	350,000	Goldcrest	600,000
European Starling	275,000	(same species)	1,500,000–3,000,000
American Goldfinch	160,000	Goldfinch	120,000
White-throated Sparrow	660,000	Chaffinch	3,000,000
Savannah Sparrow	100,000	Yellowhammer	400,000
House Sparrow	150,000	(same species)	1,400,000–2,800,000

*U.K. estimates x 0.4 [3862 squares in U.K. vs. 1682 squares]

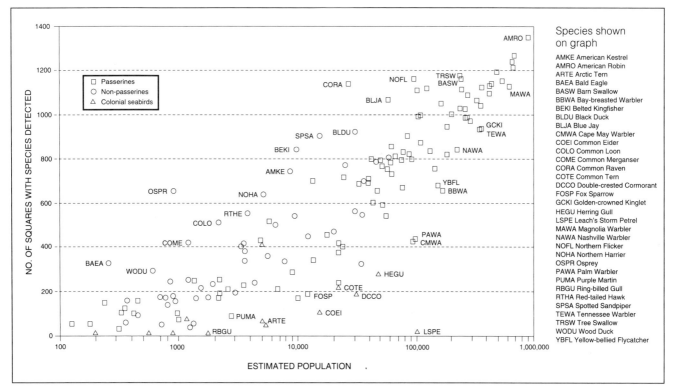

Figure 12. *Estimated population* vs. *number of squares in which species was found.*

Table 6.
Relative abundance of non-passerine *vs.* passerine & woodpecker species in the Maritimes.

Abundance class (pr)	No. species in abundance class	
	Loon to Kingfisher	Woodpeckers & Passerines
over 750,000	0	1
500,000–750,000	0	5
250,000–500,000	0	15
100,000–250,000	1	19
50,000–100,000	1	19
25,000–50,000	9	9
10,000–25,000	5	9
5,000–10,000	10	4
2,500– 5,000	8	3
1,000– 2,500	12	4
500–1,000	10	2
250–500	6	5
100–250	7	5
50–100	6	4
25– 50	6	1
10–25	4	0
under 10[a]	4	0

[a]*Excluding species not breeding annually in Maritimes.*

IX
The Maritimes Breeding Bird Atlas and species conservation

People worked on the Atlas project because it was fun. A large part of the fun was in putting our favourite form of recreation into a framework where our findings may have real value for bird conservation. The hope of being helpful as well as having fun underlies most written recording of bird sightings, but in the Atlas it is more clearly realizable than in other publications based on volunteer birding. Atlas data contribute to several aspects of conservation.

Bird conservation ensures that the existence of any species is not threatened, directly or indirectly, by human actions. Threats to birds may arise from direct exploitation of the birds themselves, or indirectly from human pressure on the vegetative cover or on the organisms on which the birds depend. Nowadays, most threats to birds arise inadvertently through failures to understand that ecosystems are made up of many interconnected elements, so that changes in one element may influence drastically the survival of other species.

Vulnerable species

"Endangered species" has been a buzz-word at government and popular levels for several decades, especially since the start of the environmental movement. Species become "endangered" or "threatened" because their numbers in an area are too few to allow local reproduction to sustain the population, and/or because they breed in so few places (as colonies or groups) that local disasters may destroy the birds there or the habitats on which they depend. All species are more or less vulnerable at the edges of their breeding ranges, which may expand or contract as a result of good or poor reproductive success in recent years. Availability of habitat acceptable to breeding birds is essential, but this is not the only factor determining where birds breed and in what numbers. What makes a particular species endangered (or threatened, or vulnerable) is a question to whose resolution the Atlas will contribute insight.

The population estimates in the Atlas provide one kind of perspective for bird conservation by allowing efforts to be focused on species that are numerically scarce as well as those that are geographically restricted. Not all scarce birds are vulnerable, as noted above. A species with 10 pairs breeding in 3 squares in the Maritimes may disappear, only to re-appear a few years

later when its populations in nearby regions expand again. But a species with 1,000 pairs clumped in only 3 squares, e.g., Atlantic Puffin, a colonial seabird, is regionally vulnerable although its numbers are not extremely low. Our Atlas, with population estimates, tells us approximately "how many birds," as well as "how many places with birds."

Many of our really scarce birds are peripheral species that reach their northern or southern limits in the Maritimes, for example Green-backed Heron or Indigo Bunting on the one hand, Semipalmated Plover or Boreal Owl on the other. We can do little to help most of these species, whose stocks rise and fall with their status in adjoining regions. Most other scarce species are associated with scarce habitats such as sand beach or salt marsh or grassland, or with scarce nest-sites in large tree cavities or on islands free of mammalian predators. If we can define the environmental factors that limit the numbers and distribution of a scarce bird, we have more likelihood of being able to improve conditions for it. Before any remedial work on a numerically scarce species is undertaken, we need to be able to measure changes in its population.

Species for which Maritimes stocks, though small, are significant in the wider picture will require actions in our region at some time, even if these species have not yet been designated as vulnerable. The long-term survival of northeastern populations of Willets and Sharp-tailed Sparrows, of Common Terns and Piping Plovers, of Northern Parula Warblers and the *bicknelli* form of the Gray-cheeked Thrush, which some argue to be a distinct species, depends on actions by the people of the Maritimes; we have more of these birds than occur in any nearby regions. For these species, protection of the birds means protection of their habitats. Systematic monitoring of numbers of designated "endangered species" is ongoing, but such monitoring also is appropriate for these other vulnerable species *before* they become endangered. In addition, other scarce species restricted to grasslands, such as Horned Lark, Vesper Sparrow, and Eastern Meadowlark, and birds with specialized niches in the forest and edge community, such as Pileated Woodpecker, Eastern Bluebird, and Northern Oriole, may also be considered as needing population monitoring in the Maritimes.

The REhabilitation of Nationally Endangered Wildlife (RENEW) program, adopted by federal and provincial governments across Canada in 1988, proposes to spend large sums on the protection and restoration of species of animals and plants that have received formal designation by COSEWIC as "endangered" or "threatened." Only a few endangered or threatened bird species occur in the Maritimes at present: Piping Plover, Peregrine Falcon (re-established after extirpation), Harlequin Duck, Roseate Tern, Loggerhead Shrike.

Bicknell's Thrush may also appear in one of these categories, when a status report is completed for it. Our Atlas data show where some of these birds breed in our area. Harlequin Ducks breed just to the north in the Gaspé peninsula, but may have bred here at one time. Loggerhead Shrikes bred here sporadically in the past, but no longer breed in the Maritimes. We should plan ahead to ensure that no more species in this area need to be added to this list.

No common pattern of threat, except human involvement, emerges for the designated bird species noted above. The decrease in Harlequin Ducks in Atlantic Canada was most plausibly attributed to hunting kill exceeding the carrying capacity of the relatively small population. The Peregrine Falcon's disappearance all across eastern North America resulted from reproductive failure brought on through accumulation of organochlorine residues from toxic chemicals, mainly DDT, sprayed in various habitats to control insect pests. The plight of the Piping Plover arose from human recreational use of the specialized beach habitat in which this species breeds. Roseate Terns, with other terns, suffered from predation by large gulls on the breeding grounds and by people in their wintering areas. Too few Loggerhead Shrikes ever bred here to allow generalizations; in other areas both habitat change and accumulation of toxic chemicals have been suggested as influential in their decline. Bicknell's Thrush habitat in New England (mountain-top conifer forest) has been lost to acid rain, but in our area this species seems mainly to use other forest habitats and has not been obviously affected by acid rain or by spruce budworm damage to forests. Only the decrease in Piping Plovers would be predictable by comparison of the species' breeding range with the occurrence of human activities in the same areas. Harlequin Ducks were shot far from their breeding areas, and Peregrines ate contaminated prey during their migrations or in wintering areas as well as in summer. Nevertheless, comparison of species' ranges with environmental factors often can give insight as to what limits particular species' numbers and the effects to be expected from future changes in the environment.

Defining the breeding distributions of birds is a necessary first step in all such assessments, and the Atlas, having done this for most species, will be a basic reference for all agencies involved in conservation of birds in the Maritimes and surrounding areas. As noted above, not all bird conservation problems arise in a species' breeding area, and other atlases, not yet begun, will be called upon in other seasonal contexts.

Distribution patterns

Another step in assessment of bird population problems to which the Atlas can contribute is the recognition of distribution patterns (see Chapter VII). Most situations where several species share a distribution pattern reflect shared use of a particular habitat or group of habitats. The deterioration or loss of such habitats reduces or eliminates the breeding opportunities for all birds and other fauna and flora that use them. Species that are restricted to an affected habitat may be extirpated, and the Atlas will assist in the recognition of species that may be vulnerable to habitat loss. Having breeding bird atlasses for adjoining regions (Maine, Quebec) makes the Maritimes Atlas more useful, as we can recognize which species with small populations here represent the edges of continuous breeding ranges, and which are true outliers; the latter are more vulnerable, and less likely to re-establish themselves if the local stocks are lost.

No bird species will become extinct as a result of environmental changes here, as the Maritimes have no endemic species. Our region lies in an area of transition between the widespread temperate, boreal, and subarctic regions, so many species that breed here also extend far to the southwest or to the north, or both. The loss of regional populations is also of concern, as these may have become adapted to the peculiar conditions that prevail here, and their loss might reduce the gene pool available to the species. One example is the Ipswich Sparrow, a subspecies of the widespread Savannah Sparrow, which is restricted in breeding season to the sand-dune grasslands of Sable Island.

Some of the species whose breeding ranges in the Maritimes are very small are here only because human actions have modified natural habitats. The impoundment of marshes on fertile sites created artificial "prairie sloughs," which are used by a number of wetland species we used to consider "western birds" (e.g., Redhead, Common Moorhen), in addition to native species. The development of towns and cities provided opportunities for species that could tolerate such environments, mainly the introduced alien birds such as Rock Dove and House Sparrow that were pre-adapted to urban areas. The provision of feeding stations for wild birds in residential areas allowed several southern species, which are largely resident where they breed and thus limited by winter foods, to become established here recently. The period of the Atlas study overlapped with the arrival and establishment of the House Finch (introduced in eastern U.S.A. in 1940), and the Atlas also shows the varying success with which the Northern Mockingbird, Northern Cardinal, and Tufted Titmouse have settled or failed to settle here. None of these species, though scarce, can be viewed as conservation problems, but their future here is unquestionably in human hands.

Resource exploitation *vs.* bird status

Scarce species attract attention in inverse proportion to their numbers, the scarcer the more noteworthy. But conservation problems are not restricted to scarce species. Human exploitation of natural resources in the Maritimes, apart from mining and excavation of sand and gravel, has been concentrated on our two major ecosystems: the seas and the forests. In the process, it has affected all animals that depend upon these ecosystems.

The direct exploitation of seabirds and their eggs for food ceased before 1900, except for local egging of gulls and terns, mostly because the more desired and more vulnerable species had been nearly extirpated by then. Threats to seabirds in the Maritimes now are less direct, but often insidious. Over-fishing of stocks of large fishes, which formerly competed successfully with seabirds for the same foods (small fish or shrimp-like invertebrates), allowed some decimated seabird stocks to recover. Currently, fishing is turning to direct use of the smaller fishes, such as capelin and sand-lance, that are eaten by seabirds. This has become a problem elsewhere, and the pattern may be repeated here. The Atlas shows that most of our pelagic seabirds, nearly all of which feed mainly on small fishes, are limited in both distribution and numbers, with little flexibility to cope with reduced food supplies. Our few primary seabirds make up trivial proportions of the total breeding populations of these species, although they were more numerous in the distant past. Their significance lies in their being among the most southerly outposts of the species; the successful re-introduction of Atlantic Puffins from Newfoundland to Maine suggests that this species' adaptation to that area may be normal variation rather than genetic specialization, but some adaptation in these primarily arctic birds is required in what to them are marginal situations here. Support systems for seabirds need to be considered in management of marine resources.

The attractions exercised on the public by remote islands and breeding seabirds, which are generally scarce here and difficult to see except at their nesting sites, combine to pose other threats to these birds. Visitation to seabird colonies is an increasing form of tourism. However, interested people may not recognize that their presence in a seabird colony, even with no hostile intent, can affect adversely the breeding success of the birds. The Atlas may even increase this problem by making public the places where these scarce breeding birds may be seen, but it also shows how very localized and potentially vulnerable such places are. We urge caution on all groups wishing, for whatever reason, to use seabird colonies as tourist attractions. With suitable care, we can have our colonies and enjoy them too; without care, they may be lost, at least for our time.

Forests in the Maritimes were originally perceived as impediments to settlement, hiding potential enemies and occupying the lands on which houses might be built and crops planted. The trees provided materials for construction from the start, and by the 18th century the forests had already become a marketable resource (see Chapter III for a fuller review). Saw-mills took their toll in the Maritimes, and pulp and paper mills later took over the use of the forests. We now have forecasts of New Brunswick, along with other parts of Canada, having a shortage of wood suitable for saw-logs and for pulp and paper manufacture, the principal industry in the province, within the next few years. How much is reduction in area and modification of forest habitats affecting the birds that depend on these areas?

Bird species requiring large tree cavities had been seriously reduced, though none was extirpated, by the loss of most of the large hardwood trees of the primaeval forest in the Maritimes. The ongoing conversion of the secondary conifer and mixed forests to vast clear-cut areas and monoculture plantations is affecting far more species. Herbicide use to suppress growth of hardwood saplings (which compete with the desired conifers for space but support greater numbers and more varied birdlife than the latter) affects still others. Shorter tree harvest cycles, now reduced from the 100 + years for growing saw-logs to 50 years for pulp-making (proposed 40-year rotations seem unlikely to be realistic here), mean fewer stands of mature trees and no old-growth forests in accessible areas. Bird use, as shown by the Atlas, needs to be considered in forest management, and programs in all three provinces are starting to include wildlife as one element in the forest planning scene. The Atlas shows which forest birds are widespread and which are scarce, a first step in multiple-use forest management.

However, the birds may be more tolerant than we realize, as they have had to be to survive until now. Less than 10,000 years ago, a very short time in the evolution of a species—though far beyond what most people can visualize—the Maritimes lay under the glacial ice-sheets. All our present breeding birds bred elsewhere, and often in other habitats, during the glacial era. Their adaptation to the present local situation is a matter of a few thousand bird generations, and a few hundred tree generations. Even the birds we consider as requiring mature conifer forests for breeding, such as Cape May, Blackburnian and Bay-breasted Warblers, use younger stands of spruce and fir as well as older forests. The crossbills and other finches, of all our birds perhaps the most intimately linked to mature seed-bearing conifers, long ago evolved a nomadic habit, breeding wherever and whenever conifers set seed, here this year and hundreds of kilometres distant next year. Even conifer-adapted birds may be able to cope, if they are not poisoned by forest spraying. Many of these species are now widespread in the Maritimes, as shown by the Atlas, and elsewhere, so their populations are not in immediate jeopardy, but all need to be considered in the planning of forest management for fibre production.

Toxic chemicals and bird numbers

The use of poisons in ecosystem management is not restricted to forestry. Agriculture uses far more toxic substances, many of them in greater quantity, than does forestry. The ultimate purpose of using such chemicals is to ensure that living things, whether plant or animal, cannot exist in an area to compete with or feed on the crop planted there. The elimination, through chemical use, of soil insects as well as foliage-eaters makes most cultivated areas unattractive to birds. The Atlas confirms what bird students knew already: the birds of farmlands now are mainly grassland birds using hay-fields and pastures rather than croplands, or else ubiquitous "edge" species. Among our grassland birds, Savannah Sparrow and Bobolink are common and widespread, but Horned Lark, Vesper Sparrow, Eastern Meadowlark, and Upland Sandpiper are among our scarcest breeding species, whose persistence here may require directed conservation efforts in the future.

Forest husbandry is much less intensive than agriculture. Despite the earlier forecasts of "Silent Spring," and occasional locally severe losses from deliberate usage of chemicals known to kill birds (e.g., phosphamidon; see Pearce et al. 1976), forest spraying usually has not affected the status of our woodland birds catastrophically. DDT, prior to its banning in Canada in 1967 and in the U.S.A. in 1973, affected populations of many birds adversely, but no species except Peregrine Falcon and Bald Eagle were extirpated over all or major parts of the Maritimes; all such birds except the Peregrine are believed to have made good their losses here subsequently. Direct, sub-lethal effects of biocides on birds (Busby et al. 1990) occur, but seldom attain levels that might be detected from Atlas data. The evidence of the Atlas, that forest birds are alive and well and living in the Maritimes, does not bear out a wholesale indictment of forest management, nor of pesticide use generally, but misuse of toxic chemicals is an ongoing problem everywhere.

Conservation initiatives involving our birds

Decreases in birds are not necessarily caused by events in their breeding areas, including the Maritimes. The increasing concern over apparent decreases in North American birds that migrate to winter in the dwindling forests of tropical America indicates that some of these birds may be limited by winter habitat. Most of "our" insectivorous birds spend little more than three months of each year here, compared to four to five months in their winter homes and another four months along their migration routes. Unlike many natural resources, birds are shared with other regions and jurisdictions. To avoid "the tragedy of the commons" (which everyone enjoyed but for which no one accepted responsibility), we have a part to play in migratory bird conservation. Our role includes using the Atlas as a baseline of information on breeding distribution of these neotropical migrants, and monitoring their numbers on the breeding grounds.

Other conservation measures, aimed at helping one species or group of species, may work to the disadvantage of other birds. The recent DUC practice of constructing impoundments for waterfowl on salt marshes destroys this habitat for breeding Sharp-tailed Sparrows and Willets as well as for migrant shorebirds and geese, as conclusively as if it had been drained for agriculture. The ongoing protection of large gulls, necessary in 1916 when these species had been decimated through direct exploitation, has allowed their populations to build over several decades to the point where they are having severe adverse effects on breeding terns, Eiders, Puffins, and storm-petrels. Ecosystems are inter-linked, and altering a few components may have unforeseen and far-reaching consequences. Sometimes the Atlas will help us to recognize where conservation action to help one bird species may hinder another.

This Atlas cannot, by itself, be a "blueprint for survival" for all Maritimes birds, but it is an essential assembly of some of the information that will be needed for future conservation efforts. This chapter is a very preliminary sketch of some of the ways in which it might be used.

Appendix A
Organization

The Maritimes Breeding Bird Atlas project began with an *ad hoc* committee that, with additions, became the Steering Committee; all other working groups were sub-committees. The Maritimes Bird Atlas Trust was established to allow receipt of financial donations to the project, rather than involving the sponsoring organizations in this matter.

Atlas Steering Committee
(All members were involved throughout the project, unless noted otherwise.) Subcommittees: M = Management, S = Data Screening, P = Publication.

Peter J. Austin-Smith	Wildlife Division, N.S. Dept. of Lands & Forests, Kentville, N.S.
William G. Caudle	N.S. Bird Society and Dept. of Supply & Services Canada, Dartmouth, N.S.
David S. Christie	N.B. Representative, Albert, N.B.
F. Rosemary Curley	P.E.I. Representative, Summerville, P.E.I.
Anthony J. Erskine (S,P)	Canadian Wildlife Service, Atlantic Region, Sackville, N.B.
Roger A. Foxall (Chairman 1985-88, M 86-88)	Halifax, N.S.
Ian A. McLaren (S,P)	Halifax, N.S.
Eric L. Mills (S 86-88)	Halifax, N.S.
Linda A. Payzant (M,P)	Waverley, N.S.
L. Peter M. Payzant (M,P)	Waverley, N.S.
Peter A. Pearce (S)	N.B. Representative, Fredericton, N.B.
Fred W. Scott (Chairman 1988-91, M 88-91, P)	Natural History Section, Nova Scotia Museum, Halifax, N.S.

Treasurer
Donald A. MacNeill	Halifax, N.S.

Staff
Judith Kennedy	Co-ordinator 1985-88
Brian Dalzell	Co-ordinator 1988-91

Maritimes Bird Atlas Trust
David A. Currie	Halifax, N.S.
Robert B. Dickie (M 88-91)	Halifax, N.S.
Fulton L. Lavender	Halifax, N.S.
Linda A. Payzant	Waverley, N.S.
L. Peter M. Payzant	Waverley, N.S.

Appendix B
Operations

Sampling framework

The sampling unit of the Maritimes Breeding Bird Atlas was the 10 km x 10 km square of the Universal Transverse Mercator grid (Figure A1). Conventional wisdom indicated that we could not expect to cover all squares in the Maritimes adequately during five field seasons with the available observers, so initially we established a 25 per cent sample of squares, uniformly distributed, termed priority squares (Figure A2), as the minimum coverage to be attempted. A few other squares, which included rare species or habitats that were scarce in a region, were accorded similar priority (special squares, Figure A2). As work progressed, coverage spread to include many other squares as well. The priority squares were expected to be worked first when a choice was available, and this was usually, but not always, done.

Standards of breeding evidence and of coverage

The categories of breeding evidence (Table A1) followed the widely accepted standards for bird atlasses, with minor variations. In the first year it became obvious that many observers did not know that some codes were inappropriate for certain species. To minimize the time needed to check on such "questionable code combinations," or QCCs, a pamphlet, setting out for each species which codes were generally acceptable vs. those which should be expected to generate an inquiry, was developed. This document is believed to be a first of its kind in any North American atlas project.

A square was considered "completed" when an observer had detected at least 75 per cent of the expectable number of species, with at least 38 per cent of the expectable number of species having evidence that confirmed breeding there. Originally it was suggested that 20 hours field-work would be a realistic minimum for completion; as the years passed, a few observers developed the aptitude for completing a square in as little as 5–10 party-hours, whereas some others who detected 75 per cent of expected species in 20 hours could not secure confirmations of 38 per cent in 100 hours.

The detection and confirmation standards were arbitrary. Many remote squares could only be visited once, and species not active then were missed. Confirmations for most species could only be obtained during visits in July. The required proportion of confirmations was made up largely of those species that were easier to confirm. Having a substantial proportion of Confirmed

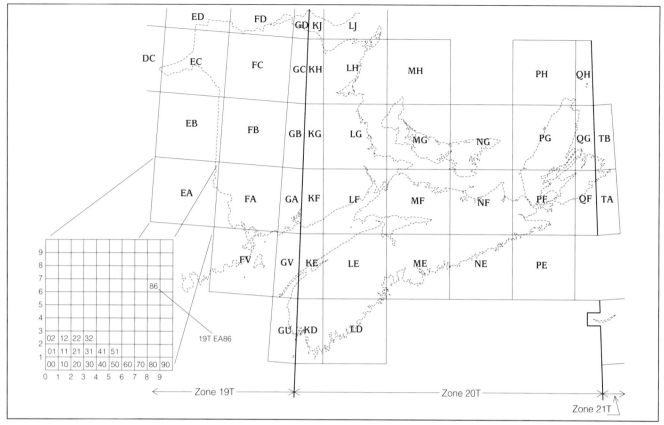

Figure A1. *The UTM zones and grid, with the numbering system for the 10 x 10km squares.*

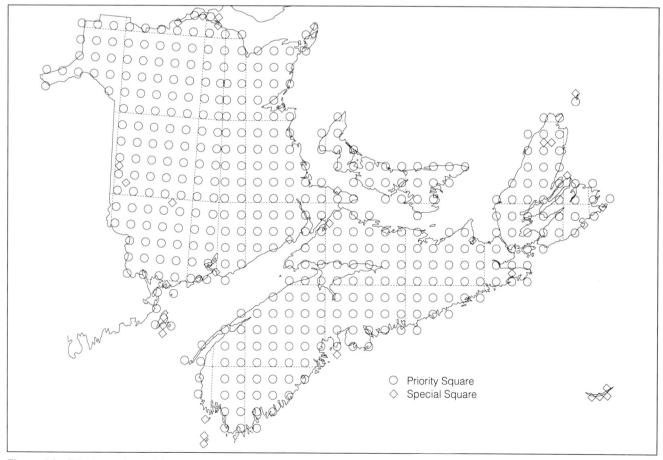

Figure A2. *Priority and special squares.*

Breeding records was seen as essential in providing
credibility for the Atlas. Surveys of other squares, especially during June when confirmation of breeding by
most species was impractically time-consuming, provided much additional coverage, and this was encouraged also, especially in the last two years. It was
accepted in the Steering Committee that obtaining the
38 per cent confirmation level, though desirable, should
not take absolute priority in all situations over securing
some coverage of poorly sampled or unsampled
squares.

Regional structure and responsibility

The Maritimes was divided into 23 regions of roughly
similar size, each with 55 to 103 squares, mostly
60–80; all coastal regions had many squares with much
less than 100 per cent land area (Figure A3). The number of regions and their boundaries were dictated partly
by geographic considerations, partly by political units
(counties), and partly by availability of observers. Each
region had a volunteer Regional Co-ordinator (RC)
(Table A2).

RCs were encouraged to recruit observers for each
square, starting with priority squares, and pass on to
them the materials needed for gathering Atlas data. At
the end of each field season, RCs assembled data cards
from the observers in their region, checked the data for
plausibility and completeness, requested additional
details where necessary, and forwarded the data to the
Atlas office.

RCs also were asked to determine the numbers of
species to be expected in each priority square in their
region and to establish a list of species for which special
documentation would be required in their region. In
addition, there were species not on the field card, for
which such documentation was required in all regions.

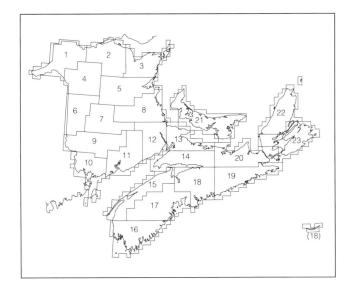

Figure A3. *Boundaries of the Atlas regions.*

Table A2.
Atlas regions and regional co-ordinators

Atlas Region (no. squares)	Co-ordinator	Period
1 Edmundston (76)	Peter Pearce	1986–90
2 Campbellton (72)	David Christie	1986–90
3 Bathurst (62)	Peter Pearce	1986–87
	Hilaire Chiasson	1988–90
4 Tobique (64)	Erwin Landauer	1986–88
	Brian Dalzell	1989–90
5 Miramichi (88)	Harry Walker	1986–90
6 Carleton (57)	Ford Alward	1986–90
7 Fredericton North (58)	Don Gibson	1986–90
8 Kent (71)	Mary Majka	1986–90
9 Fredericton South (73)	Dan Busby	1986–88
	Chris Adam	1989–90
10 Charlotte (65)	David Clark	1986
	Brian Dalzell	1987–90
11 Saint John (63)	Jim Wilson	1986–90
12 Moncton (63)	Rob Walker	1986–90
13 Border (55)	Gay Hansen	1986
	Tony Erskine	1987–90
14 Cobequid (63)	Mike Malone	1986–87
	Blake Maybank	1988–90
15 Valley (78)	Jean Timpa	1986–90
16 Yarmouth (90)	Ted D'Eon	1986–88
	Jerome D'Eon	1988–90
17 Keji (96)	Ian & Christine Ross	1986–90
18 Halifax (76)	Peter & Linda Payzant	1986–90
19 Guysborough (83)	Roslyn MacPhee	1986–90
20 Antigonish (78)	Vicky Bunbury	1986–90
21 Prince Edward Island (101)	Rosemary Curley	1986–90
22 Highlands (69)	Al Gibbs	1986–90
23 Sydney (79)	Dave Harris	1986–90

Not all RCs felt competent to carry out the last two tasks; where necessary, these were completed or revised by the Atlas Co-ordinator or other members of the Steering Committee.

In some regions, the original RCs resigned, for various reasons, and replacements were found. Additional coverage was provided by observers from other parts of the Maritimes, mainly in 1989–90, in several regions where completion of even the priority squares would not have been achieved otherwise. Several regions in northern New Brunswick were co-ordinated by non-residents and surveyed almost entirely by visiting observers.

Data collection

The first Co-ordinator, in collaboration with the Steering Committee, prepared three booklets that provided observers with the information needed to collect and report data for the Atlas. The *Atlasser Handbook* explained the purposes of the project, the sampling framework, how to go about locating a square and gathering data on breeding birds there, and how to complete the data card and species documentation forms, usually called "doc-forms," in language everyone could understand. *The Breeding Season Chart* summarized, for each species breeding regularly here, the span of dates within which eggs or flightless young were known to have occurred in the Maritimes, as documented in the Maritimes Nest Records Scheme. These two documents were available at the start of field work. French versions became available in the second year. The *Code Combinations* booklet was developed after the first field season (1986), when it became apparent that many observers did not know which codes were inappropriate for particular species. Codes which would result in a request for documentation were marked for each species. Far fewer QCC-forms were generated in succeeding years, thus easing the tasks of the Co-ordinator and Screening Sub-committee as well as of the observers. The three booklets provided the "nuts and bolts" of atlassing information for the observers.

Observers were encouraged to visit Atlas squares at all seasons so as to detect species whose breeding activity was most evident at different times. This applied mainly to squares that were close to an observer's home. Visits to more distant squares were to be timed when more species would be nesting, in late June and July. Where possible, evening or night visits to detect owls or Woodcock were encouraged; these were expected to be found most easily in April or May, but unpaved roads at that season precluded access to or minimized coverage of many suitable areas. Coverage was not often continued into August for late-nesting species such as Cedar Waxwing, American Goldfinch, and (in 1988) White-winged Crossbill; this affected confirming breeding of these species more than detecting them.

Observers were asked to enter on standard data cards (Figure A4) the highest category of breeding evidence found for each species detected in a square that year. A separate card was used for each square visited, but data for successive visits in one year to the same square by one observer or team were accumulated on the same card, with dates and durations of each visit recorded. Standard "tombstone data" covering square identification, observer(s) name(s), address(es), and atlasser number(s) appeared on each card. A French version of the card was also available.

For species detected that were not listed on the data card, and for additional species specified for each region, observers were asked to complete "doc-forms." The doc-forms served to confirm identification of improbable or confusing species, but their *principal* purpose was to indicate what evidence suggested that the species was breeding where found, rather than occurring as a vagrant. The forms used in the first years placed too much emphasis on the identification, which was seldom in question. This often-misplaced emphasis annoyed the experienced observers who contributed most of the unusual records, and tended to distract them from providing the essential details on breeding evidence, so the form was re-arranged to remedy this. The revised version (Figure A5) was in use only in the last two seasons.

In addition to the records received through people taking part in the Atlas project (the "atlassers"), efforts were made to retrieve bird distribution data collected for other purposes, mainly through surveys by government agencies. These data pertained particularly to game birds: waterfowl, American Woodcock, marsh birds encountered in waterfowl studies, and colonial seabirds. Data for these groups extracted from outside files usually were routed directly to the Co-ordinator, rather than through the RCs.

A few RCs and enterprising individuals wrote articles for local newspapers soliciting records of nests or broods of species that are seldom seen but easily identified by the public, especially raptors. This approach could have been used to advantage more often, but few people found the time needed to make the requests and to follow up on responses.

Data handling

The RCs received the data cards, with accompanying doc-forms, from the observers during September (though many were later); checked them for completeness and requested doc-forms when these were lacking; and forwarded them to the Co-ordinator. Most data were received at the Atlas office during October, but many were later; in the last two years, special efforts were made to retrieve promptly all data that RCs had received, so as not to delay subsequent processing of the information.

The Co-ordinator re-checked the data cards for completeness and entered the data into the computer. The same data were re-entered by a second person, and the two entries were cross-checked by computer to detect most entry errors (see Appendix C). The data were then run through computer programs that checked whether doc-forms were needed and whether the species and breeding code reported were compatible. The computer generated listings of records requiring doc-forms and noted whether these were in hand; the doc-form listings were sent to the RC. It also generated listings of species requiring details on QCCs, and the QCC forms were sent to the individual observers.

All doc-forms received, and QCC forms when

returned by the observers, were reviewed by all members of the Data Screening Sub-committee, as well as by the Co-ordinator, and a decision was made on each to accept or reject. Records requiring doc-forms or QCC verification for which no response was received after two years were either rejected for lack of information, or accepted if subsequent records had made the original one less unlikely than it had seemed earlier. During the final preparation of species accounts, all doc-forms were re-checked by the author, to ensure that the maps reflected both the decisions of the Screening Sub-committee and the perspective gained from subsequent records of a species.

When most data were in hand, computer print-outs of the data for each square in a region were sent to the RC for checking. These listings also indicated species for which doc-forms or QCC-forms were under review. RCs were asked to report any discrepancies between the print-outs and their file copies, using Error Report Forms (ERFs).

The results of doc-form and QCC-form reviews, and the errors reported via ERFs, were entered as amendments to the original form, each such change having a hard-copy form to back it up.

The various screening, verification, and amendment processes dragged on interminably and were further delayed by data being received late. Although RCs usually received regional detail via print-outs by January, a few details of a year's data might not be clarified before the next field season. This process was speeded up as demand for a current "clean" data-set became more urgent. Thus, 1990 data that were not received in time for processing and screening before early March 1991 were generally not used in the final revisions of the text, although a few were added before the final data base was "frozen" in May 1991 for generating the final maps.

Information feedback to observers and others

The main channel of information exchange between the Steering Committee and the observers was the Atlas Newsletter, prepared by the Co-ordinator; Linda and Peter Payzant prepared several newsletters in 1985 and 1988 when no co-ordinator was in place. Newsletters appeared four times a year from spring 1985 through winter 1990–91. Articles reported on progress of the Atlas project, highlighted deadlines for data handling, clarified methods for unfamiliar species, featured people who had made major contributions, and provided an outlet for individuals to share some of their atlassing adventures. The Newsletter mailing list included all atlassers in the Maritimes, with a large number of people and organizations in other areas who had expressed

Maritimes Breeding Bird Atlas

Form fields: Serial No. | Region | Zone | Block | Square | Year | Atlasser Number | if new address | Name | Address | Postal Code | Phone # (H) / (W) | Region Name | Identifying Feature of Square

OFFICE USE: Region | Zone | Block | Square | Year | Card # | Atlassers

Day | Month | Party Hours | Total Hours | 1st Date (D M) | Last Date (D M)

Approved by RC ____ (initials) | Keypunched | Verified | Copied

Breeding Codes

Observed
X — species identified but no indication of breeding

Possible
H — species observed or breeding calls heard in suitable nesting HABITAT

Probable
P — PAIR observed in suitable nesting habitat
T — permanent TERRITORY presumed through territorial behaviour in the same location on at least 2 occasions a week or more apart.
C — COURTSHIP behaviour between a male and female
V — VISITING probable nest-site, but no further evidence obtained
A — AGITATED behaviour or anxiety calls of adult
N — NEST-BUILDING or excavation of nest-hole by wrens and woodpeckers

Confirmed
NB — NEST-BUILDING or adult carrying nesting material: use for all species except wrens and woodpeckers
DD — DISTRACTION DISPLAY or injury feigning
UN — USED NEST or eggshells found
FL — recently FLEDGED young or downy young
ON — OCCUPIED NEST indicated by adult entering or leaving nest-site, or adult seen incubating
AY — ATTENDING YOUNG adult seen carrying food or faecal sac for young
NE — NEST with EGGS
NY — NEST with YOUNG

Comments Notes

OFFICE USE

Species Lists

(Columns for each: Breeding Evidence — O | PO | PR | CO, and A | b | c | d)

Species	Code
Common Loon	COLO
Pied-billed Grebe	PBGR
Leach's Storm-Petrel	LSPE
Great Cormorant	GRCO
D.C. Cormorant	DCCO
American Bittern	AMBI
Great Blue Heron	GBHE
Bl-Cr Night Heron	BCNH
Canada Goose	CAGO
Wood Duck	WODU
Green-winged Teal	GWTE
Am Black Duck	BLDU
Mallard	MALL
Northern Pintail	PINT
Blue-winged Teal	BWTE
Northern Shoveler	NOSH
American Wigeon	AMWI
Ring-necked Duck	RNDU
Common Eider	COEI
Common Goldeneye	COGO
Hooded Merganser	HOME
Common Merganser	COME
Red-breasted Merg	REME
Osprey	OSPR
Bald Eagle	BAEA

Species	Code
Northern Harrier	NOHA
Sharp-shinned Hawk	SSHA
Northern Goshawk	GOSH
Broad-winged Hawk	BWHA
Red-tailed Hawk	RTHA
American Kestrel	AMKE
Merlin	MERL
Gray Partridge	GRPA
Ring-N Pheasant	RNPH
Spruce Grouse	SPGR
Ruffed Grouse	RUGR
Virginia Rail	VIRA
Sora	SORA
American Coot	AMCO
Semipalmated Plover	SEPL
Piping Plover	PIPL
Killdeer	KILL
Willet	WILL
Spotted Sandpiper	SPSA
Common Snipe	COSN
American Woodcock	AMWO
Ring-billed Gull	RBGU
Herring Gull	HEGU
Gr Bl-backed Gull	GBBG
Roseate Tern	ROTE

Species	Code
Common Tern	COTE
Arctic Tern	ARTE
Black Tern	BLTE
Black Guillemot	BLGU
Rock Dove	RODO
Mourning Dove	MODO
Black-billed Cuckoo	BBCU
Great Horned Owl	GHOW
Barred Owl	BDOW
Long-eared Owl	LEOW
Short-eared Owl	SEOW
N. Saw-whet Owl	SWOW
Common Nighthawk	CONI
Whip-poor-will	WPWI
Chimney Swift	CHSW
Ruby-th Hummingbird	RTHU
Belted Kingfisher	BEKI
Yel-bel Sapsucker	YBSA
Downy Woodpecker	DOWO
Hairy Woodpecker	HAWO
Three-toed Woodp	TTWO
Black-backed Woodp	BBWO
Northern Flicker	NOFL
Pileated Woodpecker	PIWO
Olive-sided Flycatcher	OSFL

Species	Code
Eastern Wood Pewee	EWPE
Yel-bel Flycatcher	YBFL
Alder Flycatcher	ALFL
Least Flycatcher	LEFL
Eastern Phoebe	EAPH
Gr Cr Flycatcher	GCFL
Eastern Kingbird	EAKI
Horned Lark	HOLA
Purple Martin	PUMA
Tree Swallow	TRSW
Bank Swallow	BKSW
Cliff Swallow	CLSW
Barn Swallow	BASW
Gray Jay	GRJA
Blue Jay	BLJA
American Crow	AMCR
Common Raven	CORA
Bl-cap Chickadee	BCCH
Boreal Chickadee	BOCH
Red-br Nuthatch	RBNU
Wh-br Nuthatch	WBNU
Brown Creeper	BRCR
Winter Wren	WIWR
Golden-cr Kinglet	GCKI
Ruby-crowned Kinglet	RCKI

Species	Code
Eastern Bluebird	EABL
Veery	VEER
Gray-cheeked Thrush	GCTH
Swainson's Thrush	SWTH
Hermit Thrush	HETH
Wood Thrush	WOTH
American Robin	AMRO
Gray Catbird	GRCA
Northern Mockingbird	MOCK
Cedar Waxwing	CEWX
European Starling	STAR
Solitary Vireo	SOVI
Warbling Vireo	WAVI
Philadelphia Vireo	PHVI
Red-eyed Vireo	REVI
Tennessee Warbler	TEWA
Nashville Warbler	NAWA
Northern Parula	NPWA
Yellow Warbler	YEWA
Chestnut-sided Warb	CSWA
Magnolia Warbler	MAWA
Cape May Warbler	CMWA
Black-thr Blue Warb	BTBW
Yellow-rumped Warb	YRWA
Black-th Green Warb	BTGW

Species	Code
Blackburnian Warbler	BLWA
Palm Warbler	PAWA
Bay-breasted Warbler	BBWA
Blackpoll Warbler	BPWA
Black-and-white Warb	BWWA
American Redstart	AMRE
Ovenbird	OVEN
Northern Waterthrush	NOWA
Mourning Warbler	MOWA
Com Yellowthroat	COYE
Wilson's Warbler	WIWA
Canada Warbler	CAWA
Scarlet Tanager	SCTA
Rose-br Grosbeak	RBGR
Indigo Bunting	INBU
Chipping Sparrow	CHSP
Vesper Sparrow	VESP
Savannah Sparrow	SASP
Sharp-tailed Sparrow	STSP
Fox Sparrow	FOSP
Song Sparrow	SOSP
Lincoln's Sparrow	LISP
Swamp Sparrow	SWSP
Wh-thr Sparrow	WTSP
Dark-eyed Junco	DEJU

Species	Code
Bobolink	BOBO
Red-winged Blackbird	RWBL
Eastern Meadowlark	EAME
Rusty Blackbird	RUBL
Common Grackle	COGR
Br-headed Cowbird	BHCO
Northern Oriole	NOOR
Pine Grosbeak	PIGR
Purple Finch	PUFI
Red Crossbill	RECR
Wh-winged Crossbill	WWCR
Pine Siskin	PISI
American Goldfinch	AMGO
Evening Grosbeak	EVGR
House Sparrow	HOSP

Additional or Regionally Rare Species

Columns: Species | Code | Breeding Evidence (O | PO | PR | CO) | A b | UTM (1 km or 100 m) | Doc. Form ( )

Names and Addresses of Other Observers

Figure A4. *The field data card used by atlassers.*

Maritime Breeding Bird Atlas

Documentation Form for Species Not Known to Breed Regularly In Region

Atlas Card showing record: | | | | (Office use only)

Region   Zone/Block/Square   Year   Square name

What, Who, Where, How?

Full species name: _____

Species abbreviation: _____   Breeding evidence code (FL, AY, etc.) _____

Observer (s) & Atlasser # (#): _____

Location within square (UTM subdivision, & description): _____

Situation (optical equipment, distance, lighting, weather): _____

Evidence that bird was breeding where encountered

Habitat (description): _____

Date(s) observed: | | | | | | Time: _____ Duration of Observation: _____
Day Month Year

| | | | | | Time: _____ Distance: _____
Day Month Year

Behaviour observed (describe): _____

Songs or calls heard (describe): _____

Breeding evidence (additional to above): _____

Confirmation of species identity

Previous experience in identifying species as breeding bird (give years and provinces/states): _____

If no previous experience with species breeding, summarize other experience with species: _____

Differentiation from similar species: _____

Additional comments that may help in assessment of record: _____

Date of writing | | | | | | Signature of observer & address _____

Return to Regional Coordinator or to:
Maritime Breeding Bird Atlas
c/o Nova Scotia Museum
1747 Summer Street
Halifax, N.S.,
B3H 3A6

Figure A5. *The species documentation form, revised version.*

interest in the project.

Some RCs circulated regional newsletters among their observers, once to three times a year, but not in every year. These reports provided more detail, on progress and on plans for future coverage, than could be given in the Atlas Newsletter and summarized regional activities such as Atlas Days.

Press releases and articles in newspapers and magazines were used in the start-up years to bring the existence of the Atlas to public attention, and in the final years to advertise the impending publication.

Postscript on breeding evidence categories used in the Maritimes Atlas
(not all concurred in by the Steering Committee)

(a) Several years into the project, we concluded that the ranking of category AY as higher than ON (which was higher than FL) was often inappropriate under the definitions used. By then, it was thought too late to make changes, as all of these categories ranked as confirmed breeding. For future use, I suggest that the sequence be revised, with FL (young *seen* out of the nest) and ON (occupied nest *seen*, but presence of eggs or young only *inferred*) ranking higher than AY (presence of young only *inferred*). To avoid confusion, AY should be renamed CF (adult Carrying Food to *unseen* young).

(b) The lack of an agreed-on code for adult birds physically attacking predatory birds or the observer frequently perplexed the screening sub-committee. The only defined code that covered this behaviour unambiguously was A, but it was considered that attacks, wherein the adult risked its life in defense of eggs or young, confirmed the occurrence of breeding as strongly as most cases involving NB, DD, UN, and often AY. Some such cases were reported as DD or AY and in the later years accepted as such, but they do not fit the definitions established for those codes. Future projects might use a code AT for this behaviour, or include a definition clearly covering it.

(c) We did not adopt the suggested lowest Probable Breeding category based on at least 7 *separate* sightings (H; not in one or a few groups) in a square in one year, which was proposed and adopted at a North American atlas conference after our project had begun. Such evidence is somewhat stronger than a single record of Possible Breeding, but it could be erroneous in cases of delayed migration, e.g., Blackpoll Warblers may be still passing through in numbers in mid-June.

(d) Probable Breeding might be assigned plausibly in cases where a species was detected (H) in precisely the same location in successive years, when no other visits were made in either year to secure higher categories of evidence. This combination seems not to have been proposed previously and was not used in our Atlas project, although the possibility came up several times.

Appendix C
Data Processing aspects of the atlas

by Linda and Peter Payzant

Introduction

Even before the first field season began, two software professionals on the Steering Committee were planning how the data would be stored and accessed. Consultations with the Ontario and New York Breeding Bird Atlas projects were very helpful. For reasons of portability, we decided early on to keep the database and surrounding software as simple as possible—we couldn't be certain that the computer resources available to us in 1985 would still be there in 1991. Our design centred around a simple "flat file" for the database, with conventional FORTRAN software to access the database.

The Atlas database was maintained on a Digital Equipment Corporation VAX minicomputer. By the end of the project, it occupied about 1.8 megabytes of disc space—somewhat more than the capacity of a 3 1/2" diskette. About 25,500 lines of FORTRAN code (including comment lines) made up the programs used to enter and check the data and to produce the various

reports required to administer the Atlas project. Other functions such as maintaining the Atlas mailing list were also handled by the VAX.

By the end of the first year it had become obvious that it would be more efficient for us to work on our home computer whenever possible, so this computer (an AT-class machine) became increasingly important. The language used on this machine was Pascal, and about 8,600 lines of code were written. This machine was used partly for secondary data entry, but its main function was to generate plotted output. Custom-written software was producing draft-quality maps on a dot-matrix printer by the end of the first year. Output on laser printers and pen plotters followed in later years.

Data Flow

Figure A6 shows a simplified view of the major data-processing components of the atlas project. Program ENTER was used to create files containing typically a few dozen data cards. These files were termed "batch" files, and their size was arbitrary, usually being determined by the length of time that a person doing data entry could last before taking a break.

Each batch was separately processed through the path shown in Figure A6. Program RAWCHECK looked for gross errors, such as format errors, non-existent

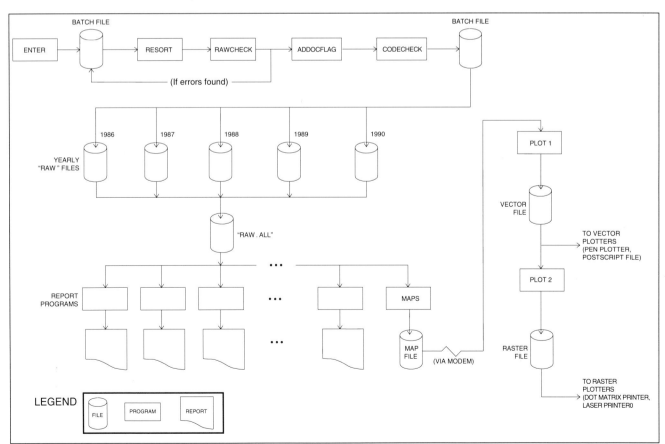

Figure A6. *Data processing flow chart.*

251

squares, erroneous species codes, etc. Program RESORT sorted the batch into order, by region, followed by UTM square designation. This was the standard order that we maintained for all of our data files. Program ADDOCFLAG looked for records requiring documentation and set the "doc" flags for these records to "#," which meant that documentation was required and not yet received. Program CODECHECK performed a similar function for records that had improbable combinations of species and breeding code. Finally, the batch file was appended to a cumulative yearly data file, which was then re-sorted into standard order. The batch file was then discarded.

After the majority of cards had been entered for each field season, reports were produced. All of the yearly data files were merged to produce the "RAW.ALL" file. This file was read by various report programs. Program MAPS created a file containing the highest category of breeding evidence (possible, probable, or confirmed) for each species for each square, and this file was further processed by plotting programs to produce maps. The RAW.ALL file was always deleted after use, ensuring that the yearly data files were the only location of valid data.

Database structure

The database consisted of the five "RAW.19xx" files. As mentioned earlier, these were "flat" files, that is, non-indexed files using only sequential access—essentially a deck of computer cards on disc. We restricted them to using only printable characters and limited the record size to 80 bytes. These files were simple to transport should this become necessary, simple to list, and easy to edit using any conventional text editor. In practice, using a text editor was the only way we ever had of manually modifying the contents of the database.

Each data card became, in the database, two or more records. The first record contained all the "header" data, including region, square, visit dates, atlasser numbers, and so on. Subsequent records consisted of up to eight 9-byte fields, each field being a single breeding record. The fields consisted of the following:

bytes 1–4:	Species mnemonic, e.g., AMRO
bytes 5,6:	Breeding code, e.g., UN
byte 7:	Abundance index, e.g., 3
byte 8:	documentation status flag
byte 9:	error flag

The "documentation status flag" field stored one of a number of single-character status flags noting the documentation status of the record, such as "!" (record requiring documentation was accepted), "*" (record was accepted with change in breeding code), and so on.

There were several obvious disadvantages to this type of database. Programs accessing the structure were necessarily inefficient owing to the sequential file access and the need for conversion between character and internal formats. Since we only used text editors to modify the database, we were somewhat exposed to human error. Nevertheless, the database was highly portable, simple, and reliable. After six years of use, we are satisfied that this structure met our needs.

Data quality control

This subject was a preoccupation with us for the entire span of the atlas project. We took every precaution possible to minimize the rate of introduction of errors into the database.

Our first line of defence was the data-entry program. It performed some rudimentary error checking, assuring that only legal species and breeding codes, for example, got into the database.

To guard against errors that the data-entry program could not catch, such as missing records, each data card was entered twice by different people, as explained in Appendix B. This revealed an average error rate (based on 1990 data) of 36 errors per 100,000 keystrokes. These were errors that the data entry program missed, i.e., those that resulted in a record that appeared to be correct but was not an accurate transcription of the data card. Typical errors were omitted records, wrong species or breeding codes, and mistakes in atlasser or square codes.

Program RAWCHECK performed as many checks as we could devise, and it saved us from a number of problems usually attributable to errors on the data card itself, typically squares incorrectly identified.

Almost every program that read the database checked some aspect of the data. For example, the report programs required the database to be sorted by region, square, and year. Rather than assuming that this was the case, most of them checked that this order was maintained and complained if they found a record out of order. This helped to guard against corrupted square designators. Similar checks were performed by most programs, and this helped us ensure that errors that crept in were promptly discovered.

Reports

Our initial suite of reports was modelled closely on those produced by the Ontario Breeding Bird Atlas. However, more were soon requested, and by the end of project we had about 20 different reports that could be produced. The more routine ones included
Raw data listing
Regional summary report (after Ontario atlas)

Square detail report (after Ontario atlas)
Matrix reports by taxonomic order and rank (after Ontario atlas)
Documentation status report
Error report
Breeding code usage report
Multiple statistics/tables

The more voluminous reports were only produced once each year. However, the smaller ones were produced on demand.

Plotting

Our original base map was digitized by the College of Geographic Sciences (C.G.S.), at Lawrencetown, Annapolis Co., N.S. They also produced a complete set of species maps after our first year. The base map file was moved from the PR1ME computer at Lawrencetown to the VAX, and then to our home computer, from which all subsequent plotting was done. This required converting floating point numbers from the PR1ME format to the format used by Turbo Pascal on our home computer.

The plotting process began with running program MAPS (see Fig. A6), to produce a file containing the highest breeding code for each species for each square. This file was down-loaded to the home computer via modem, where it was processed to produce a file containing vectors defining the map. The vector file was then processed by a second program to produce draft-quality raster images for our dot-matrix printer or for a Hewlett-Packard LaserJet II printer.

Some experiments with a pen plotter showed that drawing a visually acceptable filled disc was not a simple matter. Problems of hysteresis in the pen-positioning mechanism, varying line weight depending on pen speed, and the slowness of the whole operation made it look as if plotting the final set of maps might take up to three or four hundred hours of plotter time.

Fortunately, a casual conversation in the pre-production stages pointed out the possibility of direct digital transmission of our map data to a desktop-publishing package. A little research proved that this was indeed feasible, and in the end all of the map data were simply encoded into PostScript language files and transferred by diskette directly to the publishers.

The illustration and overlay maps were created on an Intergraph Microstation computer-aided drafting system. The original C.G.S. base map was ported from the home computer to the Intergraph system for this part of the project. Once again, this required a translation of the floating-point number format from that used by our home computer to the format required by the Microstation CAD system. As with the species maps, the illustration and overlay maps were moved to the publisher on diskette, in PostScript format.

Appendix D
Computing population estimates

by Christopher Field and Peter Payzant

This appendix provides details about the method used to derive population estimates of the number of breeding pairs from the abundance indices provided by the atlassers. These indices, ranging from 0 to 6, represent the atlasser's estimate of the number of breeding pairs of a particular species in a given square.

These estimates were coded as follows:

Index	Abundance Range
0	0
1	1
2	2–10
3	11–100
4	101–1000
5	1000–10,000
6	over 10,000

Given a particular species, it was desired to use these abundance indices to get an overall population estimate along with a standard error for the estimate. The population estimate can be computed for any geographic entity of interest, such as an atlas region, a single province, or the Maritimes as a whole.

For the species and area of interest, the following notation is used:

n = number of squares
n_0 = number of atlassed squares in which the species is not present, i.e., abundance index of 0
n_1 = number of atlassed squares with abundance indices, including the n_0 squares
n_2 = number of atlassed squares with this species reported but no abundance indices provided

The first step is to obtain a population estimate for each square based on the coded value. The number of breeding pairs in each square is assumed to follow a Poisson distribution with parameter (and mean) m, where m varies from square to square. The principle of maximum likelihood is used to get the population estimate for the square. Thus, the estimate of population is chosen to maximize the probability of getting an observation with its specific index. This leads (details at the end of the appendix) to the following estimate of population for an individual square:

Index	Abundance Range	Estimate
0	0	0
1	1	1
2	2–10	4.95
3	11–100	48.13
4	101–1000	475.74
5	1001–10,000	4751.95
6	more than 10,000	47514.06

There is a need to be cautious about isolated large index values as they can overwhelm the estimate in the aggregation. This issue is dealt with following the description of how to obtain an aggregate estimate and its standard deviation.

The population estimate is obtained as

$$M_1 = \left(M_0 + \left[M_0 \times \frac{n_2}{n_1 - n_0} \right] \right) \times \frac{n_2}{n_1 + n_0} \qquad (1)$$

where M_0 is the sum of the population estimates for the n_1 squares in the area of interest. The second term in the sum is the population estimate for the n_2 squares where species occurred but no abundance estimate was given. The fraction $n/(n_1 + n_2)$ expands the estimate to take into account the squares that were not atlassed.

As an example we will consider the reports for American Redstart (AMRE) in Region 18. Of the 79 squares in this region, 77 were atlassed. AMRE was reported from 74 squares. Abundance indices were provided for 34 squares. This gives us:

n = 79 squares
n_0 = 3 squares with species absent
n_1 = 37 squares with abundance indices (34 with species present + 3 with species absent)
n_2 = 40 squares with species present, but no abundance indices provided.

The 34 abundance indices provided were:
1 2 2 2 2 3 3 3 3 3 3 3 4 4 4 4 4
4 4 4 4 4 4 4 4 4 4 5 5 5 5 5 5

Summing the estimates corresponding to the abundance indices provided yields
M_0 = (1x1) + (4x4.95) + (7x48.13) + (16x475.74)
    + (6x4751.95)
    = 36,481.25

Extrapolating M_0 over the whole region using (1) yields a population estimate of 81,463 pairs, or about 81,500 pairs in round numbers.

Trimming

To handle possible outliers, 3 per cent of the n_1 squares were trimmed and removed from the data. This robust procedure provided protection against outliers while still giving an efficient estimate.

In the previous sample, 3 per cent of the 37 indices is 1.11, indicating that one index should be trimmed. This leaves us with 36 indices, having discarded one of the 5s. The new value for M_0 is, then,

$$M_0 = (1 \times 1) + (4 \times 4.95) + (7 \times 48.13) + (16 \times 475.74) + (5 \times 4751.95)$$
$$= 31{,}729.3$$

and the extrapolated population estimate for the region, after trimming, is 70,852 pairs.

Recoding

It was thought that, for any given species, the largest coded values were not likely to come from the whole range of the highest index, but rather would come from a region near the lower endpoint. That is, if the largest index for a species was 5, for example, this would not generally represent a range of 1001–10,000, but more likely some smaller range. We chose the following ranges as more likely to be represented by the largest index in a series:

Index	Abundance Range	Estimate
0	0	0
1	1	1
2	2–10	4.95
3	11–25	17.02
4	101–250	169.49
5	1001–2500	1694.12
6	10,001–25,000	16,941.07

The estimates were again calculated using the principle of maximum likelihood. Note that the ranges for indices 0, 1, and 2 were not changed.

Applying recoding to our (trimmed) American Redstart example yields the following:

$$M_0 = (1 \times 1) + (4 \times 4.95) + (7 \times 48.13) + (16 \times 475.74) + (5 \times 1694.12)$$
$$= 16{,}440.15$$

Extrapolating M_0 over the whole region using (1) yields a population estimate of 36,711 pairs, or about 36,700 pairs in round numbers.

Standard Error

The standard error of the estimate was computed using the bootstrap technique (Efron, B., and Tibsherani, R., *Statistical Science 1*, 1986, pp. 54–77). In brief, the technique is as follows:

We have n_1 values of population estimates, corresponding to the n_1 abundance indices. We sample n_1 values with replacement from this set of population estimates and compute the arithmetic mean of these samples. This process is repeated a large number of times (1,000 in our case). The standard deviation of all of these means, denoted by *sd*, is the estimated standard error of the population estimate for the squares with abundance estimates. It must be scaled using (1), with M_0 replaced by *sd*, to obtain the standard error for the whole region.

We can approximate 95% confidence limits using $M_1 \pm 2$ standard errors.

Population Estimate Derivation

Assigning a population estimate to a range represented by an abundance index is based on the assumption that the actual abundance, X_j, follows a Poisson distribution within the range, with parameter λ. If the range is denoted by [a, b] (e.g., [101, 1000]), then the maximum likelihood principle chooses λ to maximize

$$P_\lambda \left(a \leq X_j < b \right)$$

It can be shown that

$$\lambda^{b-a} = \frac{(b-1)!}{(a-1)!} \doteq \frac{\Gamma(b)}{\Gamma(a)}$$

maximizes this function.

A rational-fraction approximation (Kennedy, W. J., and Gentle, J. E., *Statistical Computing*, 1984, p. 90) was used to evaluate the gamma functions and hence determine the desired value of λ.

This value of λ is the population estimate used for each range covered by an abundance index.

Appendix E
Statistical Tables

Comments on Statistical Tables

Atlassers

The list of names included

(i) persons who contributed the only data for one or many squares, whether or not these were completed;

(ii) persons who, besides atlassing themselves, compiled the records of others active in the same square(s) (several of the "top 25" atlassers included observations by others (named) within some of their totals);

(iii) people who reported one or a few records to another person who was active in the atlas (such people may be unaware that they "took part" in the project);

(iv) persons who made a few excursions to a square, but most or all of their records were subsequently superseded by records at a higher level (probable or confirmed breeding) by other people.

All of these people were counted as "atlassers," although their efforts varied greatly.

Records

A "record" was a report of one species in one square in one year. The total records included about 463 (0.32 per cent) that were later rejected (for various reasons). A far larger number were superseded by records at a higher level or duplicated a level already achieved in that square. The records shown on the final maps totalled 90,643, about 63 per cent of the total entered. Squares worked intensively in several years provided most of the duplicate and superseded records, although in succeeding years some atlassers reported only records that improved on their earlier ones.

Species

In the original computer record, a species was a recognizable form for which one record (or more) was entered. The grand total thus included hybrid pairs of Mallard and Black Duck (both combinations), the Ipswich subspecies of Savannah Sparrow, and various "poultry" that escaped from captivity or were deliberately released during the Atlas period. Sightings without breeding confirmation of several other wild species, originally entered in the computer record, were later judged to represent wandering birds only, the evidence in support of possible breeding not balancing the inherent improbability of such a species breeding far outside its known range.

The number of species finally accepted for mapping as breeding was 214.

Party-hours

The overall time totals may not be far wrong, but they included lots of minor inconsistencies. All fractional time periods were rounded to whole hours; five minutes (e.g., a stop in transit) might appear as one hour, but in a square with other data it often was ignored. In squares with resident atlassers, entering the full time of every visit to an area that already had been covered repeatedly produced some unrealistic figures. Most squares with over 100 party-hours (one showed 660 hours) included time that duplicated previous coverage without new records being expectable (e.g., on trips to and from work) although squares with high party-hour totals averaged more species detected than those with less time. Nearly one-tenth of squares with data, including one with 107 species detected, showed no party-hours reported, and many others evidently reported only part of the time spent. These omissions partly balanced the "unproductive" time reported elsewhere.

Table A3
Atlasser Statistics (by year and region)

Region	Party Hours						Atlassers					
	1986	1987	1988	1989	1990	Total	1986	1987	1988	1989	1990	Total
1	208	95	59	150	274	786	5	9	8	11	14	35
2	35	38	104	336	297	810	9	9	11	16	16	38
3	176	213	263	320	312	1284	13	18	17	21	15	46
4	46	45	32	259	0	382	5	8	12	13	4	30
5	401	274	434	184	439	1732	25	14	16	24	24	64
6	720	749	585	414	239	2707	48	44	45	32	23	90
7	210	179	163	252	155	959	17	20	17	21	17	50
8	190	223	125	90	379	1007	17	22	23	12	33	65
9	276	301	198	143	319	1237	32	33	30	34	35	87
10	377	540	342	50	18	1327	20	22	25	12	9	58
11	551	466	761	251	271	2300	28	33	29	29	26	71
12	345	368	260	329	507	1809	13	28	23	34	48	85
13	583	421	675	408	269	2356	42	34	42	50	24	130
14	148	283	321	274	85	1111	13	21	14	26	18	57
15	1569	1011	652	431	481	4144	81	53	38	49	40	136
16	605	550	445	237	359	2196	29	35	25	29	29	75
17	1235	519	612	703	1025	4094	46	45	40	42	54	119
18	619	943	591	193	32	2378	35	47	40	28	34	93
19	450	384	164	465	201	1664	30	27	14	22	21	68
20	390	376	229	344	489	1828	24	27	20	26	39	74
21	868	905	731	865	620	3989	69	56	57	59	51	171
22	280	523	516	119	308	1746	17	24	18	22	15	64
23	103	55	127	318	644	1247	19	17	10	18	26	47
Totals	10385	9461	8389	7135	7723	43093	482	498	415	425	400	1120

Table A4
Record Statistics (by year and region)

Region	Records						Records w / Abundance Estimate	Rejected Records					
	1986	1987	1988	1989	1990	Total		1986	1987	1988	1989	1990	Total
1	519	377	457	1215	1494	4062	646	3	0	0	1	1	5
2	128	308	494	2826	1124	4880	2199	1	0	0	0	0	1
3	402	342	694	1475	2593	5506	1684	8	1	2	2	3	16
4	259	214	179	2801	6	3459	1927	1	0	0	0	0	1
5	957	676	1281	648	3529	7091	4306	11	3	4	2	0	20
6	2607	1824	1400	1346	851	8028	224	13	2	0	3	1	19
7	551	366	635	838	643	3033	733	2	1	0	1	0	4
8	419	391	706	318	3445	5279	2565	3	5	1	0	2	11
9	917	1012	778	1329	1627	5663	585	12	0	0	4	9	25
10	1302	2771	2240	469	201	6980	3379	35	19	12	2	0	70
11	1773	1595	1431	1373	1295	7467	3231	7	2	1	0	1	11
12	1056	1266	1172	1030	3033	7557	3592	1	1	1	1	0	4
13	1059	1508	3345	2245	309	8466	4583	8	5	8	3	4	28
14	436	610	1386	3270	772	6474	3163	3	2	5	6	0	16
15	1763	773	749	1684	1056	6025	1386	42	2	3	8	3	58
16	854	1179	1025	855	1541	5454	1125	4	9	5	3	0	21
17	1498	679	965	2094	3520	8756	2093	16	2	6	7	1	32
18	1760	1984	1811	1166	881	7602	3046	11	4	0	1	3	19
19	493	1086	790	1187	686	4242	1747	5	6	2	1	1	15
20	847	853	658	1042	2763	6163	2312	8	3	0	0	4	15
21	2403	1410	2194	4408	3232	13647	3783	20	0	9	0	3	32
22	582	902	1064	563	1450	4561	2308	4	9	6	2	3	25
23	280	419	244	1375	1926	4244	2546	7	0	0	2	6	15
Totals	22865	22545	25698	35557	37977	144642	53163	225	76	65	49	45	463

Table A5
Species Statistics

Region	Pos-sible	Prob-able	Confirmed						Total Species					
			1986	1987	1988	1989	1990	Total	1986	1987	1988	1989	1990	Total
1	10	9	87	65	67	105	104	124	110	112	104	127	139	143
2	11	14	29	9	45	102	96	123	71	83	101	145	126	148
3	6	11	72	73	76	88	129	140	116	112	125	139	149	157
4	6	6	47	26	58	129	1	132	107	85	101	143	6	144
5	5	11	71	53	93	66	128	138	124	114	134	128	146	154
6	5	13	92	94	97	117	85	137	143	139	134	142	130	155
7	12	17	70	62	79	96	76	116	123	116	121	131	125	145
8	11	20	64	88	78	48	111	128	117	117	121	112	151	159
9	13	17	87	97	82	110	126	139	140	128	136	148	161	169
10	5	11	88	135	149	102	63	163	140	161	171	134	104	179
11	11	11	119	109	112	127	111	157	153	141	145	154	150	179
12	10	12	76	92	95	92	115	146	131	134	137	140	151	168
13	12	12	91	106	119	111	84	143	131	137	145	151	129	167
14	9	15	47	75	100	116	103	129	104	120	133	142	129	153
15	7	10	127	96	96	106	111	142	142	125	130	137	141	159
16	10	16	93	98	98	90	109	135	122	139	141	127	144	161
17	2	7	113	97	111	121	119	141	133	126	133	137	136	150
18	6	6	122	118	112	104	97	146	142	140	135	140	138	158
19	8	11	75	99	76	103	84	124	122	126	114	128	123	143
20	10	9	90	108	81	105	126	141	135	131	131	138	145	160
21	14	18	95	85	92	116	116	137	133	123	129	150	152	169
22	10	7	82	98	95	87	125	137	129	125	125	124	146	154
23	8	9	66	96	67	74	121	138	106	118	114	131	146	155
Totals	**11**	**14**	**174**	**174**	**181**	**184**	**184**	**199**	**199**	**191**	**202**	**203**	**207**	**224**

Table A6
Coverage and square completion statistics

Region	No. of Squares with		No. of Squares			No. Complete			Percentage Complete		
	50% of expected spp	Adequate Sampling*	Special and Priority	Secondary and Tertiary	Total	Special and Priority	Secondary and Tertiary	Total	Special and Priority	Secondary and Tertiary	Total
1	40	32	19	57	76	19	6	25	100	11	33
2	43	31	22	50	72	22	0	22	100	0	31
3	46	30	17	45	62	17	7	24	100	16	39
4	34	21	17	47	64	17	1	18	100	2	28
5	61	31	20	68	88	19	3	22	95	4	25
6	40	30	16	41	57	15	10	25	94	24	44
7	26	21	18	40	58	18	1	19	100	3	33
8	51	23	17	54	71	17	0	17	100	0	24
9	32	25	19	54	73	18	4	22	95	7	30
10	38	28	19	46	65	19	7	26	100	15	40
11	59	59	19	44	63	19	40	59	100	91	94
12	48	39	17	46	63	17	11	28	100	24	44
13	55	55	16	39	55	16	39	55	100	100	100
14	57	46	15	48	63	15	21	36	100	44	57
15	43	37	20	57	77	20	16	36	100	28	47
16	60	43	23	67	90	15	16	31	65	24	34
17	72	56	22	74	96	21	27	48	95	36	50
18	62	49	23	56	79	23	25	48	100	45	61
19	42	30	21	62	83	21	9	30	100	15	36
20	43	33	20	57	77	20	7	27	100	12	35
21	87	77	27	74	101	27	35	62	100	47	61
22	37	34	22	48	70	19	13	32	86	27	46
23	39	29	21	58	79	19	1	20	90	2	25
Totals	**1115**	**859**	**450**	**1232**	**1682**	**433**	**299**	**732**	**96**	**24**	**44**

75% of expected species (completed or not)

Table A7
Number of times each breeding code was used (excluding rejected reports, but including duplicate and superseded records)

SPECIES	H	P	T	C	V	A	N	NB	DD	UN	FL	ON	AY	NE	NY	TOTAL
Common Loon	310	179	75	5	1	6	0	3	0	3	275	14	9	45	6	931
Pied-billed Grebe	72	6	24	0	0	0	0	4	0	1	113	5	2	6	2	235
Leach's Storm-Petrel	3	0	0	0	0	0	0	1	0	3	1	14	0	5	7	34
Great Cormorant	16	0	0	0	0	0	0	1	0	1	2	21	0	9	18	68
Double-crested Cormorant	113	2	4	0	0	0	0	4	0	0	15	73	0	11	41	263
American Bittern	297	14	106	7	2	7	0	2	0	0	37	3	0	5	3	483
Least Bittern	5	1	1	0	0	0	0	0	0	0	0	0	0	0	0	7
Great Blue Heron	397	6	14	2	0	0	0	4	0	1	16	58	1	2	51	552
Snowy Egret	0	0	1	0	0	0	0	0	0	0	0	0	0	0	0	1
Green-backed Heron	12	0	3	0	0	0	0	0	0	0	5	0	0	0	1	21
Black-crowned Night-Heron	49	2	3	0	0	0	0	0	0	0	8	3	0	1	3	69
Glossy Ibis	0	0	0	0	0	0	0	0	0	0	0	0	0	1	0	1
Canada Goose	40	21	5	0	0	0	0	0	2	0	68	13	0	10	2	161
Wood Duck	137	76	20	1	2	3	0	1	0	4	140	11	0	8	5	408
Green-winged Teal	160	152	41	2	0	3	0	0	3	0	198	1	1	2	3	566
American Black Duck	463	267	65	3	0	4	0	2	10	0	891	8	10	21	12	1756
Mallard	137	86	21	2	0	0	0	1	2	0	109	0	0	12	2	372
Mallard X Black Duck	2	3	0	0	0	0	0	0	0	0	1	0	0	0	0	6
Black Duck X Mallard	0	3	0	0	0	0	0	0	0	0	2	0	0	0	0	5
Northern Pintail	47	29	6	1	0	0	0	0	0	0	56	0	0	2	0	141
Blue-winged Teal	160	122	46	11	0	1	0	0	1	0	259	3	0	11	1	615
Northern Shoveler	22	12	3	1	0	1	0	0	2	0	22	1	0	1	0	65
Gadwall	5	11	7	0	0	0	0	0	0	0	13	0	0	1	0	37
Eurasian Wigeon	3	0	0	0	0	0	0	0	0	0	0	0	0	0	0	3
American Wigeon	80	83	24	6	0	2	0	0	2	0	115	2	0	6	1	321
Redhead	0	4	0	0	0	0	0	0	0	0	0	0	0	0	0	4
Ring-necked Duck	163	246	37	7	0	2	0	0	0	0	256	3	2	13	0	729
Greater Scaup	0	1	1	0	0	0	0	0	0	0	1	0	0	3	0	6
Common Eider	25	5	1	0	0	0	0	0	0	1	100	2	2	27	10	173
Harlequin Duck	1	1	0	0	0	0	0	0	0	0	0	0	0	0	0	2
Common Goldeneye	86	36	4	2	1	2	0	0	0	0	93	7	1	4	0	236
Hooded Merganser	76	46	0	1	0	1	0	0	0	2	69	2	0	4	1	202
Common Merganser	193	147	13	1	2	1	0	1	0	0	245	2	6	6	7	624
Red-breasted Merganser	70	48	4	5	0	1	0	0	0	0	71	0	0	16	0	215
Ruddy Duck	5	0	0	0	0	0	0	0	0	0	1	0	0	0	0	6
Turkey Vulture	7	0	0	0	0	0	0	0	0	0	0	0	0	0	0	7
Osprey	497	46	76	2	2	10	0	35	0	2	25	264	9	4	125	1097
Bald Eagle	215	20	33	3	3	2	0	11	0	1	15	59	1	14	147	524
Northern Harrier	600	119	133	11	2	9	0	5	1	0	48	6	20	9	23	986
Sharp-shinned Hawk	359	9	36	1	2	16	0	5	1	0	38	3	16	1	11	498
Cooper's Hawk	4	1	0	0	0	0	0	0	0	0	0	0	0	0	0	5
Northern Goshawk	175	7	7	0	0	3	0	1	1	0	19	12	3	0	24	252
Red-shouldered Hawk	11	1	5	1	0	1	0	0	0	0	1	0	0	0	0	20
Broad-winged Hawk	338	31	59	3	0	23	0	8	0	0	24	19	11	0	15	531
Red-tailed Hawk	471	47	77	4	0	39	0	2	0	0	55	19	9	2	25	750
Golden Eagle	4	0	1	1	0	0	0	0	0	0	0	0	0	0	0	6
American Kestrel	581	103	142	7	1	30	0	3	1	1	124	68	14	3	40	1118
Merlin	186	20	26	4	1	10	0	2	0	0	29	12	9	0	9	308
Peregrine Falcon	3	3	4	0	0	0	0	0	0	0	1	0	1	0	3	15
Gray Partridge	20	13	4	1	0	0	0	0	0	0	38	1	0	3	0	80
Chukar	0	0	0	0	0	0	0	0	0	0	0	0	0	1	0	1
Ring-necked Pheasant	128	29	47	2	0	0	0	0	0	0	130	3	3	4	3	349
Spruce Grouse	121	13	15	1	0	0	0	0	2	0	135	2	0	1	0	290
Ruffed Grouse	403	11	104	12	0	11	0	0	33	0	633	7	5	40	13	1272
Sharp-tailed Grouse	0	5	0	0	0	0	0	0	0	0	0	0	0	0	0	5

SPECIES	H	P	T	C	V	A	N	NB	DD	UN	FL	ON	AY	NE	NY	TOTAL
Wild Turkey	1	0	1	0	0	0	0	0	0	0	1	0	0	0	0	3
Yellow Rail	4	0	0	0	0	0	0	0	0	0	0	0	0	0	0	4
Virginia Rail	33	1	20	0	1	9	0	0	1	0	11	0	0	0	0	76
Sora	122	4	92	0	0	14	0	0	1	0	46	1	1	2	0	283
Common Moorhen	7	0	3	0	0	1	0	0	0	0	10	0	0	0	0	21
American Coot	21	0	3	0	0	0	0	0	0	0	7	0	0	0	0	31
Semipalmated Plover	17	1	1	1	0	2	0	0	2	0	7	0	0	2	0	33
Piping Plover	10	5	1	0	0	0	0	0	5	0	29	7	0	35	7	99
Killdeer	402	43	88	9	0	71	0	2	204	0	369	8	6	109	35	1346
Greater Yellowlegs	2	0	1	0	0	8	0	0	4	0	1	0	1	0	0	17
Solitary Sandpiper	5	0	0	0	0	2	0	0	1	0	3	0	0	0	0	11
Willet	104	7	25	5	0	52	0	0	14	2	47	14	1	10	2	283
Spotted Sandpiper	559	57	134	7	0	121	0	2	67	1	384	18	6	77	19	1452
Upland Sandpiper	9	0	7	2	0	2	0	0	0	0	6	0	0	0	0	26
Least Sandpiper	3	0	0	0	0	0	0	0	1	0	1	0	0	1	0	6
Common Snipe	556	19	296	12	2	31	0	0	26	0	79	5	1	17	6	1050
American Woodcock	424	7	141	8	0	5	0	0	16	0	111	8	0	20	8	748
Wilson's Phalarope	1	6	1	0	0	2	0	0	0	0	2	0	0	0	0	12
Laughing Gull	1	0	0	0	0	0	0	0	0	0	0	0	0	0	0	1
Bonaparte's Gull	0	0	0	0	0	0	0	0	0	0	1	0	0	0	0	1
Ring-billed Gull	4	0	1	0	0	0	0	0	0	0	1	2	0	9	8	25
Herring Gull	169	5	9	2	1	1	0	1	0	1	49	97	0	44	51	430
Great Black-backed Gull	178	12	11	2	0	3	0	1	0	2	65	109	3	52	64	502
Black-legged Kittiwake	0	0	0	0	0	0	0	0	0	0	0	4	0	1	1	6
Roseate Tern	0	1	1	0	0	1	0	0	0	0	2	0	0	4	4	13
Common Tern	135	8	16	2	1	19	0	2	0	0	31	41	14	54	26	349
Arctic Tern	24	3	1	0	0	3	0	1	0	0	10	10	6	22	9	89
Black Tern	12	3	4	0	0	1	0	0	0	0	2	2	1	3	0	28
Common Murre	2	0	0	0	0	0	0	0	0	0	1	0	0	0	0	3
Razorbill	0	1	0	0	2	0	0	0	0	0	1	3	1	1	1	10
Black Guillemot	52	2	1	0	5	0	0	0	0	0	14	37	14	6	7	138
Atlantic Puffin	1	1	0	0	0	0	0	0	0	0	0	1	2	4	0	9
Rock Dove	402	31	78	29	11	0	0	40	0	3	93	195	7	15	59	963
Mourning Dove	383	100	126	20	2	0	0	14	0	0	136	11	0	9	16	817
Black-billed Cuckoo	136	7	24	0	0	11	0	1	0	0	3	1	9	0	1	193
Yellow-billed Cuckoo	1	0	0	0	0	0	0	0	0	0	0	0	0	0	0	1
Eastern Screech-Owl	4	0	0	0	0	0	0	0	0	0	0	0	0	0	0	4
Great Horned Owl	292	20	130	1	0	4	0	2	0	0	64	13	0	1	34	561
Northern Hawk-Owl	1	0	0	1	0	0	0	0	0	0	1	0	0	0	0	3
Barred Owl	266	21	155	0	0	3	0	0	0	0	43	11	1	1	13	514
Long-eared Owl	20	1	10	0	0	0	0	0	0	0	3	1	0	0	1	36
Short-eared Owl	28	1	3	1	0	1	0	1	0	0	5	1	1	0	1	43
Boreal Owl	4	2	8	0	0	1	0	0	0	0	0	0	0	0	0	15
Northern Saw-whet Owl	115	1	61	0	1	0	0	0	0	0	10	7	0	0	5	200
Common Nighthawk	475	20	170	14	1	14	0	0	5	0	22	5	1	16	7	750
Whip-Poor-Will	35	1	33	0	0	0	0	0	1	0	1	0	0	1	0	72
Chimney Swift	381	41	69	11	10	0	0	10	0	0	10	75	0	2	13	622
Ruby-throated Hummingbird	665	100	212	41	2	29	0	14	0	6	121	12	0	4	11	1217
Belted Kingfisher	691	122	174	4	14	22	0	15	0	12	105	181	41	3	42	1426
Red-headed Woodpecker	3	0	1	0	0	0	0	0	0	0	0	0	0	0	0	4
Yellow-bellied Sapsucker	485	47	106	5	6	20	14	0	0	0	171	47	60	2	134	1097
Downy Woodpecker	599	88	143	6	7	17	14	0	0	0	162	48	81	2	57	1224
Hairy Woodpecker	617	92	125	5	4	41	13	0	0	0	138	50	52	0	73	1210
Three-toed Woodpecker	24	2	5	0	1	1	0	0	0	0	5	2	1	0	5	46
Black-backed Woodpecker	188	26	27	1	0	14	4	0	0	1	22	14	7	2	32	338
Northern Flicker	1012	108	292	12	11	29	23	1	0	0	335	154	103	5	86	2171
Pileated Woodpecker	421	50	97	1	6	4	10	0	0	3	34	37	9	0	29	701
Olive-sided Flycatcher	528	24	186	2	1	62	0	7	0	0	29	5	53	4	6	907

SPECIES	H	P	T	C	V	A	N	NB	DD	UN	FL	ON	AY	NE	NY	TOTAL
Eastern Wood-Pewee	665	21	313	3	0	42	0	7	0	1	39	8	89	4	13	1205
Yellow-bellied Flycatcher	503	12	174	1	0	37	0	1	0	1	31	1	83	7	7	858
Willow Flycatcher	15	0	8	0	0	1	0	0	0	0	1	0	1	0	0	26
Alder Flycatcher	809	15	401	3	1	53	0	11	0	1	56	5	162	11	9	1537
Least Flycatcher	611	16	257	4	1	68	0	8	0	2	47	8	110	11	15	1158
Eastern Phoebe	147	7	26	0	1	2	0	10	0	19	13	31	14	16	41	327
Great Crested Flycatcher	95	13	40	1	2	6	0	6	0	0	4	10	10	1	9	197
Eastern Kingbird	517	99	150	5	2	58	0	23	0	0	118	54	109	26	39	1200
Horned Lark	41	1	9	1	0	0	0	1	0	0	16	1	5	2	4	81
Purple Martin	30	5	1	0	2	0	0	5	0	0	15	57	12	5	30	162
Tree Swallow	694	63	101	20	21	10	0	64	0	2	271	556	118	31	212	2163
Northern Rough-winged Swallow	5	1	1	0	0	0	0	0	0	0	1	1	0	0	0	9
Bank Swallow	334	11	34	2	14	2	0	35	0	11	69	619	57	12	59	1259
Cliff Swallow	165	9	23	0	3	2	0	95	0	14	66	357	41	6	129	910
Barn Swallow	518	45	71	5	21	10	0	127	0	28	273	550	108	64	333	2153
Gray Jay	378	39	63	3	1	9	0	2	0	0	487	1	3	1	2	989
Blue Jay	1008	31	220	9	1	36	0	33	0	0	509	10	16	2	16	1891
American Crow	1025	34	223	13	4	31	0	97	0	3	629	43	9	15	79	2205
Common Raven	1024	59	188	26	2	17	0	49	0	3	549	36	1	6	63	2023
Black-capped Chickadee	882	45	259	6	8	40	0	28	1	0	502	29	237	16	44	2097
Boreal Chickadee	495	24	120	1	1	26	0	4	0	0	267	8	101	5	10	1062
Tufted Titmouse	0	1	0	0	0	0	0	0	0	0	0	0	0	0	0	1
Red-breasted Nuthatch	693	54	177	4	5	38	0	15	0	0	332	22	142	0	25	1507
White-breasted Nuthatch	121	18	37	1	1	6	0	1	0	0	44	4	16	0	6	255
Brown Creeper	248	8	61	4	0	18	0	0	0	0	59	5	31	4	4	442
House Wren	3	0	4	0	0	0	2	0	0	0	0	0	0	0	0	9
Winter Wren	664	4	273	0	2	71	1	0	4	0	76	0	43	1	1	1140
Marsh Wren	9	0	1	0	0	0	3	0	0	1	0	1	2	0	0	17
Sedge Wren	3	0	0	0	0	0	0	0	0	0	0	0	0	0	0	3
Golden-crowned Kinglet	540	45	139	3	0	24	0	3	0	0	385	4	148	1	1	1293
Ruby-crowned Kinglet	796	44	317	1	0	75	0	5	1	0	205	4	218	0	2	1668
Blue-gray Gnatcatcher	0	0	0	0	0	0	0	0	0	0	1	0	0	0	0	1
Eastern Bluebird	37	17	5	0	5	0	0	0	0	0	30	12	6	4	27	143
Veery	657	10	270	0	1	47	0	1	0	0	84	4	124	24	8	1230
Gray-cheeked Thrush	50	2	16	0	0	10	0	0	0	0	6	1	16	1	3	105
Swainson's Thrush	872	10	349	1	1	76	0	7	6	2	112	7	249	22	11	1725
Hermit Thrush	836	11	338	3	1	53	0	3	3	0	151	6	219	55	30	1709
Wood Thrush	117	5	51	0	1	19	0	3	2	0	7	0	10	4	3	222
American Robin	697	56	144	2	0	128	0	76	0	28	426	108	883	108	222	2878
Gray Catbird	463	31	158	3	1	112	0	10	9	3	75	21	195	12	32	1125
Northern Mockingbird	62	5	47	3	2	5	0	7	0	0	16	5	13	6	13	184
Brown Thrasher	28	1	7	0	0	2	0	0	0	0	4	0	4	0	0	46
Bohemian Waxwing	1	0	0	0	0	0	0	0	0	0	0	0	0	0	1	2
Cedar Waxwing	950	161	211	76	1	15	0	106	0	6	203	36	68	34	49	1916
Loggerhead Shrike	2	0	0	0	0	0	0	0	0	0	0	0	0	0	0	2
Starling	427	13	48	2	6	0	1	73	0	3	549	119	437	7	183	1868
Solitary Vireo	665	21	217	2	0	108	0	10	1	2	96	4	202	2	11	1341
Warbling Vireo	65	0	35	0	0	6	0	1	0	0	1	0	14	0	1	123
Philadelphia Vireo	103	0	28	0	0	15	0	3	0	0	14	0	19	0	2	184
Red-eyed Vireo	973	48	366	5	0	103	0	20	2	1	104	14	216	15	17	1884
Tennessee Warbler	695	27	273	2	0	50	0	6	2	0	101	2	216	6	3	1383
Nashville Warbler	613	18	178	1	0	54	0	7	7	0	100	1	165	0	0	1144
Northern Parula Warbler	766	64	310	3	2	43	0	20	2	1	133	9	233	5	6	1597
Yellow Warbler	665	89	251	8	2	59	0	33	5	3	175	38	401	41	47	1817
Chestnut-sided Warbler	475	37	194	2	0	38	0	6	0	0	69	2	236	5	5	1069
Magnolia Warbler	744	48	275	3	0	71	0	8	8	2	218	8	441	10	10	1846
Cape May Warbler	279	22	117	1	0	12	0	2	1	0	31	0	61	2	0	528
Black-throated Blue Warbler	301	24	64	0	0	20	0	0	1	1	23	1	65	0	0	500

SPECIES	H	P	T	C	V	A	N	NB	DD	UN	FL	ON	AY	NE	NY	TOTAL
Yellow-rumped Warbler	754	106	216	6	1	52	0	34	2	0	316	10	425	8	15	1945
Black-throated Green Warbler	696	47	249	1	0	43	0	14	1	1	152	8	282	4	4	1502
Blackburnian Warbler	526	72	194	3	0	37	0	7	0	0	121	0	206	0	1	1167
Pine Warbler	4	1	6	0	0	0	0	0	0	0	3	0	2	0	1	17
Palm Warbler	222	19	59	2	0	25	0	2	3	0	101	2	136	2	7	580
Bay-breasted Warbler	435	41	154	0	0	24	0	0	2	0	68	1	165	0	2	892
Blackpoll Warbler	160	15	57	0	0	6	0	1	0	0	32	0	43	0	1	315
Black-and-white Warbler	754	75	217	3	0	49	0	9	4	0	177	3	315	2	8	1616
American Redstart	827	144	257	6	0	104	0	23	4	2	252	24	507	14	40	2204
Ovenbird	815	19	368	5	1	104	0	11	28	0	116	17	168	17	13	1682
Northern Waterthrush	469	6	166	1	0	69	0	4	11	0	35	10	76	0	4	851
Louisiana Waterthrush	1	0	0	0	0	0	0	0	0	0	0	0	0	0	0	1
Mourning Warbler	356	20	110	0	0	58	0	3	8	0	42	5	170	2	1	775
Common Yellowthroat	772	90	263	3	0	160	1	12	41	0	207	19	615	15	18	2216
Wilson's Warbler	258	8	79	0	2	45	0	2	2	0	31	0	73	1	1	502
Canada Warbler	479	35	166	3	0	114	0	3	3	0	60	4	198	2	0	1067
Scarlet Tanager	151	40	41	1	0	11	0	3	0	0	8	0	22	0	1	278
Northern Cardinal	6	6	4	0	0	0	0	1	0	0	5	0	1	0	1	24
Rose-breasted Grosbeak	680	145	229	5	1	53	0	26	1	1	155	11	153	10	14	1484
Indigo Bunting	27	3	11	0	0	3	0	0	0	0	3	0	2	1	0	50
Dickcissel	1	0	0	0	0	0	0	0	0	0	0	0	0	0	0	1
Rufous-sided Towhee	4	0	1	0	0	0	0	0	0	0	0	0	0	0	0	5
American Tree Sparrow	0	0	0	0	0	0	0	0	0	0	0	0	1	0	0	1
Chipping Sparrow	649	28	212	16	1	41	0	32	5	0	328	14	315	28	49	1718
Clay-coloured Sparrow	0	0	6	0	0	0	0	0	0	0	0	0	0	0	0	6
Field Sparrow	1	0	3	0	0	0	0	0	0	0	0	0	0	0	0	4
Vesper Sparrow	55	2	20	0	0	2	0	0	0	0	13	0	15	2	0	109
Savannah Sparrow	553	28	195	4	1	89	0	4	7	0	174	19	329	32	32	1467
Ipswich Sparrow	0	0	0	0	0	0	0	0	0	0	0	0	1	0	0	1
Sharp-tailed Sparrow	119	6	76	0	2	21	0	2	2	0	30	1	28	1	4	292
Fox Sparrow	117	1	34	0	0	19	0	0	1	0	22	2	17	3	0	216
Song Sparrow	796	29	305	8	1	157	0	32	23	1	374	28	530	42	68	2394
Lincoln's Sparrow	359	7	101	0	0	94	0	3	17	0	105	3	177	1	2	869
Swamp Sparrow	461	20	174	2	0	97	0	5	8	0	124	6	200	9	7	1113
White-throated Sparrow	873	48	340	10	0	200	0	18	15	0	401	23	398	34	31	2391
Dark-eyed Junco	713	77	177	3	1	114	0	26	6	1	478	15	281	55	58	2005
Bobolink	501	188	203	8	8	52	0	9	14	1	123	28	315	12	17	1479
Red-winged Blackbird	561	171	173	10	2	181	0	39	19	0	264	62	312	80	48	1922
Eastern Meadowlark	18	1	7	1	0	1	0	0	0	0	3	0	7	0	1	39
Rusty Blackbird	238	38	50	2	0	43	0	4	0	0	100	2	75	2	5	559
Common Grackle	702	68	130	9	1	38	0	44	0	3	372	32	563	38	97	2097
Brown-headed Cowbird	534	174	96	39	8	0	0	0	0	0	162	1	0	22	13	1049
Northern Oriole	135	40	41	1	1	14	0	24	1	29	21	26	27	2	16	378
Pine Grosbeak	254	89	27	2	0	4	0	5	0	0	41	1	13	0	1	437
Purple Finch	737	340	230	44	1	18	0	25	0	0	286	7	122	7	4	1821
House Finch	20	13	10	0	0	0	0	0	0	0	8	0	0	0	1	52
Red Crossbill	114	26	21	3	0	0	0	2	0	0	23	0	2	0	0	191
White-winged Crossbill	364	96	82	25	1	2	0	11	0	0	108	2	5	0	2	698
Common Redpoll	3	0	0	0	0	0	0	0	0	0	0	0	0	0	0	3
Pine Siskin	668	47	134	29	0	16	0	8	0	0	173	3	34	0	5	1117
American Goldfinch	842	474	230	25	4	11	0	86	0	1	144	13	58	20	18	1926
Evening Grosbeak	596	318	100	14	1	4	0	12	0	0	192	3	46	1	1	1288
House Sparrow	367	64	76	8	18	1	0	68	0	0	198	192	178	3	106	1279
Number of rejected records	5	8	271	0	57	31	4	9	3	45	3	5	6	4	12	463

Appendix F
Names of living things mentioned in the text

Animals

List does not include birds for which species accounts, in any category, were provided

Ants	Formicidae
Beaver	*Castor canadensis*
Beetles	Coleoptera
Spruce Budworm	*Choristoneura fumiferana*
Bumblebee	*Bombus* sp.
Cat (domestic)	*Felis catus*
Cattle (domestic)	*Bos taurus*
Common Sandpiper	*Tringa hypoleuca*
Crane-flies	Tipulidae
Dipper (in Europe)	*Cinclus cinclus*
Dog (domestic)	*Canis familiaris*
Dragon-flies	Odonata
Fish	Pisces
Frogs	Anura
Grasshoppers	Orthoptera
Great Auk	*Pinguinnis impennis*
Snowshoe Hare	*Lepus americanus*
Herring	*Clupea harengus*
Horse (domestic)	*Equus caballus*
Jay (in Europe)	*Garrulus glandarius*
Labrador Duck	*Camptorhynchus labradorius*
Little Egret	*Egretta garzetta*
Man (including woman)	*Homo sapiens*
MacGillivray's Warbler	*Oporornis tolmiei*
Moth	Lepidoptera
Muskrat	*Ondatra zibethicus*
Mussels	*Mytilus* (usually *M. edulis*)
Rat	*Rattus* (probably all *R. norvegicus*)
Red-bellied Woodpecker	*Melanerpes carolinus*
Atlantic Salmon	*Salmo salar*
Sand-eel or Sand-lance	*Ammodytes maritimus*
Sea-Urchin	*Strongylocentrotus droebachiensis*
Snails	Gastropoda
Tent caterpillar,	*Malacosoma* spp.
Eastern	*Malacosoma americanum*
Forest	*M. disstria*
Speckled Trout	*Salvelinus fontinalis*
Meadow Vole	*Microtus pennsylvanicus*
White-faced Ibis	*Plegadis chihi*

Plants

Alder	*Alnus* (usually *A. rugosa*)
Apple	*Malus pumila*
Aspen	*Populus tremuloides* or *P. grandidentata*
Bayberry	*Myrica pennsylvanica*
Beard-lichens	*Usnea* spp.
American Beech	*Fagus grandifolia*
Birch,	*Betula* spp.
White	*B. papyrifera*
Yellow	*B. allegheniensis*
Blueberry	*Vaccinium* spp. (usually *V. angustifolium* or *V. myrtilloides*)
Bulrush	*Scirpus* spp.
Giant Bur-reed	*Sparganium eurycarpum*
Cat-tail	*Typha* (usually *T. latifolia*)
White Cedar	*Thuja occidentalis*
Elm	*Ulmus* (usually *U. americana*)
Balsam Fir	*Abies balsamea*
Grasses	Gramineae
Eastern Hemlock	*Tsuga canadensis*
Larch	*Larix laricina*
Leatherleaf	*Chamaedaphne calyculata*
Maple,	*Acer* spp.
Sugar	*A. saccharum*
Manitoba	*A. negundo*
Moss	Bryophyta
Pine,	*Pinus* spp.
Jack	*P. banksiana*
White	*P. strobus*
Poplar	*Populus* spp.
Potato	*Solanum tuberosum*
Raspberry	*Rubus idaeus*
Sedges	Cyperaceae; *Carex* spp.
Spanish moss	*Tillandsia* spp.
Spruce,	*Picea* spp.
Black	*P. mariana*
Sweet-gale	*Myrica gale*
Thistle	*Cirsium* spp.
Wild rice	*Zizania aquatica*
Willow	*Salix* spp.

References

[Note: The list includes references (i) cited in chapters and species accounts, and (ii) to breeding studies of species with field work partly or wholly conducted in the Maritimes.]

1. Adamus, P.R., comp. 1987. *Atlas of breeding birds in Maine, 1978–1983.* Augusta, Me.: Maine Dept. Inland Fish. Game. 366 pp.
2. American Birding Association. 1975. *A.B.A. Checklist: Birds of continental United States and Canada.* Austin, Tex.: Foxx Printers. 64 pp.
3. American Ornithologists' Union (A.O.U.). 1983. *Check-List of North American birds.* 6th ed. Washington, D.C.: American Ornithologists' Union.
4. Andrle, R.F., and J.R. Carroll, eds. 1988. *The atlas of breeding birds in New York State.* Ithaca, N.Y.: Cornell University Press. 551 pp.
5. Anonymous 1988. *Prince Edward Island field check-list of birds.* 4th ed. Charlottetown: P.E.I. Dept. Tourism & Parks, Prince Edward Island National Park and Island Nature Trust.
6. Austin-Smith, P.J., and G. Rhodenizer. 1983. Ospreys, *Pandion haliaetus*, relocate nests from power poles to substitute sites. *Can. Field-Nat.* 97:315–319.
7. Bartlett, C.O. 1960. American Widgeon and Pintail in the Maritime Provinces. *Can. Field-Nat.* 74:153–155.
8. Benkman, C.W. 1990. Intake rates and the timing of crossbill reproduction. *Auk* 107:376–386. [And references therein]
9. Bent, A.C., and collaborators. 1968. *Life histories of North American cardinals, grosbeaks, buntings, towhees, finches, sparrows, and allies.* (Comp. & ed. by O.L. Austin, Jr.). United States National Museum Bulletin 237, parts 1–3. 1889 pp. [And earlier issues in the Bent Life histories series, listed in the introduction to the 1968 work]
10. Bird, D.M., ed. 1983. *Biology and management of Bald Eagles and Ospreys.* Ste. Anne de Bellevue, Que.: Macdonald Raptor Research Centre, McGill University. 325 pp. Pertinent papers by Greene et al. (pp. 257–267), Seymour & Bancroft (pp. 275–280), Stocek & Pearce (pp. 215–221), all on Ospreys in Maritimes.
11. Black, W.A., and J.W. Maxwell. 1972. Resource utilization: Change and adaptation. Pp. 73–136 in *The Atlantic Provinces*, edited by A.G. Macpherson. Toronto: University of Toronto Press.
12. Blacquiere, R.; S. Griesbach, and M. Morris. 1986. Some details of the nesting of the American Robin in New Brunswick. *N.B. Nat.* 15(2): 55–59.
13. Bondrup-Nielsen, S. 1984. Vocalizations of the Boreal Owl, *Aegolius funereus richardsoni*, in North America. *Can. Field-Nat.* 98:191–197.
14. Boyer, G.F. 1959. Hand-reared Mallard releases in the Maritime Provinces. *Can. Field-Nat.* 73:1–5.
15. Boyer, G.F. 1972. *Birds of the Nova Scotia–New Brunswick border region.* 2nd ed. (with addendum by A.J. Erskine and A.D. Smith). Canadian Wildlife Service. Occasional Papers no. 8. 46 pp.
16. Brookes, I. 1972. The physical geography of the Atlantic Provinces. Pp. 1–45 in *The Atlantic Provinces*, edited by A.G. Macpherson. Toronto: University of Toronto Press.
17. Brown, R.G.B. 1986. *Revised atlas of eastern Canadian seabirds. I. Shipboard surveys.* Ottawa: Canadian Wildlife Service. 111 pp.
18. Busby, D.G., L.M. White, and P.A. Pearce. 1990. Effects of aerial spraying of fenitrothion on breeding White-throated Sparrows. *J. Appl. Ecol.* 27:743–755.
19. Cadman, M.D., P.F.J. Eagles, and F.M. Helleiner, comps. 1987. *Atlas of the breeding birds of Ontario.* Waterloo, Ont.: University of Waterloo Press. 617 pp.
20. Cairns, W.E. 1977. Breeding biology and behaviour of the Piping Plover (*Charadrius melodus*) in southern Nova Scotia. M.Sc. thesis, Dalhousie University, Halifax. 115 pp.
21. Cairns, W.E., and I.A. McLaren. 1980. Status of the Piping Plover on the east coast of North America. *Am. Birds* 34:206–208.
22. Cannell, P.F., and G.D. Maddox. 1983. Population change in three species of seabirds at Kent Island, New Brunswick. *J. Field Ornithol.* 54:29–35.
23. Carter, B.C. 1958. *The American Goldeneye in central New Brunswick.* Canadian Wildlife Service, Wildlife Management Bulletin, ser. 2, no. 9. 47 pp. [Based on M.S. thesis, University of Maine]
24. Christie, D.S. 1979. Changes in Maritime bird populations, 1878–1978. *J. N.B. Museum* 1979:132–146.
25. Dickerman R.W. 1987. The "Old Northeastern" subspecies of Red Crossbill. *Am. Birds* 41:189–194.
26. Dixon, C.L. 1978. Breeding biology of the Savannah Sparrow on Kent Island. *Auk* 95:235–246.
27. Dobell, J.V. 1977. Determination of Woodcock habitat changes from aerial photography in New Brunswick. *Proc. Woodcock Symp.* 6:73–81,
28. Drury, W.H. 1973–74. Population changes in New England seabirds. *Bird-Banding* 44:267–313; 45:1–15.
29. Elderkin, M.F. 1987. The breeding and feeding ecology of a Barred Owl, *Strix varia* Barton, population in Kings County, Nova Scotia. M.Sc. thesis, Acadia University, Wolfville, N.S. 268 pp.
30. Eliason, B.C. 1986. Mating system, parental care, and reproductive success in the Blackpoll Warbler (*Dendroica striata*). Ph.D. diss., University of Minnesota, Minneapolis. 132 pp.
31. Erskine, A.J. 1967. Range extension of Willets in eastern Canada. *Can. Field-Nat.* 81:147.
32. Erskine, A.J. 1968. Northern birds summering in eastern Canada. *N.S. Bird Soc. Newsletter* 10:128–130.
33. Erskine, A.J. 1970. Starlings nesting in eastern Nova Scotia. *N.S. Bird Soc. Newsletter* 12:33–36.
34. Erskine, A.J. 1971. Some new perspectives on the breeding ecology of Common Grackles. *Wilson Bull.* 83:352–370.
35. Erskine, A.J. 1972a. *Populations, movements and seasonal distribution of mergansers in northern Cape Breton Island.* Canadian Wildlife Service Report Series no. 17. 35 pp.
36. Erskine, A.J. 1972b. *The Great Cormorants of eastern Canada.* Canadian Wildlife Service, Occasional Papers no. 14. 21 pp.
37. Erskine, A.J. 1977. *Birds in boreal Canada: Communities, densities and adaptations.* Canadian Wildlife Service Report Series no. 41. 71 pp.
38. Erskine, A.J. 1978. *The first ten years of the co-operative Breeding Bird Survey in Canada.* Canadian Wildlife Service Report Series no. 42. 59 pp.
39. Erskine, A.J. 1979. Man's influence on potential nesting sites and populations of swallows in Canada. *Can. Field-Nat.* 93:371–377.
40. Erskine, A.J. 1980. Estimates of species populations from census and atlas data. Pp. 254–263 in *Bird census work and nature conservation*, edited by H. Oelke. Dachverbandes Deutscher Avifaunisten.
41. Erskine, A.J., ed. 1987. *Waterfowl breeding population surveys, Atlantic Provinces.* Canadian Wildlife Service, Occasional Papers no. 60. 80 pp.
42. Erskine, A.J., and W.D. McLaren. 1972. Sapsucker nest holes and their use by other species. *Can. Field-Nat.* 86:357–361.
43. Erskine, A.J., and S.M. Teeple. 1970. Nesting activities in a Cliff Swallow colony. *Can. Field-Nat.* 84:385–387.
44. Erskine, A.J., M.C. Bateman, R.I. Goudie and G.R. Parker. 1990. *Aerial surveys for breeding waterfowl, Atlantic Region, 1985–89.* Canadian Wildlife Service Tech. Rep. no. 85. 13 pp.

45. Erskine, A.J., B.T. Collins, and J.W. Chardine. 1990. *The co-operative Breeding Bird Survey in Canada, 1988.* Canadian Wildlife Service Progress Notes no. 188. 15 pp.

46. Findlay, D.D. 1969. Nesting Sharpshins. *N.S. Bird Soc. Newsletter* 11:70–71.

47. Flemming, S.P. 1988. Osprey *Pandion haliaetus* L. social organization and foraging ecology in northern mainland Nova Scotia. M.Sc. thesis, Acadia University, Wolfville, N.S. 182 pp.

48. Forbes, M.R.L. 1983. The nesting ecology and breeding behaviour of the Pied-billed Grebe, *Podilymbus podiceps* (L.), at a national wildlife refuge in Nova Scotia. B.Sc.(Hons.) thesis, Acadia University, Wolfville, N.S. 86 pp.

49. Forsythe, B. 1986. Problems of nesting Chestnut-sided Warblers. *N.S. Birds* 28(1): 58–59. [Problems with Cowbirds]

50. Gates, A.D. 1975. *The tourism and outdoor recreation climate of the Maritime Provinces.* Environment Canada, Atmospheric Environment Service, Publications in Applied Meteorology, REC-3-73. 133 pp.

51. Germain, P., and S. Tingley. 1980. *Population responses of songbirds to 1979 forest spray operations in New Brunswick.* Report to Forest Protection Ltd. by Avifauna Ltee/Ltd. 32 pp. + 7 pp. app.

52. Gibbon, R.S. 1964. Studies and observations of the Black-backed Three-toed Woodpecker near Stewiacke. *N.S. Bird Soc. Newsletter* 6(3): 5–10.

53. Gibbon, R.S. 1966. Observations on the behaviour of nesting Three-toed Woodpeckers, *Picoides tridactylus*, in central New Brunswick. *Can. Field-Nat.* 80:223–226.

54. Gibbon, R.S. 1970. The breeding biology and food of the Yellow-bellied Sapsucker in New Brunswick. M.Sc. thesis, York University, Downsview, Ont. 126 pp.

55. Gilliland, S.G. 1991 Predator prey relationship between Great Black-backed Gull and Common Eider populations on the Wolves archipelago, New Brunswick: a study of foraging ecology. M.Sc. thesis, University of Western Ontario, London. 96 pp.

56. Gittens, E.F. 1968. A study on the status of the Bald Eagle in Nova Scotia. M.Sc. thesis, Acadia University, Wolfville, N.S. 205 pp.

57. Godfrey, W.E. 1954. *Birds of Prince Edward Island.* National Museum of Canada Bulletin 132:155–213.

58. Godfrey, W.E. 1966. *The birds of Canada.* National Museum of Canada Bulletin 203. 418 pp.

59. Godfrey, W.E. 1986. *The birds of Canada.* Revised ed. Ottawa: National Museum of Natural Science. 595 pp.

60. Goudie, R.I. 1989. Historical status of Harlequin Ducks wintering in eastern North America—a reappraisal. *Wilson Bull.* 101:112–114.

61. Grubb, T.C., Jr. 1971. Olfactory navigation in Leach's Petrel and other Procellariiform birds. Ph.D. diss., University of Wisconsin, Madison. 144 pp.

62. Halliday, W.E.D. 1937. *A forest classification for Canada.* Canada Department of Mines and Resources, Forestry Service Bulletin 89.

63. Hansen, G.L. 1979. Territorial and foraging behaviour of the Eastern Willet *Catoptrophorus semipalmatus semipalmatus* (Gmelin). M.Sc. thesis, Acadia University, Wolfville, N.S. 174 pp.

64. Harris, R.N. 1979. Aggression, superterritories, and reproductive success in tree swallows. *Can. J. Zool.* 57:2072–2078.

65. Hawkesley, O. 1950. A study of the behavior and ecology of the Arctic Tern *Sterna paradisaea* Brunnich. Ph.D. diss., Cornell University, Ithaca, N.Y. 228 pp.

66. Hébert, P.N. 1989. Decline of the Kent Island, New Brunswick, Herring Gull, *Larus argentatus*, colony. *Can. Field-Nat.* 103:394–396. [Based on M.Sc. thesis, University of Manitoba, Winnipeg]

67. Hickey, T.E. 1980. Activity budgets and movements of Black Ducks (*Anas rubripes*) in Prince Edward Island. M.Sc. thesis, McGill University, Montreal. 95 pp.

68. Hogan, G.G. 1979. Breeding parameters of Great Cormorants (*Phalacrocorax carbo carbo*) at mixed species colonies on Prince Edward Island. M.Sc. thesis, Brock University, St. Catharines, Ont. 99 pp.

69. Howard, D.V. 1967. Variation in the breeding season and clutch-size of the Robin in the northeastern United States and the Maritime Provinces of Canada. *Wilson Bull.* 79:432–440.

70. Hughson, W.B. 1971. Habitat preferences of breeding Black Ducks (*Anas rubripes* Brewster) in Nova Scotia. M.Sc. thesis, Acadia University, Wolfville, N.S. 114 pp.

71. Hunter, R.E. 1967. *Purple Martin survey in New Brunswick.* A centennial project. Moncton, N.B.: Moncton Publishing Co. 25 pp.

72. Huntington, C.E. 1963. Population dynamics of Leach's Petrel, *Oceanodroma leucorhoa.* Proc. Int. Ornithol. Congr. 13:701–705.

73. Huntington, C.E., and E.H. Burtt. 1972. [Abstract] Breeding age and longevity in Leach's Petrel (*Oceanodroma leucorhoa*). Proc. Int. Ornithol. Congr. 15:653.

74. Johnson, B.C. 1969. Home range, movements and ecology of the Gray Partridge, *Perdix perdix perdix* L., in a selected area of eastern Kings County, Nova Scotia. M.Sc. thesis, Acadia University, Wolfville, N.S. 127 pp.

75. Keppie, D.M. 1982. A difference in production and associated events in two races of spruce grouse. *Can. J. Zool.* ˙60:2116–2123.

76. Kerekes, J.J. 1990. Possible correlation of summer common loon (*Gavia immer*) population with the trophic state of a water body. *Verh. Internat. Verein. Limnol.* 24:349–353.

77. Kimball, D. 1989. The Eastern Bluebird—More common than we believe? *N.B. Nat.* 16:109–110.

78. Kirkham, I.R., and D.N. Nettleship. 1987. Status of the Roseate Tern in Canada. *J. Field Ornithol.* 58:505–515.

79. Knudsen, B.M. 1987. Reproductive success and behavior in Herring Gulls breeding in adjacent cliff and flat habitats. Ph.D. thesis, University of Manitoba, Winnipeg. 195 pp.

80. Lee, S.C. 1988. Third-egg neglect in the Herring Gull *Larus argentatus.* M.Sc. thesis, University of Manitoba, Winnipeg.89 pp.

81. Linton, A. 1979. The food and feeding habits of Leach's Storm-Petrel (*Oceanodroma leucorhoa*) at Pearl Island, Nova Scotia, and Middle Lawn Island, Newfoundland. M.Sc. thesis, Dalhousie University, Halifax. 126 pp.

82. Lock, A.R. 1973. A study of the breeding biology of two species of gulls nesting on Sable Island, Nova Scotia. Ph.D. thesis, Dalhousie University, Halifax. 135 pp.

83. Lock, A.R. 1987. Recent increases in the breeding population of Black-legged Kittiwakes, *Rissa tridactyla*, in Nova Scotia. *Can. Field-Nat.* 101:331–334.

84. Lock, A.R. 1988. Recent increases in the breeding population of Ring-billed Gulls, *Larus delawarensis*, in Atlantic Canada. *Can. Field-Nat.* 102:627–633.

85. Lock, A. 1989. A brief history of terns in Nova Scotia. *N.S. Birds* 31(3):59–62.

86. Lock, A.R., and R.K. Ross. 1974. The nesting of the Great Cormorant (*Phalacrocorax carbo*) and the Double-crested Cormorant (*Phalacrocorax auritus*) in Nova Scotia in 1971. *Can. Field-Nat.* 87:43–49.

87. Loucks, O.L. 1962. A forest classification for the Maritime Provinces. *Proc. N.S. Inst. Sci.* 25:85–167.

88. Mackinnon, C.M. 1988. Population size, habitat preferences and breeding ecology of the Leach's Storm-Petrel *Oceanodroma leucorhoa* (Vieillot) on Bon Portage Island, Nova Scotia. M.Sc. thesis, Acadia University, Wolfville, N.S. 181 pp.

89. MacLellan, C.R. 1959. Woodpeckers as predators of the Codling Moth in Nova Scotia. *Can. Entom.* 91:673–680.

90. Majka, C.G.; B.L. Roscoe; and M.V. MacKinnon. 1976. The first nest record of the Greater Yellowlegs (*Tringa melanoleuca*) in Nova Scotia. *Can. Field-Nat.* 90:200–201.

91. Martin, K., and D. Guignon. 1983. Canada Goose numbers, daily movements and foraging patterns on Prince Edward Island. *Proc. N.S. Inst. Sci.* 33:107–114.

92. McAloney, R.K. 1973a. Brood ecology of the Common Eider *Somateria mollissima dresseri* in the Liscombe area of Nova Scotia. M.Sc. thesis, Acadia University, Wolfville, N.S. 92 pp + 7 pp. app.

93. McAloney, K. 1973b. The breeding biology of the Great Blue Heron on Tobacco Island, Nova Scotia. *Can. Field-Nat.* 87:137–140.

94. McAlpine, D.F., M. Phinney, and S. Makepeace. 1988a. New Brunswick breeding of Wilson's Phalarope, *Phalaropus tricolor*, confirmed. *Can. Field-Nat.* 102:77.

95. McAlpine, D.F.; J. Finne; S. Makepeace; S. Gilliland; and M. Phinney. 1988b. First nesting of the Glossy Ibis, *Plegadis falcinellus*, in Canada. *Can. Field-Nat.* 102:536–537.

96. McAlpine, D.F, S. Makepeace, and M. Phinney. 1988c. Breeding records of the Greater Scaup, *Aythya marila*, in New Brunswick. *Can. Field-Nat.* 102:718–719.

97. McAlpine, D.F., J. Finne, M. Phinney, S. Gilliland, and S. Makepeace. 1988d. Breeding records for the Gadwall (*Anas strepera*) in New Brunswick.*Naturaliste Canadien* 115:95–96.

98. Mendall, H.L. 1958. *The Ring-necked Duck in the northeast.* University of Maine Studies, Second Series, no. 73. 317 pp.

99. Mengel, R.M. 1964. The probable history of species formation in some northern wood warblers (Parulidae). *Living Bird* 3:9–43.

100. Merriam, C.H. 1894. The geographic distribution of animals and plants in North America. Pp. 203–214 in *U.S. Department of Agriculture Yearbook.* Washington, D.C.

101. Meyer, K.D. 1987. Sexual size dimorphism and the behavioral ecology of breeding and wintering Sharp-shinned Hawks (*Accipiter striatus*). Ph.D. diss., University of North Carolina, Chapel Hill. 184 pp.

102. Miller, E.H. 1977. Breeding biology of the Least Sandpiper, *Calidris minutilla* Vieill., on Sable Island, Nova Scotia. Ph.D. thesis, Dalhousie University, Halifax. 320 pp

103. Milton, G.R. 1977. A population census and nesting habitat study of the Sora (*Porzana carolina*) and the Virginia Rail (*Rallus limicola*) on the Tintamarre National Wildlife Area. B.Sc.(Hons.) thesis, Mount Allison University, Sackville, N.B. 110 pp.

104. Milton, G.R. 1983. The winter ecology of the Common Crow (*Corvus brachyrhynchos* Brehm) in eastern Kings County, Nova Scotia. M.Sc. thesis, Acadia University, Wolfville, N.S. 231 pp.

105. Milton, G.R., and P.J. Austin-Smith. 1983. Changes in the abundance and distribution of Double-crested (*Phalacrocorax auritus*) and Great Cormorants (*P. carbo*) in Nova Scotia. *Colonial Waterbirds* 6:130–138.

106. Minot, E.O. 1980. Tidal, diurnal and habitat influences on Common Eider rearing activities. *Ornis Scand.* 11:165–172. [Based on M.S. thesis, University of Maine]

107. Morris, M.M.J., and R.E. Lemon. 1988a. American Redstart nest placement in southwestern New Brunswick. *Can. J. Zool.* 66:212–216.

108. Morris, M.M.J., and R.E. Lemon. 1988b. Mate choice in American Redstarts: by territory quality? *Can. J. Zool.* 66:2255–2261.

109. Nettleship, D.N., and T.R. Birkhead. eds. 1985. *The Atlantic Alcidae. The evolution, distribution and biology of the auks inhabiting the Atlantic Ocean and adjacent water areas.* London: Academic Press. 574 pp. [Including Nettleship, D.N., and P.G.H. Evans. pp. 53–154]

110. Newell, R.B. 1985. Nesting ecology of Arctic Terns *Sterna paradisaea* Pontoppidan at Machias Seal Island. M.Sc. thesis, Acadia University, Wolfville, N.S. 167 pp.

111. Palmer, R.S., ed. 1962. *Handbook of North American birds.* vol. 1. *Loons through Flamingos.* New Haven: Yale University Press.

112. Palmer, R.S., ed. 1976. *Handbook of North American birds..* vols. 2,3. *Waterfowl.* New Haven: Yale University Press.

113. Palmer, R.S., ed. 1988. *Handbook of North American birds.* vols. 4,5. *Diurnal birds of prey* . New Haven: Yale University Press.

114. Paynter, R.A., Jr. 1954. Interrelations between clutch-size, brood-size, prefledging survival, and weight in Kent Island Tree Swallows. *Bird-Banding* 25:35–58, 102–110, 136–148.

115. Pearce, P.A., D.B.Peakall, and A.J. Erskine. 1976. *Impact on forest birds of the 1975 spruce budworm spray operation in New Brunswick.* Canadian Wildlife Service, Progress Notes, no. 62. 7 pp.

116. Pettingill, O.S. 1939. The bird life of the Grand Manan archipelago. *Proc. N.S. Inst. Sci.* 19:293–372.

117. Philipp, P.B., and B.S. Bowdish. 1919. Further notes on New Brunswick birds. *Auk* 36:36–45.

118. Phillips, J.C. 1923–26. *A natural history of the ducks.* 4 vols. Boston.

119. Preston, W.C. 1968. Breeding ecology and social behavior of the Black Guillemot *Cepphus grylle*. Ph.D. diss., University of Michigan, Ann Arbor. 136 pp.

120. Prévost, Y.A., R.P. Bancroft, and N.R. Seymour. 1978. Status of the Osprey in Antigonish County, Nova Scotia. *Can. Field-Nat.* 92:294–297. [From M.Sc. thesis (by Prévost), McGill University]

121. Prince, H.H. 1965. The breeding ecology of Wood Duck (*Aix sponsa* L.) and Common Goldeneye (*Bucephala clangula* L.) in central New Brunswick. M.Sc. thesis, University of New Brunswick, Fredericton. 109 pp.

122. Quinney, T.C. 1979. Reproductive success, growth of nestlings and foraging behaviour of the Great Blue Heron, *Ardea herodias herodias* L. M.Sc. thesis, Acadia University, Wolfville, N.S. 143 pp.

123. Quinney, T.E., and P.C. Smith. 1980. First breeding record of Black-crowned Night Heron in Nova Scotia. *Can. Field-Nat.* 94:463.

124. Redmond, G.W., D.M. Keppie, and P.W. Herzog. 1982. Vegetative structure, concealment, and success at nests of two races of spruce grouse. *Can. J. Zool.* 60:670–675.

125. Reed, A., ed. 1986. *Eider ducks in Canada.* Canadian Wildlife Service Report Series no. 47. (including Erskine, A.J., and A.D. Smith. pp. 20–29; Lock, A.R., pp. 30–38).

126. Rigby, M.D. 1982. Clutch-size and prefledging survival in Red-winged Blackbirds at Williamstown Lake, New Brunswick. *Wilson Bull.* 94:569–571. [From B.Sc.(Hons.) thesis, University of New Brunswick]

127. Robbins, C.S. 1966. *Birds of North America.* New York: Golden Press. 340 pp.

128. Robbins, C.S., D. Bystrak, and P.H. Geissler. 1986. *The Breeding Bird Survey: Its first fifteen years, 1965–1979.* U.S. Fish Wildlife Service, Resource Publication no. 157. 196 pp.

129. Roland, A.E. 1982. *Geological background and physiography of Nova Scotia.* Halifax: Nova Scotia Institute of Science and Nova Scotia Museum. 311 pp.

130. Ross, R.K. 1973. A comparison of the feeding and nesting requirements of the Great Cormorant (*Phalacrocorax carbo* L.) and Double-crested Cormorant (*P. auritus* Lesson) in Nova Scotia. M.Sc. thesis, Dalhousie University, Halifax. 82 pp.

131. Rowe, J.S. 1959. *Forest regions of Canada.* Canada Dept. Northern Affairs and Natural Resources, Forestry Branch Bulletin no. 123. 71 pp. (rev. ed. 1972)

132. Sabean, B.C. 1972. Breeding biology of the American Eider (*Somateria mollissima dresseri* Sharpe) on Tobacco Island, Nova Scotia. M.Sc. thesis, Acadia University, Wolfville, N.S. 110 pp.

133. Saunders, R.M. 1935. The first introduction of European plants and animals into Canada. *Can. Hist. Rev.* 1935 (Dec.): 388–406.

134. Seymour, N.R. 1977. Social aspects of reproductive behaviour with Black Duck (*Anas rubripes*) in eastern Nova Scotia. Ph.D.

thesis, McGill University, Montreal. 111 pp.

135. Sharrock, J.T.R., comp. 1976. *The atlas of breeding birds in Britain and Ireland*. Tring, U.K.: British Trust for Ornithology. 477 pp.

136. Simmons, R.E. 1983. Polygyny, ecology and mate choice in the Northern Harrier *Circus cyaneus* (L.). M.Sc. thesis, Acadia University, Wolfville, N.S. 177 pp.

137. Squires, W.A. 1952. *The birds of New Brunswick*. Monograph Series no. 2. Saint John: New Brunswick Museum. 164 pp.

138. Squires, W.A. 1976. *The birds of New Brunswick*. 2nd ed. Monograph Series no. 7. Saint John, N.B.: New Brunswick Museum. 221 pp.

139. Stewart, P.A. 1971, Persistence of remains of birds killed on motor highways. *Wilson Bull.* 83:203–204.

140. Stobo, W.T., and I.A. McLaren. 1975. The Ipswich Sparrow. *Proc. N.S. Inst. Sci.* 27 (2nd suppl.). 105 pp.

141. Stocek, R.F. 1988. Notes on Tree Swallow breeding biology—III. Nest building and success in Tree Swallows. *N.B. Nat.* 16:50–51.

142. Stocek, R.F. 1990. The status of the Common Loon in New Brunswick. *N.B. Nat.* 17:54. [Abstract from a lengthy unpublished report to World Wildlife Fund]

143. Stocek, R.F., and P.A. Pearce. 1981. Status and breeding success of New Brunswick Bald Eagles. *Can. Field-Nat.* 95:428–433.

144. Thompson, G.R. 1979. The habitat and population dynamics of the Ruffed Grouse (*Bonasa umbellus togata* L.) in Prince Edward Island. M.Sc. thesis, Acadia University, Wolfville, N.S. 172 pp.

145. Tuck, L.M. 1972. *The Snipes: A study of the genus* Capella. Canadian Wildlife Service Monograph Series. no. 5. 428 pp

146. Tufts, R.W. 1962. *The Birds of Nova Scotia*. Halifax: Nova Scotia Museum.

147. Tufts, R.W. 1965. Probably a "first" for North America. *N.S. Bird Soc. Newsletter* 7(3):10–11. [It wasn't.]

148. Tufts, R.W. 1968. Banding Ravens. *N.S. Bird Soc. Newsletter* 10:18–19.

149. Tufts, R.W. 1986. *Birds of Nova Scotia*. 3rd ed. (with revisions by members of the Nova Scotia Bird Society under the coordination of Ian A. McLaren). Halifax: Nimbus Publishing and Nova Scotia Museum. 478 pp.

150. Voous, K.H. 1960. *Atlas of European birds*. London: Thomas Nelson and Sons Ltd. [Translated from Dutch]

151. Warkentin, J., ed. 1968. *Canada: A geographic interpretation*. Toronto: Methuen. 608 pp. [Including Erskine, D.S., The Atlantic region. Chapter 9. pp. 231–280]

152. Welsh, D.A. 1971. Breeding and territoriality of the Palm Warbler in a Nova Scotia bog. *Can. Field-Nat.* 85:31–37. [An earlier version was in *N.S. Bird Soc. Newsletter* 11:115–118, 1969, including population data on other species.]

153. Welsh, D.A. 1975. Savannah Sparrow breeding and territoriality on a Nova Scotia dune beach. *Auk* 92:235–251. [From M.Sc. thesis, Dalhousie University]

154. White, G. 1788. *The natural history of Selborne*. Reprint. London: Everyman's Library Ed. J.M. Dent & Sons, 1906, 1949.

155. White, H.C. 1953. *The Eastern Belted Kingfisher in the Maritime Provinces*. Fisheries Research Board of Canada Bulletin no. 97. 44 pp.

156. White, H.C. 1957. *Food and natural history of mergansers on salmon waters in the Maritime Provinces of Canada*. Fisheries Research Board of Canada Bulletin no. 116. 63 pp.

157. Wilbur, H.M. 1969. The breeding biology of Leach's Petrel, *Oceanodroma leucorhoa*. *Auk* 86:433–442.

158. Winn, H.E. 1950. The Black Guillemots of Kent Island, Bay of Fundy. *Auk* 67:477–485.

159. Young, A.D. 1985. Costs and benefits to Red-breasted Mergansers nesting in tern and gull colonies. M.Sc. thesis, McGill University, Montreal. 55 pp.

Index

This index contains the common, French, and scientific names of birds for which there are species accounts. In the common name entries, the page number of the species account is given in boldface type.

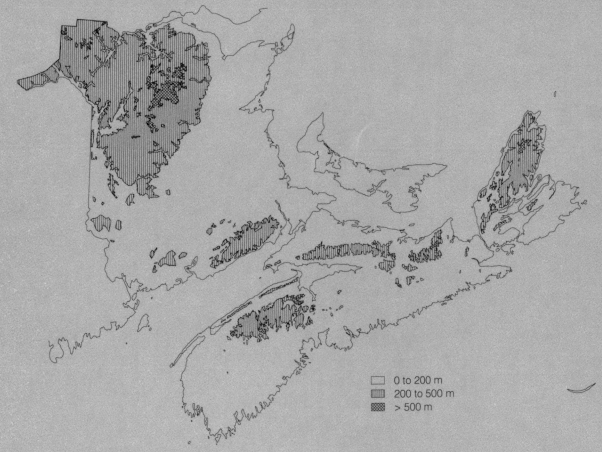

0 to 200 m
200 to 500 m
> 500 m

Elevation contours

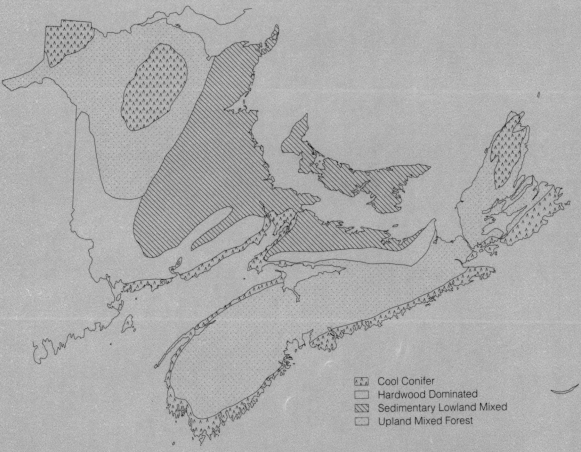

Cool Conifer
Hardwood Dominated
Sedimentary Lowland Mixed
Upland Mixed Forest

Main forest types

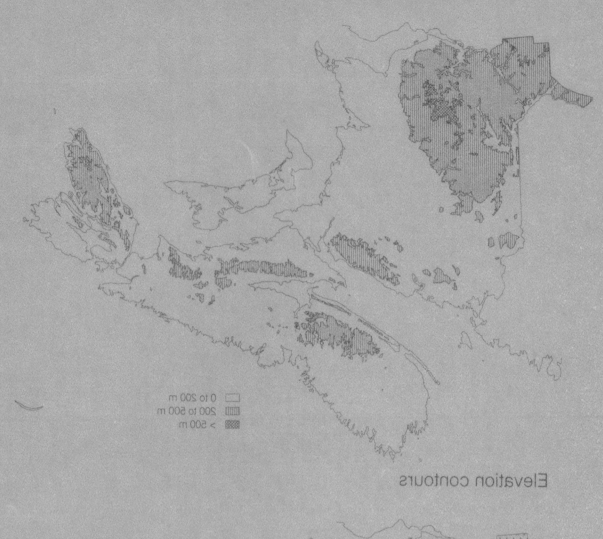

0 to 200 m
200 to 500 m
> 500 m

Elevation contours

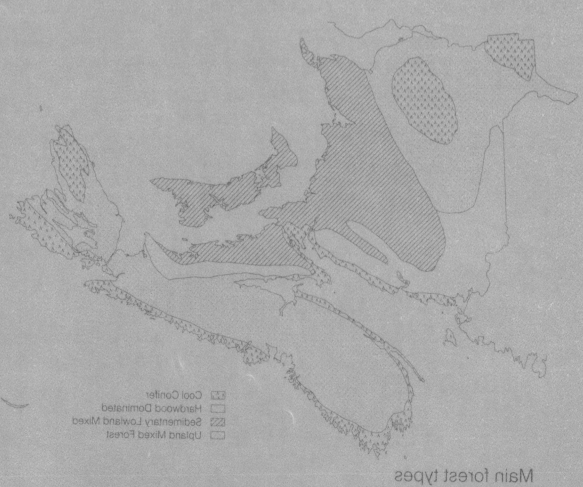

Cool Conifer
Hardwood Dominated
Sedimentary Lowland Mixed
Upland Mixed Forest

Main forest types

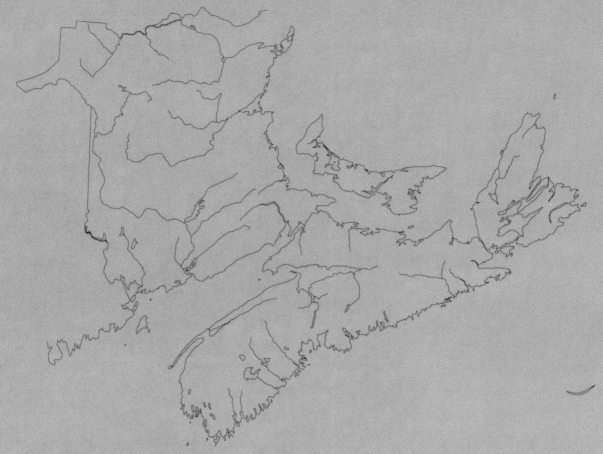

Major rivers and lakes

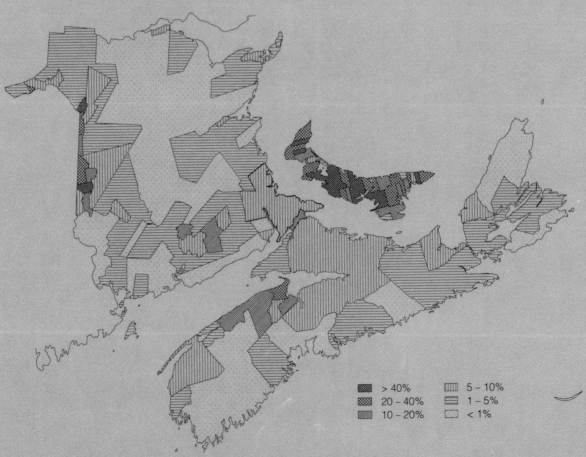

▨ > 40%	▥ 5 – 10%
▧ 20 – 40%	▤ 1 – 5%
▩ 10 – 20%	░ < 1%

"Improved" farmland as a percentage of total land areas

Major rivers and lakes

▦ > 40%	▥ 5 - 10%	
▨ 20 - 40%	▤ 1 - 5%	
▥ 10 - 20%	▢ < 1%	

"Improved" farmland as a percentage of total land areas

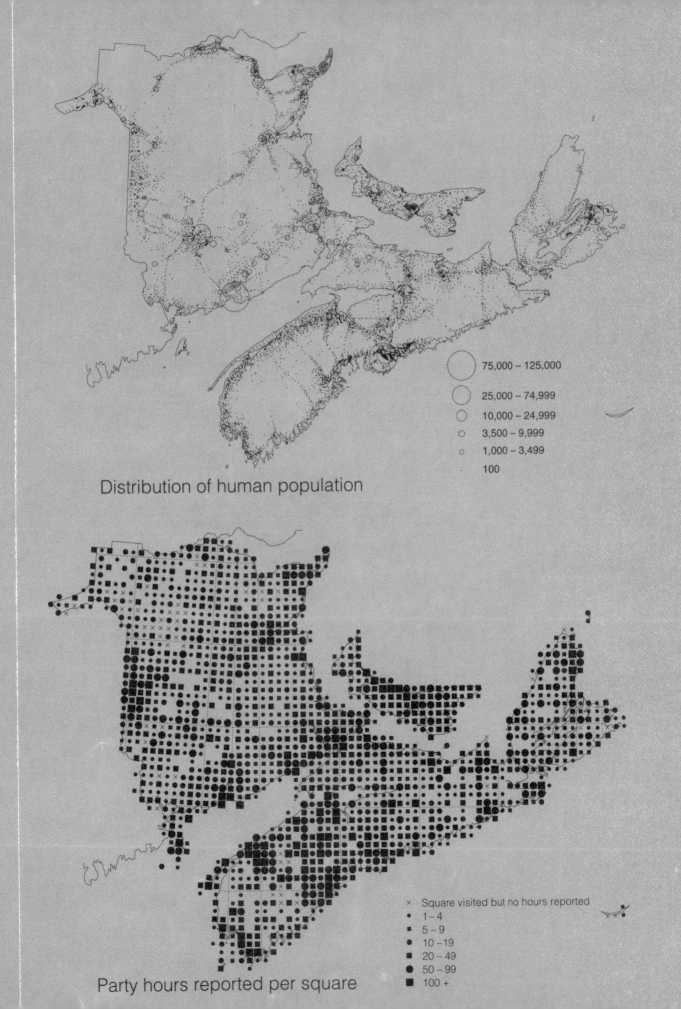

Distribution of human population

Symbol	Population
◯	75,000 – 125,000
◯	25,000 – 74,999
○	10,000 – 24,999
○	3,500 – 9,999
○	1,000 – 3,499
·	100

Symbol	Party hours
×	Square visited but no hours reported
•	1 – 4
■	5 – 9
●	10 – 19
■	20 – 49
●	50 – 99
■	100 +

Party hours reported per square

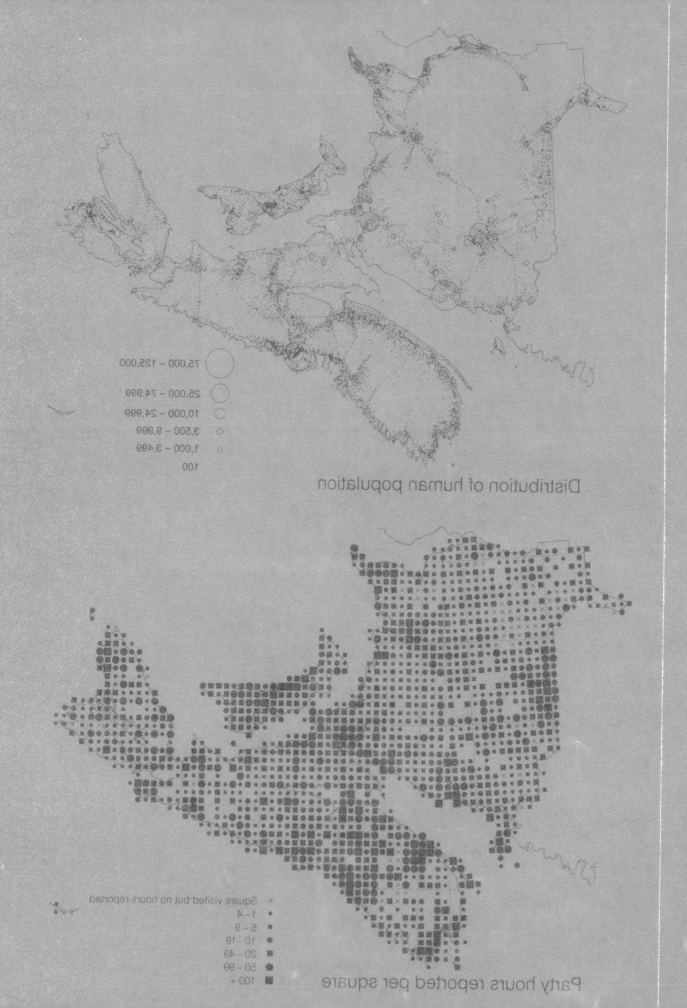

Distribution of human population

100
1,000 – 3,499
3,500 – 9,999
10,000 – 24,999
25,000 – 74,999
75,000 – 125,000

Party hours reported per square

Square visited but no hours reported
1 – 4
5 – 9
10 – 19
20 – 49
50 – 99
100 +

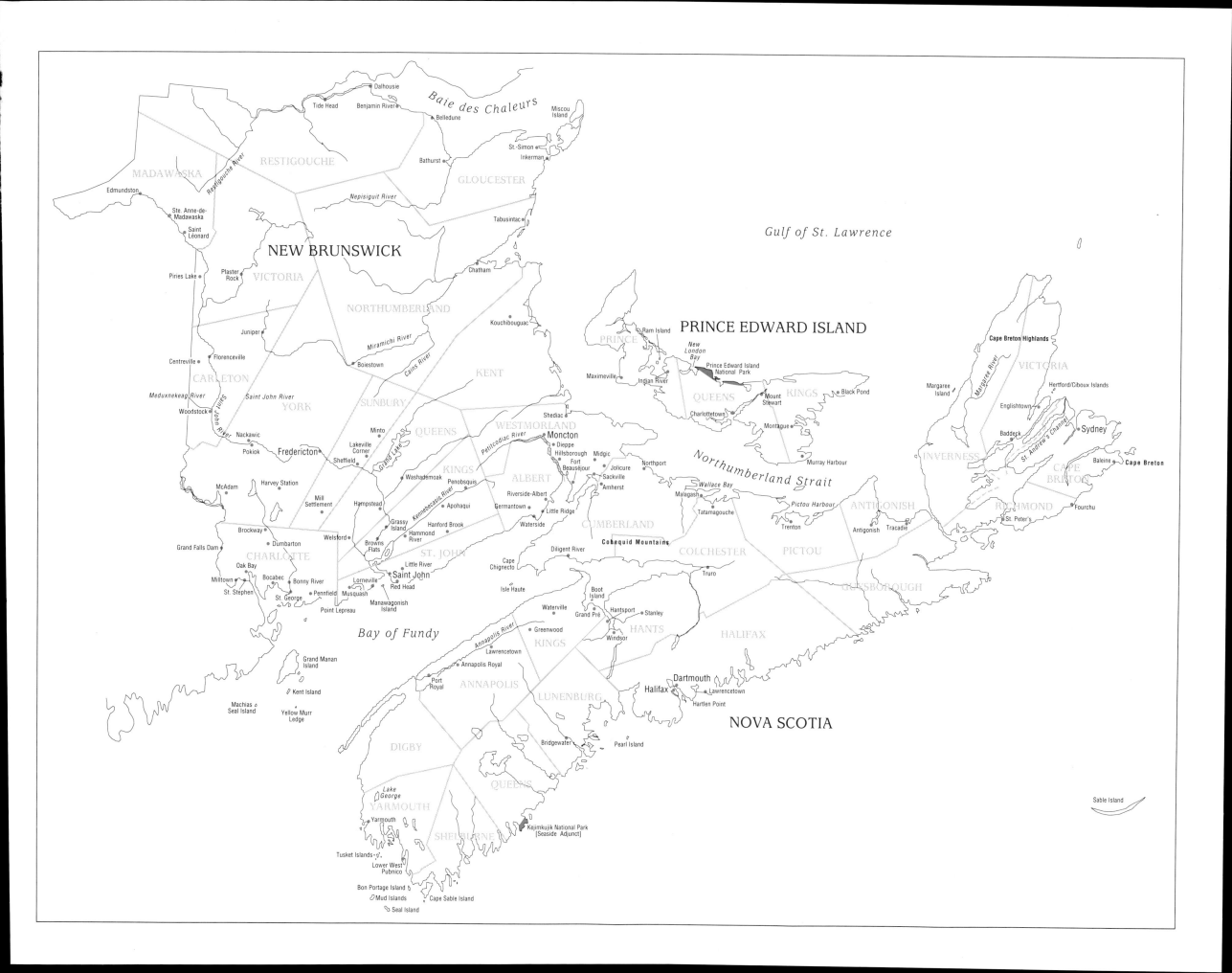